WHY
SEX
MATTERS

REVISED EDITION

WHY SEX MATTERS

a darwinian look at human behavior

~~~~~~~~~~~~~~~~~~~~~

BOBBI S. LOW

PRINCETON UNIVERSITY PRESS
PRINCETON AND OXFORD

Published by Princeton University Press, 41 William Street,
Princeton, New Jersey 08540
In the United Kingdom: Princeton University Press, 6 Oxford Street,
Woodstock, Oxfordshire OX20 1TW

press.princeton.edu

Jacket image © Jezper/Shutterstock.
Jacket design by Lorraine Betz Doneker.

Library of Congress Cataloging-in-Publication Data

Low, Bobbi S.
    Why sex matters : a darwinian look at human behavior / Bobbi S. Low. —
Revised edition
        pages cm
    Summary: "Why are men, like other primate males, usually the aggressors
and risk takers? Why do women typically have fewer sexual partners? In Why
Sex Matters, Bobbi Low ranges from ancient Rome to modern America, from
the Amazon to the Arctic, and from single-celled organisms to international
politics, to show that these and many other questions about human behavior
largely come down to evolution and sex. More precisely, as she shows in this
uniquely comprehensive and accessible survey of behavioral and evolutionary
ecology, they come down to the basic principle that all organisms evolved to
maximize their reproductive success and seek resources to do so, but that
sometimes cooperation and collaboration are the most effective ways to
succeed. This newly revised edition has been thoroughly updated to include
the latest research and reflect exciting changes in the field, including how our
evolutionary past continues to affect our ecological present"— Provided by
publisher.
    Includes bibliographical references and index.
    ISBN 978-0-691-16388-8 (paperback)
    1. Human evolution. 2. Sex role. 3. Nature and nurture. 4. Sociobiology.
I. Title.
    GN281.4.L68 2015
    304.5—dc23   2014027259

British Library Cataloging-in-Publication Data is available

This book has been composed in Minion Pro and Helvetica Neue
Printed on acid-free paper. ∞

Printed in the United States of America

10 9 8 7 6 5 4 3 2 1

FOR MY SON, MICHAEL,

and his family, and for my growing academic lineage, many of whom work in the "real" world, tackling the problems I raise in chapter 15.

I now have grandchildren in both lineages and am grateful to all generations, up and down.

May the lineages increase!

Contents

Preface

WHEN I WROTE the first version of *Why Sex Matters*, I suggested that sex differences are central to our lives, wherever and whenever—however—we live. They still are. And no matter our age or stage, we still think about them. Sex differences present us with a fascinating tangle: What do the widely acclaimed (and equally widely denied) differences between men and women mean in terms of the ways in which men and women use resources, take risks, make war, and raise children? Which differences are lasting, which are ephemeral? If we follow the differences through time, across space, and into different environments, what do they mean in today's societies?

Fourteen years ago was an exciting time to study sex differences, but it is even more exciting now. The general recognition was that, although environmental conditions can exaggerate or minimize differences in male and female behaviors, under most conditions each sex has been successful as a result of very different behaviors. The results are more complex—and more interesting—than we knew. We humans must solve the same ecological problems as all other species. In biological terms the currencies are, in the end, reproductive; that is, who survives and who reproduces best? This principle seems so simple that it is hard to imagine that diverse and complicated behaviors could arise from it. Yet behaviors vary greatly, because the ecological (including social) conditions that shape our lives' success vary so widely.

There are three themes in this work, as in the first edition. First, resources are useful in human survival and reproduction; like other living things, we have evolved to wrest resources from the environment for our benefit, and today we know much more about how this operates. Second, the two sexes tend to differ in how they can use resources most effectively to accomplish survival and reproduction. Third, how each sex accomplishes these ends relies not only (and not obviously) on differences in genes but also on differences in environment—there are no identified genes specific for polygyny, for example, but in many environments the trends for male mammals to profit from trying to be polygynous are strong.

These intertwined themes of resource utility, sex differences, and environmental constraints soon lead us to consider other problems: for example, status striving and risk taking. They help us understand why homicide is largely a male endeavor, and why men and women are jealous about different things. What happens to populations is the aggregate of what men and women do as they grow, mate, raise children, and endow those children as well as they can. What can we do about the fact that many of today's problems have relatively straightforward technical solutions but that many of our strategies to get both men and women to "do the right thing" are ineffective? Is prosociality—the tendency to be social—enough?

So much has been explored, and today we understand more deeply some issues that were new and explained too simply at first. Peter Richerson, Rob Boyd, Richard McElreath and others have deepened our understanding of cultural evolution; Ernst Fehr and colleagues have extended our understanding of fairness; Joseph Henrich and colleagues have demonstrated how variable such concepts can be cross-culturally. Anthropologists such as Michael Gurven have taken analyses of a traditional society to depths not previously reached (and more) by anyone except perhaps Kim Hill and Magdalena Hurtado, spurred in part by the pioneering work of Napoleon Chagnon and Bill Irons.

I have not changed the structure of the book greatly, but much information is new, and in chapter 15 I try to link our evolutionary past with our ecological future. I will begin with the basic arguments and assumptions of behavioral and evolutionary ecology: selfish genes, conflicts of interest, and why two (and not more or fewer) sexes have specialized to reproduce through different behaviors (chapters 1–3). Then I ask, how do these basic sex differences, whose theory we understand, actually play out in primates, including humans (chapters 4–6)? Next, I take an empirical glance at the diverse ways in which both traditional and transitional societies make a living, how men's and women's roles and lives diverge, and how even marriage is affected by ecology and resources (chapters 7, 8). The complexity of these patterns leads us back to basic theory, to explore how conflicts of interest are mediated, literally from the level of genes in genomes to whole societies (chapters 9, 10). Sex differences and conflicts of interest help us predict why there are so few women warriors or high-roller politicians in many societies—and the kinds of societies in which they are likely to occur (chapters 11–14). Finally, I ask, How does our heritage from our evolutionary past interact with current global population and resource-consumption problems (chapter 15)? Does sex still matter? Have we, in creating novel environments, changed the rules so that now it may even be detrimental to "strive" to our utmost abilities? Have we gotten ourselves into a bind in which the behaviors we have evolved to do, and do ever more efficiently, are now the behaviors that threaten our very existence?

I draw cross-cultural examples from traditional, historical, and modern societies and from physical, physiological, demographic, and behavioral data (very occasionally, a little genetics will sneak in). I'll share my interests and my concerns, but I cannot offer cut-and-dried solutions. My purpose is to ask, here is a puzzle, a conundrum—what ideas, old and new, can we use to solve it? I would like to reach both general readers and scholars in the traditional human disciplines with concepts that may be new.

Many people helped me greatly. Thanks first to Rob Kurzban, whose kind words about an interview convinced Princeton that a revised edition, when fields are moving so fast, might be a good idea.

Thanks so much to friends and colleagues who took time from their own over-busy schedules and sometimes responded from around the world and who provided advice, papers, and figures: Lee Cronk, Jim Wilson, Kathleen Welch, K. G. Anderson, Bill Irons, Martin Daly, Margo Wilson, Ray Hames, Rachel Smolker and Andrew Richards, Kim Hill (who also corrected a couple of mistakes!), Bill Irons, Napoleon Chagnon, John Mitani, Laura Betzig, and Rebecca Sear.

Thanks to Monique Borgerhoff Mulder and the great group she and Sam Bowles got together to look at how parents try to skew resources in favor of their offspring, in societies from hunter-gatherers to modern developed nations, and for showing how universal it is. Thanks also to the following people for discussions, joint work on problems leading to parts of this manuscript, or for providing data: Monique Borgerhoff Mulder, Elizabeth Cashdan, Napoleon Chagnon, Martin Daly, Mary Anne Fisher, Justin Garcia, Alice Clarke, Kim Hill, Deborah Judge, Elinor Ostrom, Richard Connor, and Rachel Smolker. All remaining errors and blunders are, sadly, mine alone.

Many thanks to Whitney Conard for logistic help well above and beyond the call of duty. Alison Kallett, Quinn Fusting, Alison Anuzis, Leslie Grundfest, and Gail Schmitt of Princeton University Press actually made the manuscript come together in publishable form. Thank you!

The best-ever reciprocator in the world, Carl Simon, has provided expertise, humor, and a sense of perspective (and put up with me in the process). Thanks to my son, Michael, once a reason to expend ridiculous amounts of parental effort; now, he and his wife Ryoko Imai (thanks!) are expending amazing amounts of parental effort in their turn.

It is my hope that general readers will have fun discovering how similar, and how different, the sexes are and how modern times differ and are similar to the past. I hope that experts in other fields will find themselves saying: "I know a way to tackle that problem; my field can contribute something here although it's not the sort of problem I usually analyze." My sense is that many crucial problems haven't been solved because of disciplinary approaches and that no one will be able to solve them alone in an attempt to use interdisciplinary approaches. But a number of us, reaching across boundaries with tolerance and patience, might make more progress in getting answers.

WHY
SEX
MATTERS

1

Introduction

Probably a crab would be filled with a sense of personal outrage if it could hear us class it without ado or apology as a crustacean, and thus dispose of it. "I am no such thing," it would say: "I am MYSELF, MYSELF alone."
—William James *Varieties of Religious Experience* (1902)

To the same natural effects we must, as far as possible, assign the same causes.
—Newton

"**W**HY CAN'T A WOMAN be more like a man?" wailed Professor Henry Higgins in *My Fair Lady*, the musical derived from George Bernard Shaw's *Pygmalion*. Certainly in many societies, across time, there have been women who were "more like a man." Think of Joan of Arc, who was burned at the stake (on the minor charge of wearing men's clothing); or of George Sand, of whom Elizabeth Browning said, "You are such a large-brained woman and a large-hearted man." Yet in part, we remember such women *because* they are singular, whether we envy their ability to break free or whether we imagine that they missed a lot. What really contributes to the patterns we see—and to their exceptions?

Here I want to explore sex differences from a relatively unusual perspective, one that is often misunderstood. Understanding and explaining human behavior is a central concern for all of us. But doing so—especially when sex differences are the issue—presents a real dilemma. We are complicated, highly social beings. We live in a staggering array of environments, both ecological and social. Our families, lovers, and friends are not exemplars or prototypes, but unique, particular individuals. None of us wants to be "reduced" to some formula.

For, like William James's crab, we *know* we are above crude analysis. Even the name we give ourselves, *Homo sapiens*, reflects both the value that we give to understanding and the fact that we feel ourselves to be special. Like that crab, many people may be appalled at the approach I will use here, that is, to assume that, once we understand environmental and social pressures, we humans are as predictable as other animals in our behavior and are governed by the same rules. And I want to begin with *simple* rules, no less.

Many of us assume that humans operate under rules that are different from these other species, that our rules are based entirely on culture rather than on biology, or some combination. I will ask: what can we learn if we begin without assuming that this were true? I want to begin by exploring what a biologist would predict if he or she knew only that here was a smart, upright-walking, highly social primate and *nothing more*. I will explore the ecology of being male and female, beginning with simple rules and with what I can discern about the environments with which the evolutionary rules interact.

The approach I use, behavioral ecology, is an evolutionary approach with roots in Charles Darwin's work. It focuses on the question, how do environmental conditions influence our behavior and our lifetimes? This approach has proved profitable in exploring other realms of human behaviors.[1] Behavioral ecology and its intellectual relatives seek to understand how comparatively simple operating rules interact with historical accidents, and with temporal and spatial specifics, to yield a rich diversity of patterns. There is no doubt that genes influence not only our physical structure and physiology but also our behavior; there is no doubt that historical accident often plays a role, nor is there any doubt that cultural and social pressures can influence behavior. But where lies the balance? Perhaps by beginning with very simple rules and assumptions, and finding where they fail to predict well, we can gain some insight.[2]

VAMPIRE STORIES AND BEYOND

Humans have always sought to explain the patterns they see. In fact, one of the strongest selective forces on human behavior has been to understand pattern—not only in order to deal with environmental variation but also to be the first in one's tribe who is able to predict events and when they will occur (imagine being the first human to predict a solar eclipse). But creating stories that more or less match our observations is not science but folklore.

Consider vampires. From Bram Stoker through Anne Rice, from Bela Lugosi to Tom Cruise and Leslie Nielson, vampires have always fascinated us: aristocratic, sexy, dangerous, and (almost) invincible. Vampire folklore provides a wonderful example of how our need to explain can drive us to spin stories that *seem* to explain something that we see, and that can be hard to refute, but that nonetheless do not reflect what actually happens.

The folklorist Paul Barber, in a delightful examination of vampire myths around the world, notes that the ways people in preindustrial societies interpreted phenomena associated with death and the decay of corpses are "from our perspective, quite wrong. What makes them interesting, however, is that they are also usually coherent, cover all the data, and provide the rationale for some common practices that seem, at first, to be inexplicable."[3]

The variety of myths and legends about vampires all begin simply: death—especially unexpected or unusual death—brings more death. If someone died, "why" was usually unknown, and epidemics leading to death and more death were once far more common than today. Once people were buried (often without coffins), not all corpses had the decency to stay below ground. In a prebacteriology culture, people weren't likely to see a "flailing" corpse as the natural by-product of bacterial decay, but rather as the will of the dead person or as the rejection of the corpse by Mother Earth. Since death brings death, those first to die, as in an epidemic, were dangerous and somehow had to be disarmed so that they could not continue to bring death. Only when all "changing" ceased, and ashes or bones alone remained, was the corpse neutral, inactive, and no longer dangerous.

People thus began with a repeatable observation: that death brings death. This applies not only to vampires but also to the general idea that dead people call to their relatives and friends and must be propitiated to protect those still alive. Because they had no knowledge of disease transmission, people imbued the corpse with dangerous properties. Not a bad idea, particularly in times of plague, when unexpected deaths were frequent and vampire fears were heightened—but an idea that led to a misinterpretation of the normal signs of decomposition.

In folklore, in a variety of societies around the world, vampires are described as undecomposed; they have a ruddy or dark complexion, do not suffer rigor mortis, are swollen or plump, have blood at the mouth and/or in internal cavities, and grow new skin or nails after burial. In an interesting twist, people suspected of being vampires were frequently buried differently, in ways that made a diagnosis of vampirism *more*, not less, likely. Suspected vampires were often buried face down, so that if they tried to claw their way out of the earth to torment others, they would dig themselves in deeper. But because blood settles into the lowest capillaries after death, face-down burial meant that the body's face (rather than the back) would be dark and ruddy. And a ruddy face was believed to be a sign of a vampire. Some putative vampires were buried with lime to hasten their decomposition—but lime in fact retards it. Thus, someone who died in an unusual way and was in danger of becoming a vampire was likely to be buried face down, covered with lime—and thus to have a ruddy face, to decompose more slowly, and, on exhumation, to be confirmed as a vampire. Such practices reinforced mistaken beliefs.

Folklore about vampires arises from an entirely sensible and consistent desire to explain something in the absence of complete information. But what people *say* about what they see and do can be a rotten path to explanation. Although observers called the corpses undecomposed, they described unmistakable signs of rot (e.g., a stench). Descriptions like "ruddy" or "swollen," which were used to assert failure to decompose, are in fact signs of the ordinary (but variable) process of decomposition.

It is important to separate carefully what people *describe*, as they see, hear, and smell what happens, from the *causes they attribute*. The observation that a corpse stinks is, in fact, consistent with the contention that decay is occurring, but not that it is failing to decompose. It is important to avoid this kind of muddle at all times, not only when we are no longer likely to believe in something like vampirism. Being led astray by "vampire myths" that sound reasonable but are untested is most likely to occur when a behavior is complicated and we want to believe the stories we tell.

Explaining Behavior without Folklore

Other species' behavior can be more complex than we realize. An excellent example of such complexity generated by the interaction of operating rules (genes), environment, and historical accident is the biologist Bernd Heinrich's work on food sharing in ravens, *Corvus corax*. Heinrich saw a curious behavior on a hike in Maine late one October: a group of ravens that were feasting on a dead moose were giving a distinctive, loud, and high-pitched "yell."[4] Because other ravens were attracted to the yelling, the result was that there were more competitors at the kill, and thus less food for each of the yellers. Why didn't the raven who first found the kill just keep quiet? And why weren't the ravens now fighting over the kill?

If we saw humans share like this, or if we extended our often untested social perceptions about humans to ravens, we would probably think how kind all this indiscriminate sharing is. In fact, that would constitute a vampire story; behavioral ecologists find true genetically costly altruism to be so rare, as I will explain below, as to be a fluke in nonhumans. The ravens who shared seemed to be doing it at a cost. One of Heinrich's first questions was whether the "called" birds were related to the callers, for sharing with individuals who have at least some genes in common can help copies of one's own genes. Or were they reciprocators? After much work marking and recapturing ravens in the field, Heinrich was able to eliminate the possibility that ravens were summoning their kin or their reciprocators. Where next?

Heinrich made a series of careful observations, comparing the behavior of different ravens under different circumstances. His summary begins with eleven clues and proceeds from simpler to more complex deductions.[5] Without giving away the whole plot, I can say that one of his major findings was that *adult* ravens are territorial, controlling access to any carcasses in their territory and driving off any juveniles found on a carcass. When a *juvenile* found a carcass, it was likely to "yell," attracting other juveniles (the largest group at a carcass was about 150). When enough juveniles were present, the resident territorial adults could not drive all of them off. So the cost of additional juvenile competitors could be offset by the benefit of attracting a group large enough to stay on the carcass even if adults were nearby. Clearly the costs and benefits of yelling would vary in different circumstances. Heinrich did not simply create a plausible vampire or just-so story about the juvenile behavior he observed. Instead, he observed, made hypotheses, and tested them to discover the most probable functional reason for the ravens' behavior.

Heinrich could not know what role any gene plays in this behavior, so he used a technique called the "phenotypic gambit" to make testable predictions: starting from what he saw—the phenotype—he made certain assumptions. He assumed that, whatever the relationship between genes and the set of behaviors he saw, enough time had passed for the system to come to equilibrium and thus what he saw represented the outcome of competing strategies.[6] The phenotypic gambit is a powerful tool (although sometimes controversial), and I'll return to it below.

For ravens as for humans, both ecological and social conditions can change the costs and benefits of any action. Heinrich, of course, had to observe the ravens without their cooperation; he had to concentrate on what the ravens actually did. Sociologists, psychologists, and anthropologists find it useful to interview people, and this tells us something about what people themselves imagine they are doing (and, of course, speech itself is a behavior).

Because we can make decisions consciously, often we assume that unless a behavior is consciously considered, it is of no interest. Yet many other species routinely learn and behave in complicated ways without (so far as we can tell, at least) consciousness—or at least without the ability to share abstract sentiment through speech. Furthermore, as I noted above, it is presumptuous to assume that people's conscious attributions of their behavior are analytically helpful, and if we do assume that, it can cause real trouble.

Behavioral ecologists cannot interview ravens about why they call to other ravens when they find a carcass (who knows what reasons a raven might give, anyway?), and they don't know the genetics of the situation (is such calling the result of a single gene's action?). For these reasons, behavioral ecologists concentrate on what happens—on what behaviors show up under what conditions. If we take the same approach in looking at human behavior, we will lose some information about people's intentions, but we won't get distracted by our human reports of conscious reasons. And such lack of distraction may prove useful, for what people say is often not consistent with what we observe them doing.[7]

Perhaps new connections will appear as we look past what we imagine behavioral causes to be and as we look beyond what people say about why they act a certain way—if we instead examine carefully *what sorts of behaviors we see in particular environments*. Without requiring consciousness or rationality (or even speculating on their existence), we can ask what behaviors will be profitable under specific environmental conditions. Then we can ask explicitly how conscious, cultural influences can influence the costs and benefits of these behaviors.

Kinds of "Why" Questions

Heinrich's analysis of raven behavior highlights an important distinction. To understand "why" we do things, to explain both the behaviors that seem almost universal or unvarying and those that vary greatly, we can seek answers in different ways. "Why" questions have two principal complementary forms in biology: "proximate" and "ultimate" explanations.[8] Why do birds migrate? One answer might be "changing day length causes hormonal changes, triggering migration." Both changing day length and changing hormones are *proximate* triggers, or cues. If we could interview birds, we might have another set of proximate causes, the equivalent of our reasons: "I really hate the cold," "it makes me feel good," "that way I get to see my relatives."

However, proximate answers are no help in explaining why one species migrates while others don't, why not all individuals in this species migrate (costs and benefits may differ for older, younger, weak, or healthy individuals), or why day length rather than some other cue, or a combination of cues, has become the trigger.

The *ultimate* cause of migration (and other behaviors) always concerns reproductive success. Seasonal better-versus-worse geographic shifts in foraging and nesting areas mean that individuals who seek the better areas, shifting seasonally, leave more descendants than those who remain in one area. When day length is the most reliable predictor of these seasonal shifts, individuals who use it as a cue will fare better than those that use some other proximate cue or who fail to migrate. Thus, we would predict migration patterns triggered by day length for birds that are (for example) insectivores or nectar-eaters in northern temperate regions; their food disappears seasonally. We expect variation in which individuals will migrate when: the benefits and costs of migration (in terms of survival and reproduction) differ for older, prime-age birds compared with yearlings in a particular species, for example. Proximate cues and ultimate (selective) causes tell us very different things.

It is useful to ask questions about both "proximate" triggers and the "ultimate" selective cause, and it is important to understand that these two approaches are not alternatives but that they complement each other. Proximate triggers, the *mechanisms that release behaviors*, are sometimes also called "causes." They tell us what kinds of environmental factors are reliable. The ultimate cause of a behavior's existence, in evolutionary terms, is always its impact on family (or genetic lineage) persistence through time. We seldom think of such matters, perhaps because few of us can now trace our ancestry in the same way a certain schoolteacher from Cheddar, England could: a preserved "bog man" from ancient times was found, through DNA analyses, to be a clear direct ancestor of his.[9] Nonetheless, the persistence of genes through time, and the clustering of genes in family lines, are real.

Sometimes it is important to ask "why" at the proximate level. Suppose we wish to ask about variation in human fertility. Reduced fertility could have the proximate "cause" of later marriage age and an ultimate selective cause of greater lineage success through fewer, better-invested children than through more numerous, but less able, children. We humans would naturally think about the first of these, but seldom about the second.

We could interview people about their conscious reproductive decisions, the proximate causes. If you were interested in manipulating what people will do, this would be the appropriate level at which to ask the question.

Behavioral ecology, in contrast, seeks to discover which behaviors in which particular environments result in greater success (more about definitions of success in a moment). It starts with a bias toward "ultimate" questions, although it seldom can ignore proximate correlates. We can profit from disaggregating—teasing apart—behavior patterns in a population: who does what, under what circumstances. Consider the following: we might discover that, in a particular society, men who marry younger women have more children over their lifetimes than men who marry older women; we would not then be surprised to find a "proximate" social preference for youth in wives, nor would we be terribly surprised to find that older wealthy men in this society marry younger women more often than do poor men. The behavioral ecologist is more interested in the first question: Does marrying a young wife affect a man's lifetime reproduction? A cultural anthropologist, on the other hand, would be more interested in the ways women and men make marital decisions in this society.

Answers to both kinds of "why" questions are informative. And certainly any human society can make decisions to foster behaviors that are counterproductive in terms of ordinary natural selection, though no proximate "cause" is likely to remain common for more than a few generations if it does not serve an ultimate selective cause. For example, the Shakers are a religious group that imposes celibacy on all its members. That's certainly a cultural rule with biological implications: the few (perhaps three?) Shakers remaining today are not being replaced.

Proximate mechanisms can enrich our understanding. The usefulness of specific mechanisms depends on (1) what best predicts important changes in the external environment and (2) what internal devices already exist in the organism. When our primary concern is intervention (as in medicine or family planning), these particulars of proximate mechanisms become important. To understand the ultimate evolutionary purpose, we are more likely to study the correlations between organismal traits and environmental conditions. For example, when predictably timed periods of very cold weather alternate with food-rich moderate-temperature times, we expect trees to lose leaves (to conserve water), nectar-eating birds to migrate, and so forth.

Simple Rules, Complex Outcomes

Although humans are more complicated than other species in many ways, the exercise of asking questions in the same way about ourselves as we do about other animals may be instructive for two reasons. First, other species, like ravens, are often more complex than we realize—and we learn much from studying their behavior. Second, even as we tout our human complexity, we have sometimes offered remarkably simplistic explanations about human culture and behavior. Recent work applying the same standards of repeatability and hypothesis testing to our own behavior as we do to that of other species suggests that we can gain important new insights.[10]

My explorations here assume that humans are indeed animals, even if elegantly complex ones, and that they are therefore subject without special exemption to the general rules of natural selection, the rules that govern behavior and life history among living things. We are still exploring the gene-behavior links that are likely to be important, but it is nonetheless clear already that genes are a "currency" to be maximized in various behavioral equations. We can explore to see what we can learn about human behavioral patterns by considering genes alongside more standard currencies like status or money. The philosophy of keeping underlying assumptions as simple as possible is sometimes called "Occam's razor" in the sciences. To paraphrase Einstein, "Keep things as simple as possible, but no simpler." That is, seek the simplest model that captures the important patterns, whether the model is verbal or formal. Then, step by step, add real-world complexities to the model.[11]

Humans as Critters

Despite our cultural complexity, we humans must solve the same ecological problems as all other organisms in order to survive and to reproduce. That is, in any environment, individuals must extract sufficient resources to survive and to reproduce alone or in groups, competing and/or cooperating with others, both among our own and different species. Perhaps because of the scope of our actions, we seldom think, except in the most personal terms, about the impact of our behavior on our genetic lineage. Yet when one family lineage dies out, it is replaced by other competing lineages. This is why we can trace *Homo sapiens* all the way back to Mitochondrial Eve, for example. More colloquially, a colleague of mine tells an old story about a farmer who is thinking of buying a bull and asks the seller about the bull's potency. "Well," drawls the seller, "he comes from a long line of fertile ancestors."[12] The same is true for each of us.

I will apply to humans, at least for the purposes of generating hypotheses, the central paradigm in biology: What would it mean if humans, like other living organisms, have evolved to maximize their genetic contribution to future generations through producing offspring and assisting nondescendant relatives such as nephews, nieces, and siblings? How will the particular strategies that accomplish such maximization differ in specific ways in different environments? And, just as for other mammals, how will these strategies typically differ between the sexes?

This is a complicated endeavor, at best. We change our own environments probably more often and more completely than any other organism. Further, history contains not only "selective" events, but events that are random with regard to fitness: when Mount Vesuvius erupted, the evolutionarily fittest Pompeian died as well as the least fit. Such histories complicate our problem: most of the time, we can expect the emergence of strategies that produce, compared to other strategies, the largest increase in genetic contribution—but sometimes sheer historical chance can alter what we see.[13]

Since we typically know so little about how much effect any particular genetic locus has on any specific behavior, behavioral ecologists assume that behaviors are the product of the *interaction* of genes and environment—not the result purely of genetics *or* environment. By using the phenotypic gambit, we assume that, when we look at behavior, we are seeing the result of gene-environment interactions over time and that the most common behaviors in an environment are working well compared to available alternatives. We ask when and how environmental conditions (including social conditions) change individuals' genetic costs and benefits. If we understand how particular conditions are likely to affect behavior, and if we are cautious, we can predict the kinds of behavior we are likely to see.[14]

It is important to note that predictions are not absolute but statistical. We do not predict that genetically costly behaviors never arise, only that they will not become and remain common. Of course, there is a catch. Although we probably know more about the genes of *Homo sapiens* than of any other species except perhaps fruit flies, some yeasts, and some prokaryotes, we know the specific genetics of relatively few behaviors.[15] Historical accident can present problems as well. The Pompeian tragedy probably changed human population genetics, at least locally. Thus we can get unexpected and interesting results from simple rules and historical accident.

Despite such complications, the phenotypic gambit is a good place to start— it works in many cases, helping us simplify and clarify what we see, as well as highlighting those behaviors that are more complicated than we had thought. When it doesn't work, we have learned something valuable; when we do not find what we expect, we look for usually more complex alternative explanations. In animal behavior, this has proved to be an extremely powerful technique, as in the example of Heinrich's ravens.

The rules may be simple, but rules never operate in a vacuum, and environments are varied and complex. Both physical and social/cultural environments are major determinants of what strategies will succeed. Humans are remarkably complicated and flexible organisms, and human environments, with their elaboration of social and cultural rules, are multifaceted. There is little that tells us, in most cases, how important the various possible influences are or what the relative role of genes versus individual experience is. But if we ask about human behavior without assuming that humans are qualitatively different from other animals, perhaps we can get rid of the false dichotomy that has persisted between "biological" and "social" causes (both defined narrowly, with only proximate mechanisms considered as hypotheses).

Earlier "biological" hypotheses were typically concerned with, for example, whether sex differences could clearly be related to hormonal or brain lateralization differences. This is of interest but not related to (and not contradicting) questions about the evolution, or selective importance, of sex differences. "Biological determinism" has often been inferred from such observations that, for example, a behavior occurs in all cultures and/or a behavior occurs at a typical age. But to do this ignores the possible differences in the ecology of succeeding as a male or female mammal and simply makes assumptions based on analogy. It is surely misleading to construct a dichotomy between some sort of "biological/genetic determinism," which is assumed to be fixed and immutable, and "social" causes of behavior, which are assumed to have no correlation with genotype. Most biologists now think that *all* behavior is likely to be the result of interactions between genes and environment and that experience is important for many species, not just humans.[16]

Both verbal and quantitative behavioral ecology differ from older approaches to behavior in two crucial regards: (1) currencies to be maximized are not simply economic or social but also genetic; and (2) following from this, an individual will treat others differently on the basis of what those others can do for that individual's genetic representation, for example, treating kin and reciprocators better than others.[17] We predict some widespread biases, and we can test for them. Thus, for example, "society" is not our primary concern as we dispense social and economic largesse;[18] we typically leave our wealth to our children and nondescendant relatives unless we have none, or unless we have so much wealth that we can take care of our kin as well as endow foundations and chairs in universities (with, as a colleague noted, our family's name attached).

With all these complexities, what then does behavioral ecology suggest about a view of "human nature"? It suggests that some traditional approaches have previously ignored an important currency: genes. It suggests that we do indeed look a lot like calculators but that we are not necessarily more conscious in calculation than other species, which may forage as optimally as if they carried Hewlett-Packard calculators in their cheek pouches, for example.[19] It suggests that ecological constraints are important in setting limits to the strategies that will, and will not, "work," and that human social complexity cannot be ignored.[20]

Although genes are a currency, we seldom know the actual genetic influence on any particular trait, as I already noted. Using the approach I outline here, this lack of information need not keep us from testing hypotheses. The questions we *are* able to answer, both in modeled systems like genetic algorithms and in empirical behavioral ecological studies of many species, including humans, are these: *In any particular environment what is the success of individual variants with different traits in reproducing?* and *How strong is the parent-offspring correlation in traits?*

Genes are more important in this view as a currency to be conserved and multiplied than as behavioral dictators, because external environment, development, and genes interact in a complex way. As the geneticist Theodosius Dobzhansky once put it: "Everything in the organism's the result of the interactions of all genes, subject to the environment to which they are exposed. Inheritance is particulate, but development is unitary."[21] That is, although we conceptualize the effects of "a" gene as though it were separable, almost no gene acts alone; it is embedded along with other genes in a particular organism, which develops in a particular environment—and all this affects how the genetic influence plays out.

Many different particular internal mechanisms may be called upon to create complex behavioral responses. If we search too hard for the mechanism in each particular case, or if we ignore development and ecology, we may miss the forest as we stumble about in the trees. Consider this metaphor. The link between genes and bits of body or behavior is rather like the link between a cake recipe and the resulting cake. There is little one-to-one mapping. One cannot pick up this crumb and match it to that word or phrase in the recipe. Rather, the words of the recipe, like the genes in the chromosomes, together comprise a set of instructions for carrying out a process: development. In most cases, changing a word or phrase in the recipe will not change a particular crumb; more likely, it will subtly change the cake's characteristics. Changing "baking powder" to "yeast" will change the cake considerably, but not in a particulate way. So will a recipe "mutation" to a sharply different oven temperature.[22]

To extend the metaphor a little further, perhaps the reason we initially find this complexity confusing is that there are some well-known one-gene-one-trait correlations that seem particulate—like changing "walnuts" to "pecans"—and these have become quite famous. For example, the disease phenylketonuria arises from an individual getting two alleles for this disease at a single locus; this results in disturbed metabolism of phenylalanine (an important amino acid; the condition is diagnosed by a peculiar odor in the urine as phenylpyruvic acid is excreted) and leads to mental retardation.[23] So here is a case of a "one-word" (single locus) change that *is* particulate.

We are discovering other examples, and they make the front page of the news about once a month. But these dramatic single-gene effects can distract us from the ordinary, more subtle paths, and it is among these more subtle and complex interactive paths that I think we must look for the important links among environmental conditions, gene persistence, and observed behavior.

Hence, my focus is not on allelic specifics or precise models, but on more general problems that we're still learning to model precisely. As a classic text on behavioral genetics concluded, for the majority of behaviors studied so far, there is clear evidence of substantial genetic influence, though seldom any evidence of a really particulate single-gene-single-trait relationship.[24]

Here, I focus on questions about the current utility, in selective terms, of different behaviors, with special interest in sex differences: What advantage, or disadvantage, accrues to an individual by virtue of having this trait in this environment?[25] This isn't always simple. The *process of optimization* (in each particular population, better strategies displace and replace inferior ones and the best *available* strategies prevail) is different from the *state of optimality* (the best imaginable fit between strategy and environmental conditions). Because selection acts only on existing variants, optimization is always local and often incomplete (these variants in this environment—some prevail, some disappear); as a result, we will see variety—perhaps a lot of rather similar, pretty-good varieties, not necessarily a settled, singular strategy.

Another difficulty in asking about utility or optimality is that one is asking about tradeoffs, and the "phenotypic correlation" can hide them. For example, suppose I hypothesize that, for an individual, what is spent on housing cannot be spent on transportation. So I would expect a negative relationship between housing and transportation expenses. But when I measure, I find a positive relationship. Does that mean that no trade-off exists? No, it probably means that I have compared quite incomparable individuals. If I compare not only graduate students, or hunter-gatherers in a particular society, but instead lump them together with professors and millionaires (who have enough money to own both a mansion and a Porsche), the range of variation obscures the trade-offs.[26]

As we look at species, including our own, in a variety of environments—some quite new—we will find variety. We might find an excellent fit between trait and environment because of a long evolutionary history of unchanging selection. For example, the fact that Arctic fish can die of heat prostration at temperatures cold enough to freeze humans to death reflects for the fish a long selective history of constant cold. Or, we could see a trait that is currently advantageous, like running away from a fast-approaching truck. Since there were no trucks in the Pleistocene, the evolved rule was probably something like "run away from large, fast-approaching things," and trucks, though relatively new, fit the same general category as dangerous fast predators of the Pleistocene. Because the process of optimization is complex and few traits can be easily isolated from developmental and historical constraints, we may rarely see a really fine-tuned "fit" between any single trait and environmental conditions.[27] Finally, humans seem to me to be at least as likely as any other species to show interpopulation differences not only as a result of natural selection and adaptation (perhaps sickle-cell anemia allele frequencies) but also because of historical—and cultural-historical—events (e.g., lactase distribution in humans; see chapter

10). There are very real difficulties, but I hope defining the problem as one of current utility may help avoid some of the less useful controversy.

This approach, I hope, creates natural linkages—to empirical fields of human behavior such as anthropology, psychology, and sociology on the one hand, and to ecology, genetics, behavioral genetics, and population theory on the other. Scholars in each of these (and other) fields have information and perspective on constraints and on how to consider human behavior. The "current utility" approach links us to what we know about other species as well.

We know something about the ecological and genetic components of behavior and about what behaviors become common under what circumstances, but our knowledge is still unconnected across disciplines; my work daily leads me to conversations with colleagues with whom I have more shared interests than shared knowledge. Now we need to reach across disciplines, and I hope experts in other fields will read this not as a postulated expert disquisition in their field but as an invitation to contribute what they know to solving the questions I raise.

2

Racing the Red Queen

SELFISH GENES AND THEIR STRATEGIES

Now here, you see, it takes all the running you can get to do, to keep in the same place. If you want to get somewhere else, you must run at least twice as fast as that!
—The Red Queen in Lewis Carroll's *Through the Looking Glass*

ALICE HAD SOME TROUBLE following the Red Queen's logic: that one has to run as fast as one can just to stay in place because everything else in the landscape is running as well. Biologists, however, find the image an apt one. Consider Matt Ridley's engaging book on the origins of sexual selection, which he chose to call *The Red Queen* in recognition of the problem that the sexes continually change each other's costs and benefits. In a way, much of biology is a record of such selective arms races.[1] Ecology is rife with examples: if faster rabbits escape coyotes, tomorrow's rabbits are faster than today's—but once this is true, fast rabbits put pressure on coyotes, so that faster or sneakier coyotes become the only successful ones. For us humans, our families, friends, and rivals are forces to be reckoned with. Such social selection (table 9.1) is surely a good example of the Red Queen's problem: the goal you seek is situated in a moving landscape, and it may always be moving away from you.

At the core of behavioral ecology rests the notion that Richard Dawkins aptly called the "selfish gene," the idea that genes that get themselves copied into more and more individuals will be the genes that prevail and persist through time. This measure of success is a modern version of the simple logic first employed explicitly by Darwin.[2] Genes compete for locations on the chromosome, and groups of genes make what the biologist Leigh Van Valen called a "parliament" (they interact to produce complex effects); individuals housing effective parliaments survive and reproduce relatively better than those who don't.

At some level, this competition among genes is like the old adage, "I don't have to outrun the bear; I only have to outrun you";[3] that is, no guarantees exist that the chance events of mutation, recombination, and drift, combined with the filtering of natural selection by the environment, have generated the best possible combinations. It is only true that, at any moment, relatively more effective combinations do better than others. Individuals thus never represent the "best conceivable" combinations of genes plus environment, only the "currently most effective"—which may be superb or less than wonderful and likely to disappear if a better alternative appears.

Although the idea of selfish genes is simple, great complexity results. On the one hand, genes can affect more than one trait; on the other, groups of genes can cooperatively affect a single trait. Some genes regulate the activity of other genes; some switch others on and off. Genes are carried about by individuals, yet genes in one individual may affect the success of genes in other individuals. Some information goes from generation to generation through the cytoplasm, not the genes, and in humans and in some other species, information is transferred across individuals through culture (chapter 9). Individuals differ genetically, and they live in varied physical and social environments. Thus, while the rules are simple and individual strategies for survival and reproduction are all-important, the analysis of either may be complicated.

We know a great deal about the evolution and ecology of resource use in other species: the costs and benefits; the impact of various environmental conditions; the evolution of sex differences.[4] Genes, history, and environment interact, but the basic patterns are clear. Recent empirical tests suggest that the relative power of chance, selection, and history can differ under various circumstances. Nonetheless, for traits strongly correlated with fitness, even when chance is great in the environment, natural selection is still powerful.[5]

If we humans, like other species, evolved simply to get resources and to survive long enough to get duplicates of our genes into the next generation, why are we so complicated about the process? Even in our life history (chapter 6), we humans are unusual in the pace of our maturation, growth, and reproduction. Socially, we not only live in families (common in many species) but also cluster together in villages, cities, and heterogeneous nations. We ally ourselves with one another in more complicated ways than most species, and our groups are based on more than simple reciprocity. We have society-wide recognized third-party roles (police, legal systems, realtors). We have formal trade agreements and schooling, art galleries, transportation networks, and so on.

This complexity and diversity seem a far cry from any simple set of strategies. Yet despite our complexity, this simple observation is true: those of us alive today are the descendants of those who successfully survived and reproduced in past environments. Historical accidents can happen, but the rules still are true. This means that genetically profitable behaviors, those that enhance an individual's total genetic representation, are always favored by ordinary natural selection.

Why doesn't the favoring of genetically selfish behavior always result in bloody outright battles? The short answer is (1) sometimes it does, (2) but when it does not, it is because the costs of attempting bloody battles outweigh the potential gains. Rules operate under environmental pressures. While all living things have evolved to acquire and use resources to survive and reproduce, the ways they do so are constrained by ecological conditions. Individuals that use the most effective and efficient resource strategies in any particular environment are those that tend to survive and reproduce; but there is no reason to suspect that what works in the desert will work in the river, or that what works among small kin-based societies will work in nation-states.

Further, fertility is complex. Although a simplistic interpretation might imply that the best strategy is to produce as many offspring as physically possible as soon as possible, this is seldom, in fact, a winning strategy, even for relatively nonsocial animals. In some environments, only "superkids" survive and reproduce at all (chapter 15); the result is that *fertility responds to the cost of parental investment to make successful offspring*, so that high per-child investment correlates with small families.[6] Life-history theory is, in fact, largely the study of tradeoffs: size against number of offspring in nonhumans, for example. Finally, in social animals, other individuals create some of the most important environmental pressures, and rampant short-term self-interest will often fail (more on this in later chapters). Clearly, rich diversity is likely.

The starting point is that, other things being equal, individuals that use efficient strategies produce more successful offspring for the next generation than their competitors. But other things are often *not* equal. Are we sure that we are measuring costs and benefits correctly? At first glance, many behaviors appear counterproductive. For example, infanticide occurs commonly in many species, including lions, ground squirrels, and a number of primates.[7] To determine whether infanticide is an evolutionary "mistake" or an effective strategy, we must ask, Who commits infanticide, and under what circumstances? Who profits from it? Most infanticide is committed by reproductive competitors; its evolutionary logic seems clear. Sometimes, however, parents kill their own children. How shall we view this? Surely parental infanticide is an evolutionary mistake that decreases, rather than increases, reproductive success? Indeed, evolutionary mistakes are possible. But across species, except for rare pathologies, infanticide is found under specific ecological and social circumstances, and its impact in these circumstances is greater lineage success for the killer—even when the killer is a parent.[8]

How can this be? In species in which a successful offspring requires considerable parental effort, there are circumstances in which terminating a particular investment pays off—for example, a deformed offspring or a mother in poor health or with no support system. In these circumstances, parents win who discriminate by investing more in offspring that are more likely to be successful, and less in weak or deformed offspring—even in the extreme case of infanticide. Other similar puzzles of apparently maladaptive, yet common, behaviors include lethal conflict (when does it pay to risk getting killed?), delayed reproduction (how can waiting to reproduce increase one's reproductive success?), and sterility (the ultimate in nonreproduction).[9] Consider honeybees, in which all females except the queen are sterile. How can a (female) worker bee's genetic representation possibly be increased by remaining sterile and devoting her life to caring for the queen's, her sister's, eggs? Darwin worried how a trait like sterility could be inherited. He understood that honeybee colonies were somehow special but left the solution for the future researchers.[10]

Perhaps the most blatant examples of hard-to-explain phenomena fall under the heading Darwin called "sexual selection." He understood that anything that helped you survive would be "favored," but unless you also reproduced, that trait would disappear. Part of the reason Darwin treated sexual selection separately from "ordinary" natural selection was that successful strategies in so many species were also often dangerous, life-threatening strategies, which were usually associated with male-male competition. Darwin struggled to understand how such behaviors could be favored by selection. It took much observation to determine that (1) sometimes the most effective thing you can do is take a huge risk, and (2) in a sexual species, reproducing means that you must face both the competitors of your own sex and the preferences of the opposite sex you wish to win. So, Darwin argued, others of your species could "select" just as effectively as the horse or cattle breeders who change the genetics of future racehorses and milk cows by selecting who gets to breed. Sex differences, how they arose, and how and when they are maintained are at the heart of my exploration, and I will return to these themes repeatedly.

WHOSE GENES COUNT, AND WHY? KIN SELECTION

Though the concept of selfish genes is a simple one, it has been repeatedly misinterpreted, just like Darwin's original formulation. How horrible: "Nature red in tooth and claw"! How could we possibly believe such a noxious idea, when we can see generosity all about us in many species. It remained for the biologist W. D. Hamilton to quantify and formalize some of the most important genetic costs and benefits of behavior—and his formula, though simple, is to much of biology what $E = mc^2$ is to physics.

The first general rule is that a behavior will become common only if its genetic benefits outweigh its costs—if $b > c$. Hamilton pointed out that, since "nondescendant" relatives of any individual ("ego"), such as nieces, nephews, and siblings, share genes with ego, helping these relatives (even if it has some direct cost to ego) can help ego's genes. To be favored, there simply must be a net genetic benefit. Relatives share only some genes with the helper, and this varies with r, the degree of relatedness; your full sister has more genes (higher r) in common with you than with your fifth cousin. So the benefit must be discounted, and only the help that goes to identical genes counts. Hamilton pointed out that "giving" behavior should evolve whenever (in Alan Grafen's reformulation)

$rb > c$; that is, whenever the benefit to the recipient b times the degree of relatedness r is greater than the cost to the doer—for example, whenever the benefit to one's sibling, who shares on average one-half of one's genes, exceeds twice the cost of one's act, or when the benefit to each of two siblings exceeds the cost of the act.[11] Notice that this also means that not all help to kin will be favored: if the cost to self exceeds the (benefit × relatedness of kin), the behavior should disappear. And sometimes being positively mean to one's kin pays—when $b_{self} > rc_{rel}$ (simply reorganizing the above equation).

This concept had been recognized informally for some time. The mathematical biologist J.B.S. Haldane had noted that, while he would not give his life to save his brother (who would, on average, share half his genes), he would die for two brothers or for eight cousins.[12] Hamilton applied this concept of *inclusive fitness maximization*, or *kin selection*, to Darwin's puzzling problem of sterile castes in social insects (surely it is a great sacrifice not to reproduce and to spend one's life caring for a relative's offspring). He could show why Darwin was correct in maintaining that the self-sacrificial behavior persisted only because it occurred in the familial context and that it would not persist in other sorts of groups.[13]

The point is that a behavior can be genetically profitable even if, to the casual observer, it appears to be costly and of benefit to others (table 2.1). This distinc-

Table 2.1. Categorization of the Impacts of Behaviors on Phenotypic (Apparent) and Genetic Condition.

Behavior	Apparent Effect	Genotypic Effect
Overt competition	PROFITABLE} ("selfish")	PROFITABLE ("selfish")
Parenting, nepotism, reciprocity	COSTLY ("altruistic")	PROFITABLE ("selfish")
BECAUSE NATURAL SELECTION FAVORS ONLY GENETICALLY PROFITABLE BEHAVIORS, UNDER NATURAL SELECTION, BEHAVIORS ABOVE THE box SHOULD BE COMMON; BELOW THE box, RARE.		
??	PROFITABLE ("selfish")	COSTLY ("altruistic")
Mother Teresa?	COSTLY ("altruistic")	COSTLY ("altruistic")

Source: Modified from Alexander 1974.
Notes: If we only look at superficial, apparent ("phenotypic") impact, we miss crucial differences.
?? = this category is so very rare that I have trouble imagining a noncontroversial example: Perhaps a rich miser (phenotypically selfish, since he is a miser) who disinherits his family, leaving an anonymous gift to a home for unwed mothers (genetically altruistic, since he hurts his relatives in order to help genes not identical by descent). See chapters 9, 10 for further exploration.

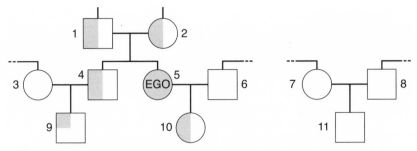

Figure 2.1. Hamilton's (1964) theory of inclusive fitness maximization suggests that behavior will evolve as a result of its costs and benefits not only directly to the doer, but also to those to whom the behavior is directed, depending on the degree of relatedness. Helping behavior will be favored as it helps more genes identical by descent, and less favored as its effects are diluted by genes not identical; thus the proportions of shared genes are important to calculate. Here, Ego (5) receives half her genes from her father (1) and half from her mother (2). Her daughter (10) receives half her genes from Ego. Because her brother (4) is a full sibling, with the same mother and father, he will, on average, be genetically half identical to Ego; his daughter (9) will be exactly half like him, and on average, one-quarter like Ego. Individuals 3 (brother's wife), and 8, and 11 (relatives of Ego's mate) share no genes identical by descent. (From Williams 1992a.)

tion between what seems to be true and what a behavior does for relative genetic representation is a thorny one, but not one we can ignore. Measuring both costs and benefits correctly is crucial.

How many genes we are likely to share with another individual (r, the degree of relatedness) is one key to how much we will profit by helping. As humans, we are likely to have, on average, half of our genes in common with a sister or brother. All adult humans have two alleles at each locus, or location, on every chromosome (with some special conditions for the X and Y sex chromosomes); these two alleles can be duplicates (homozygous; perhaps both are the allele we call a), or different (heterozygous, perhaps a and A). We received one set from our father, one set from our mother. Mother's egg and father's sperm are produced in a lotterylike process called "meiosis"; each resultant egg or sperm has only one set of chromosomes. An egg or sperm has some chance of getting either allele at any location. Suppose my mother is heterozygous and has aA at the locus I am interested in. My sister and I might both receive an a or A from Mom, or one of us might have a and the other A. In the first case, we would be 100% alike at that locus through Mom; in the second, we would be zero alike. This pattern is true for each genetic location, and true for alleles from our mother and our father. As a result, I share on average one-half of my genes with my full sister, and one-fourth of my genes with her child (fig. 2.1). The same is true for my full brothers. Your own child, of course, shares exactly half your genes.

Hamilton suggested that striking phenomena follow from the proportion of genes identical by descent. To demonstrate, he used the extreme case of social insects such as bees and wasps, which have the peculiar genetic arrangement (haplodiploidy) in which mothers produce sons by laying unfertilized eggs. Because males have no father and get only one set of alleles from their mother, they have and pass on exactly the same genetic material in each sperm. So in honeybees, for example, full sisters share identical genes through their father. They can never be less than one-half alike if they have the same father, and they are more closely related to each other on average (3/4) than they are to their mother (1/2). Sterile female workers, Hamilton suggested, were not paying, but gaining, genetically by raising their three-fourths-alike sisters rather than half-alike daughters.[14]

Hamilton's statement of the theory of kin selection was general and not dependent on the odd case of haplodiploidy (e.g., termites and naked mole rats are also eusocial, with one breeding "queen," who is fertile, and many workers, who are sterile (sometimes through pheromonal suppression by the queen). Queen bees occasionally mate multiple times, so not all daughters are full sisters. It is an important part of the general theory of natural selection. Hamilton's summary hypothesis makes a strong, testable prediction: "The social behavior of a species evolves in such a way that in each distinct behavior-evoking situation the individual will seem to value his neighbor's fitness against his own according to the coefficients of relationship appropriate to that situation."[15]

Thus, other things being equal, we expect individuals to treat their kin more gently than strangers and to treat close kin more gently than distant kin—and a large body of empirical data upholds this expectation.[16] Even though we expect no organism to be able to calculate its relatedness to others, we expect them to *act* as though they could. In fact, of course, what we see is that organisms treat those with whom they grow up differently from others, because under most circumstances in most species, individuals grow up among relatives, not strangers (a good proximate cue). Hamilton described an individual's strategy for making effective use of proximate cues of relatedness.[17]

In many species this proximate cue of nearness works because the individuals with whom one grows up typically *are* one's relatives. This is why, for example, a researcher can fool ground squirrel mothers into adopting unrelated babies before her own children emerge from the nest to forage above ground, but not after, when there is opportunity for mothers to make costly mistakes, because then youngsters occasionally blunder down the wrong burrow. Research on a wide variety of species has produced a wealth of empirical data that support Hamilton's prediction.[18] Each individual has reproductive interests, but these interests are shared to a predictable degree by others who also share common genes—genes identical by descent. Overtly selfish behavior, "nature red in tooth and claw," is thus only one route to enhanced genetic representation (table 2.1, fig. 2.1). Individually risky or energetically costly behaviors and helping one's kin can also be genetically profitable. We expect organisms, including humans, frequently to engage in activities that benefit relatives. Further, highly selfish behaviors will not be genetically profitable in the longer run if they harm relatives too much.[19] Human history has a wealth of examples: the impact of family on death rates in crisis situations, alliances and internecine warfare in Icelandic and English history, and others.[20]

How much relatives profit from helping or harming one another will depend on the degree of relatedness as well as on the costs and benefits of the act. Helping relatives, even at some cost to oneself—and helping friends who will reciprocate, for example—can be genetically profitable, and many behaviors that *appear* to cost the performer are actually profitable in terms of genetic representation (table 2.1). Here, it is the actual effect on reproduction and genes that we care about. Reciprocal coalitions can be complex and quite elaborate in humans (think of politics and of warfare), and they affect men and women differently; I will return to them in chapters 10–14.

Summing Up the Basics: Assumptions and Objections

The forces of selection, including kin selection, explain much of the variation in the behavior of individuals in other species. To ignore them completely when it comes to humans would be absurd. Imagine how much credence you would give to me if I told you that gravity exists, making all other animals behave in certain ways (walking on the ground, being able to glide only a short distance, or expending considerable effort to fly), and then went on to explain that humans, on the other hand, are exempt because we have cultural phenomena such as airplanes. We have more complexity and more variation in *how* we can travel on the ground or in the water, and because of our cultural transmission and elaboration (including inventing and sharing technology), we are not bound to the ground; we can clearly fly in ways unknown among other species.[21] And when we fly, it is because of our cultural innovations and personal desires, not because we "need" to. Nonetheless, you would probably still bet that gravity influences all of our travel. While we can circumvent gravity in sophisticated ways, it never disappears as a force to be reckoned with.

This is not the place for a lengthy review of behavioral basics; some excellent reviews already exist.[22] But because the behavioral outcomes I explore here involve not only rules but also environments and history, I want to be explicit about what I am and am not suggesting about our human sex differences. Behavioral ecology asks whether, if we know about environmental conditions (including social conditions) and about how those conditions change actors' (genetic) costs and benefits, we can predict the kinds of behavior we are likely to see. To explore this possibility requires certain assumptions. These include the following:

1. Organisms are generally well suited to the environments in which they live; they achieve success in any environment by getting resources that enhance their survival and reproduction. The strategies we see have had time to compete against

other strategies, and "what you see is what you get"—it is what has worked best in the past (the phenotypic gambit).

2. Only heritable variation is appropriately considered in testing predictions about changes in gene frequencies over time. An individual can assist its genes to spread by (a) reproduction, the most direct method; or (b) assisting individuals carrying copies, identical by descent, of its genes (kin selection); or (c) helping individuals that do not carry identical copies of its genes if such assistance is returned in genetically effective ways (reciprocity). This implies that individuals who help reproductive competitors without any reciprocation will lose descendant representation in existing lineages. In short, people like them will cease to exist except perhaps for an occasional mutation.

3. Organisms that are more efficient in getting resources in any environment will survive and reproduce better than others. In the evolutionary history of all species, there have been important proximate correlates of reproductive success, including resources (food, territory), rank (status or power), wealth, and, in highly social species such as humans, social reputation.

4. Further, no organism, including humans, has evolved to perceive or assess directly the spread of genes; rather, organisms behave as though the proximate correlates were their goal. Thus, species may find themselves in novel evolutionary environments, and individuals' behavior may be currently maladaptive; when this is true, the assumption of the phenotypic gambit is violated.

5. In their "deep" objectives—in what they evolved to do—humans are not *qualitatively* different from other living organisms. Like other living things, they evolved to get and use resources to survive and enhance the spread of their genes. To evolutionary biologists, this is parallel to the argument above that humans, while they can make airplanes and fly, for example, are still subject to the laws of gravity. Yet because we are highly intelligent and highly social, humans are the likeliest of species to be in novel environments, making it a complex task to make assumption 4 and to distinguish evolutionary history from current utility.[23]

Perhaps because the study of human social and sexual behavior has in the past "belonged" to fields in which only humans were studied, this is a novel approach. A behavioral ecologist would answer the query "How do I love thee? Let me count the ways" far less poetically than Elizabeth Barrett Browning. For several reasons this approach may not be palatable to all. It may also be controversial; certain assumptions are simply not accepted by everyone. Many think, for example, that it is inappropriate to use the same general principles to examine human behavior as to study the behavior of other species, even to test hypotheses; and many feel that not all behaviors can be reasonably analyzed.

A widely held discomfort with any evolutionary approach to human behavior is the one reflected in the following (perhaps apocryphal) story, which I first heard attributed to Bishop Wilberforce's wife. When Darwin presented his theory of natural selection, hers was a typical response as she leaned over to say to her friend, "My dear, have you heard Mr. Darwin's theory that we are all descended from apes? Let us hope that this not be true; and, if true, let us hope that it not become generally known."[24] It is easy to agree with the bishop's wife, or to fear that knowledge of behavioral ecology will lead us to make bad or cynical policy. This confusion of "is" and "ought" is sometimes called the "naturalistic fallacy." Because evolution is simply genetic change over time and thus amoral, we need analytic rather than normative approaches. Evolution has no bearing on human moral decisions (expect, perhaps, to predict why some exhortations will be more easily followed than others). To observe that something is true does not constitute a moral endorsement.[25]

Human complexity and flexibility raise special concerns. Simply because a trait looks useful, we cannot assume that it evolved through the action of selection. As George Williams put it long ago, it is an onerous task to infer adaptation. It is important to articulate the proposition one is using to generate hypotheses and to state's one's assumptions—and find clearly relevant data.[26]

In relatively stable environments, we expect selective pressures to remain similar over time. Thus, for example, constant cold leads to stumpy limbs and cold-tolerant physiology or to the Arctic fish with very narrow temperature tolerance. This does not mean that any particular observation can offer strong support; the camel's hump, for example, might have come about for nonselective reasons, because of natural selection, or due to some combination; and humans are a particularly difficult entity to study. But if we accumulate a series of a priori predictions, they can grow into a significant number of directional results. Special insulating hair and feathers are found in numerous species—but only in Arctic and Antarctic environments; this suggests selection. For complex phenomena in complex species, such as social and sexual behavior in humans, this teasing apart is an issue to which we must pay careful attention.

What about anecdotes and exceptions, what a colleague of mine calls the Bongo Bongo argument? "This is how the world works, and you can see I'm right, for among the Bongo Bongo they . . ." Or the reverse: "Your view of how the world works is obviously wrong, for among the Bongo Bongo, they are absolutely altruistic to everyone." Or: "Gravity cannot be a natural force, for birds and humans can fly." Hardly persuasive. Behavioral ecological predictions are statistical, arguing that at any moment, behaviors that, in a particular environment, get genes passed on effectively will spread compared to other behaviors and that (employing the phenotypic gambit, statistics, and modeling) we are likely to be able to detect this trend. Thus, if we find that the Bongo Bongos are truly genetically altruistic, this means only that the Bongo Bongo are likely to decline over time, to be replaced by competing peoples—and this is a testable prediction.

All of the arguments in this book, in the context of natural selection, are statistical propositions that, other things being equal, individuals with certain traits will be, or will become, more common than competitors in particular environments. None is a statement of absolutes. Consider: Many people would deem Mother Theresa a genetic altruist, helping nonrelatives for no genetic payback. If I argue that genetic selfishness is favored, I am not suggesting that we will never see or hear of a Mother Theresa—only that, over time, her genetic lineage will likely decrease compared to others, and she and others who are *genetically* altruistic will remain a rarity. We are making predictions about what a statistician would call "central tendencies" rather than rare exceptions.

Novel Evolutionary Environments: Can the Principles Still Hold?

A very large question remains: How far can we usefully explore our current behavior? Environments that are new and novel in an evolutionary sense introduce significant complication. After all, it might be easy to see how reciprocity and discrimination are favored in people living in small bands, interacting daily with the same few people, and protecting some resource against outsiders. But today our societies are large and complex. We may interact with literally hundreds of people; we do business daily with people we have never seen before; we have information about what happened today across the world to strangers we may never see. History also complicates our problem: we know of many nonselective accidents in our history that are equal to the destruction of Pompeii.

Evolutionary novelty poses a great difficulty for studying human behavior; through our cleverness, we constantly create environments that are novel in selectively important ways for ourselves and other species.[27] The behaviors that helped hunter-gatherers in the savanna may not be useful in the suburbs of Chicago (chapter 15). But we combine study of, say, urbanites, with the study of people in traditional societies, our best window to our past.

Novel evolutionary events influence the behavior and demography of other animals as well as humans. For example, a male chimpanzee in a well-studied group gained dominance status by banging together empty metal containers instead of the more traditional branches. In Great Britain, great tits (*Parus major*) began to feed out of milk bottles, which were evolutionarily novel, although the bird's probing behaviors had evolved to forage on bark and twigs and their digestive systems certainly did not evolve to deal with milk. On the East Coast of the United States, gulls, which evolved as generalist feeders, showed marked increases in population density as a result of an increase in the number of garbage dumps, whereas other seabirds declined in abundance due to gull predation. In experiments with monkeys, when two individuals were asked to do a task but one was offered a rich reward and the other a less desirable reward, the less-rewarded monkey was likely to refuse; and monkeys can also show the "endowment effect"—refusing to give up something after they had gotten it, unless they got a much larger payment for it than they had spent to get it (humans do this also).

Now we are changing the environment rapidly (chapter 15). If global climate change continues, the water in some areas will become warmer, and whether any fish in a particular species still retain the genes for warmth tolerance is a toss-up. For some Arctic fish, decline, even extinction, is likely, because over many past fish generations, cold tolerance was profitable and the ability to tolerate warmth was not (and likely had some cost), so a narrow cold-tolerance evolved; why pay the cost of developing heat tolerance if you never, ever, encounter heat? In contrast, a coral species (*Acropora hyacinthus*) is doing well off the Samoan coast as waters warm (up to 95°F); such tolerance to great changes typically reflects great variance in the organism's past environment.[28]

The impact of novel environments is heightened by the fact that no organism, including humans, has evolved to be mindful of ultimate selective effects; their awareness is only of proximate cues. Selection acts so that what enhances our survivorship or reproduction—forming friendships, having sex—tends to be perceived as pleasurable; and acts that typically detract from our survivorship or reproduction—for example, getting burned—are unpleasant or painful. But this relationship can change when the environment changes.

Consider a simple example of novelty. In nature, sweet foods are seldom harmful, and sour and bitter tastes are often correlated with the presence of harmful alkaloids. Thus a preference for sweet tastes (a good proximate cue to nutrient-rich, safe food) became widespread in omnivores, including humans. In most past environments, it was difficult to obtain enough sugar to create problems of obesity. Once we humans invented technologies for refining and concentrating sugar, we created foods that had enormous concentrations of sugar, breaking the selective link between sweet taste, the proximate cue, and a good food source that had previously led to enhanced survivorship and reproduction. But proximate cues drive the system, and selection simply acts as a passive sieve. So we retain a preference for the sweet taste that can make us fat and fill our teeth with cavities.[29]

Because we humans can modify our environments so extensively, and because our cultural transmission can respond more quickly than genetic intergenerational transmission, we are frequently in novel environments. It is surely fair to ask how far we can really expect to see selectively advantageous behavior in our current environments. As an example, consider the following: In other species and in preindustrial human societies for which we have data, males who have more resources typically have more offspring (usually because they have more mates) than others (chapters 4, 7, 8). Now, in much of the world, effective contraception has broken the link between resource accumulation and fertility (the Pill is so extraordinary that *The Economist* included it in a list of the Seven Modern Wonders of the World). For the first time in all of history, men and

women could be on equal sexual footing; sex could be without parental consequences for either. Certainly this is a novel environment. But we humans, like
other species, operate on proximate cues, not any awareness of selection; the existence of the Pill, breaking the link between sexual pleasure and parenthood,
doesn't mean that sex is no longer fun. So do wealthier men today, as in the past,
have more children than others? As I explain in chapter 15, often not, because of
novelties such as birth control. But they do have more sex, if they want it.[30]

MORE THAN ANTS OR PEACOCKS: LIFETIMES, CULTURE, ECOLOGY, AND VARIATION

My explanation of the basics of selfish genes does not yet come close to being
useful in looking at human behavior in all its diversity and complexity. Three
important ingredients remain to join the play: the social effects of sexual reproduction (chapters 3–5) and how these play out in human lifetimes (chapter 6);
the impacts of external environmental influences (chapters 7, 8); and the intense pressures of group living (social influences), which elaborate sociality as
a force beyond anything we see in other species (chapters 9, 10).

Three phenomena—kin selection, reciprocity (and its relatives: chapter 9),
and sexual selection, or how we interact with family, friends, and mates—lie at
the heart of why we behave as we do in many circumstances. Phenotypic plasticity, the ability to be different in different environments, even with the same
genes, facilitates this. The basics give us a perspective on complex phenomena,
and I will try to weave together the themes of the ecology of resource consumption and sex differences in different ecological situations. My central task here
is to ask why we behave as we do, especially about resource issues, and why the
sexes differ so consistently in some areas and not at all in others.

I began by assuming that we humans share some constraints with other animals (and plants, too, for that matter): we must get resources to survive and
reproduce; parents and offspring are more like each other than like strangers;
and what is effective in one environment won't necessarily work in another.
Although only heritable variation is important, in complex social animals like
ourselves, cultural transmission is one kind of heritability, and interaction between cultural and genetic transmission can certainly complicate analysis (chapters 9, 10).

The advantage to this approach is that we can frame questions in ways that
have some rigor and repeatability. That is, we can go beyond convenient just-so
stories, such as "How the Camel Got Its Hump" or what causes vampires. It's
not easy. In its crudest formulation, natural selection theory sounds like a circular argument (what works, works, so if you see it, it must be working).[31] Of
course, so do the principles of physics. We also are tackling complex and inter-
related phenomena. The trick is to figure out for *which* traits we wish to predict

the direction, along with *which* environmental forces we predict will favor this versus that version of the trait. Then we can predict the direction before we go out to measure it: "If the environment is A, then version X of our trait should increase over time; if the environment is B, version Y should increase." Then, if we see what we expect, we have support (but not proof) that we are likely to be at least partially right; and we know that if we see something else (e.g., version Y runs riot in environment A), we must be wrong or else there is a factor missing from our analysis.

The philosopher Helena Cronin developed the apt imagery of the ant and the peacock to suggest how kin selection (ants) and sexual selection (peacocks) influence the lives of living things.[32] I wish to explore here not only the similar ways in which selection has acted on us (to what extent we are all ants and peacocks), but also the specific ways in which conditions in the external environment are good predictors of differences in the ways men and women approach resources. Finally, I will discuss the impacts of today's evolutionarily novel environment.

3

The Ecology of Sex Differences

Of those who were born as men, all that were cowardly and spent their life in wrongdoing were transformed at the second birth into women. . . .

Such is the origin of women, and of all that is female.
—Plato

We acknowledge a biological difference between men and women, but in and of itself this difference does not imply an oppressive relation between the sexes.

The battle of the sexes is not biological.
—Editorial Collective, *Questions féministes* (1977)

It is theory that determines what we can see.
—Albert Einstein

A FAVORITE CARTOON of mine shows two deer, a buck and a pretty annoyed-looking doe standing on a hillside. The buck is tilting his head, saying, "So I like rutting—so sue me." In the genetic gambling casino, success depends not only on individual strengths and weaknesses but also on environmental conditions and whether or not there are groups to contend with. The buck, however, highlights an influence on all sexual species. From fish to flying squirrels, from Hanuman langurs to humans, males and females of most species experience different costs and benefits in reproducing, and these differences influence both lifetimes and social behavior. Professor Higgins's plaintive cry could be universal among males of sexual species (as could the converse: why can't males be more like females?).

Sexual reproduction seems something of a puzzle: it means a loss in genetic representation, because half your offspring's genes are identical to someone else's, not to yours. So, why is there sex? is an important question.[1] Sexual reproduction is in fact common among vertebrates, suggesting that the loss of genetic representation is compensated for—but how? Most current hypotheses argue in some form that unpredictably changing conditions may make it so valuable for an individual to produce variable offspring that the genetic cost becomes worthwhile. For example, W. D. Hamilton argued that, in the face of rapidly evolving parasites, producing offspring exactly like yourself, as in asexual or some parthenogenetic cases (even if you are maximally fit and resistant), is futile and costly, because your parasites, having many generations during your lifetime, can always evolve new strategies faster than you can respond, so offspring identical to you will be vulnerable.

Sexual reproduction is far more diverse than you would think from looking at humans: In many turtles, crocodilians, and some fish, an individual's sex is determined by environmental conditions (primarily temperature, sometimes pollutants). There are multisex species (like the lowly slime mold), female-only species (some fish and lizards), bisexual species (many plants, snails), environmentally induced sexual or asexual species (aphids), and environmentally induced sex changers like the blue-headed wrasse (*Thalassoma bifasciatum*). In this coral-reef fish, most individuals begin life as a female. Large females make more eggs than small females, but even small females can make some eggs and have no trouble getting them fertilized. Male-male competition is severe and risky; only very large males get mates, but they are highly successful—so size has a very different impact on probable reproductive success for males versus females.[2] If one could choose, and blue-headed wrasses in some sense can, it would be reproductively more profitable to be a female when one is small and to become a male only when one is very large and highly competitive. So in blue-headed wrasses, the sequence of sexes is not fixed but triggered by changes in the social environment.

How do big males come to exist? Within any group, if an experimenter acts like a predator and removes the large male, the largest female switches sex, becoming a bright, yellow-and-violet-colored male, in a process that takes about two weeks. If that male dies or if an experimenter again removes it, the next-largest female switches to become a male. And if two females are close in size, they may jockey back and forth, changing and rechanging sex until one of them wins the battle to become the supermale. Sex is not genetically, but, rather, socially and ecologically determined in this species. A very few individuals follow an alternate strategy: they are born male and remain male, though they are forever small and inconspicuous and must follow a sneaker's strategy, darting in to deposit sperm when a large male courts a female. These sneaker males are rare, for their success is low.[3]

Here I want to explore, for sexual species, the interplay between sexual strategies and environmental conditions, to begin asking when and how the optimal strategies of males and females differ—the ecology of sex differences, if you will.

Sex and Strategies

At the heart of all life history strategies lies a single problem: How can effort (calories and time spent, risks taken) be spent optimally? An organism faces a central constraint: what is spent for one purpose cannot be spent for something else; effort is a constant-sum game. An organism can spend effort on maintaining its body, or it may spend effort in reproducing: getting a mate or parenting offspring. The relative payoffs for these competing activities differ with age, competitors, external conditions—with many things. The central problem facing an individual is how best to spend effort to replicate its genes by allocating effort sufficiently well enough to survive and reproduce relatively better than its competitors. This is the key to being relatively "fit." Surviving and reproducing was the original meaning of "fitness" as used by Darwin, though there is a plethora of additional uses today.[4] To analyze the contribution to fitness of particular strategies, such as age at maturity in particular environments, we test specific predictions about what behaviors should succeed best in specific environments.

First we must back up and ask a series of questions: Why is there sexual reproduction in the first place?[5] Why does the number of sexes almost always reduce to two, not three or more? The answer is not immediately obvious. Some slime molds, for example, have about a dozen "sexes." There are also single-sex, all-female (parthenogenetic) species. In some of these, like whiptail lizards (*Cnemidophorous*), some females behave like males in mating, though no sperm or eggs are exchanged.[6] In other species, like *Poeciliopsis* fish, females mate with males of other species. They use the sperm to start the physical process of egg development and then throw out the male genetic material. Some species, such as the water flea *Daphnia,* reproduce parthenogenetically so long as the environment is stable; females produce diploid daughter eggs without mating. When conditions begin to deteriorate (e.g., as when temporary ponds dry up in temperate latitudes), the females, again without mating, produce diploid daughters and haploid males (from unfertilized eggs). These mate and thus sexually produce daughters—who are physically different from their parthenogenetically produced mothers and are ready to overwinter.[7]

There are also plants and animals in which each individual has male and female parts: they are the hermaphrodites. Most of these do not "self"—they mate with another hermaphrodite, producing more variable offspring than they could by selfing. These observations lend support to the general argument that the evolution of sexual reproduction is the creation of variable offspring in unpredictably changing environments.[8] In some hermaphroditic species, individuals are first one sex and then the other (small maple trees are male: pollen is cheaper than seeds; blue-headed wrasses are female first). In other species (usually low-density, relatively sessile species), individuals may be both sexes simultaneously.

Even though blue-headed wrasses are unusual in changing sex, they are typically in another way. Like most sexual species, there are only males and females—two sexes. One clue lies in the observation that reproducing in gametic sexual species requires two quite different sorts of effort: getting a mate (mating effort: striving to gain resources or status, attracting a mate) and raising healthy offspring (parental effort: e.g., feeding, protecting, and teaching offspring).

Imagine a population of something like jellyfish floating in the ocean and reproducing by releasing into the sea haploid gametes, each carrying half the adult number of chromosomes. These can combine with gametes from other individuals to make zygotes—new offspring. This is a basic and simple form of sexual reproduction. Since in this example any gamete can recombine with any other gamete, even "sibling" gametes from the same individual, there is nothing so specialized as eggs or sperm. But the scene is set for a conflict of interest that becomes important: to make a zygote, a gamete requires genes from another gamete. Genes are in the nucleus, but cells also have some genetic instructions in the cytoplasm (the rest of the cell). And while the gamete needs the genes (in the nucleus) from another gamete, it is better off without the extra instructions (in the cytoplasm) from the other gamete. I'll return to this in a moment.

First, imagine that there is variation in the size of gametes released: from very small to extremely large. Remember the two tasks to be accomplished in contributing to a successful zygote: hitching up with another gamete and making a sturdy, well-endowed zygote. Simple physical laws lead us to conclude that the tiny gametes, those cheapest to produce and most likely to move farthest in currents, will have the greatest chance of bumping into other gametes. The largest gametes have resources to live longer and also have the most to contribute to producing the best-endowed zygotes. Over time, because the smallest and largest gametes are favored for the two distinct tasks and middle-sized gametes lose out to both, the system reduces to individuals that specialize in making either large or small gametes (fig. 3.1).

The only advantages to a small gamete are that it travels faster and is energetically cheap; the interests of small gametes are promoted by behaviors like traveling far and searching for another gamete. The only advantage to a large gamete is its contribution to a healthy, well-endowed, competitive zygote, and any risks, such as those of roaming far, may be counter to its interests. These gamete traits come to influence the behavior of the entire organism. So typically, it is more profitable for a single individual to make—and promote the success of—only one of the two gamete types. This pattern, *anisogamy* (unlike gametes), is so ubiquitous that, without thinking about it, we tend to call small gametes "sperm" and small-gamete makers "males" and to call large gametes "eggs," and large-gamete makers "females."[9]

Any behavior by the gamete carrier—the organism—that enhances the advantage of either the small gamete or the large gamete is likely to be favored. If one carries small gametes (whose only advantage is meeting other gametes),

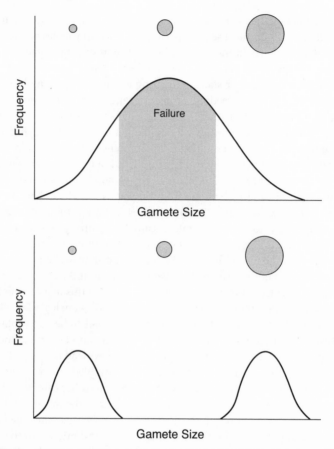

Figure 3.1. Two feats are required for a gamete to contribute to a successful zygote: it must find another gamete, and the resulting zygote must be well endowed and healthy. Small gametes do the first task well; large gametes are best at the second. Middle-sized gametes lose out to both extremes. This disruptive selection leads to anisogamy (unlike gametes) and a bimodal size distribution: small gametes (sperm) and large gametes (eggs). See text for further discussion and for the role of cytoplasmic conflict in development of anisogamy.

then traveling far and wide, spreading those sperm about, will be a better strategy than sitting alone in a safe place, meeting only others who do travel. And so we associate sets of behavioral characteristics with each of the sexes: The risk-taking travelers we call males; the risk-averse nurturers, females.[10] And we're almost always right—the number of sex-role-reversed species is minuscule.

This observation may help us puzzle out why males and females diverge so strikingly in size, weapons, and behavior in so many species like deer, seals, and sea lions[11] and what conditions lead to the sexes behaving similarly in species like Canada geese (*Branta canadensis*). Within the single species *Homo sapiens*, the variety of behaviors between the sexes is extraordinary, rivaling the variety in the European bird, the dunnock (*Prunella modularis*). Perhaps stepping back and looking first at the general rules of sexual reproduction in other species will help us to tease out why the general pattern exists and what conditions make exceptions profitable.[12]

Once specialization into small- or large-gamete making exists, any subsequent changes in *either* the gamete *or* the carrier that enhance these specific advantages will be favored by natural selection. That is, if ever a proto-sperm is produced that is not round but ovoid (hydrodynamically superior) and has a tail, it will be favored in the race to the egg. Proto-sperm-making individuals who travel farther, thus finding other gamete carriers over a larger area, will be favored. Proto-egg makers who invest less and less in other functions, and more in big healthy future offspring, will be favored. Thus, starting from the simple physics of gametes bumping into other gametes, a strong bias toward differentiation is built up. Both physical and behavioral characteristics are affected, for the benefits and costs of searching and endowing differ. The rarity of simultaneous hermaphroditism in animals is related to this phenomenon, for it is hard to search and to endow, both maximally, at the same time.[13]

The nuclear genes in the gametes had to fight for their spot—in meiosis, the competition is among genes within the individual. But in gamete-sexual species like humans that produce eggs and sperm, the cytoplasm in the gamete also brings along information and "interests," and there is still the problem that each gamete would be better off not having to deal with the other's cytoplasmic material. Much of our understanding of the species that do have more than two sexes arises from the biologist Lawrence Hurst's examination of this cytoplasmic "war of all against all."[14] If a zygote is formed by the fusion of two gametes, there can exist a strong conflict of interest (mother's versus father's genes and cytoplasmic material).

Typically, the egg (or egg-producer) has mechanisms to keep out the cytoplasmic material from the sperm.[15] In contrast, when sex consists not of fusion between two gametes but of conjugation—transferring just a nucleus across a "pipe" between two exchanging cells—there is no conflict between the cytoplasmic materials, and there can be any number of sexes or mating types.[16] There is even a "hypotrich" ciliate, a microbe, that has both sorts of reproduction; in fusion sex, it behaves as if it had only two sexes, but in conjugation sex, it has many. The slime molds with thirteen mating types have a complicated hierarchical sort of fusion sex. Sex 13 always contributes the organelles, whomever it mates with. Sex 12 contributes the organelles if it mates with 11, 10, etc., but not if it mates with 13. And so on.

While both conjugation sex and gametic sex are options for very simple species, complex multicellular species (including humans, of course), are stuck

Figure 3.2. (A) Mating effort has a large "fixed cost," compared to parental effort: much effort must be spent before any success is achieved. Parental effort "levels off" much earlier than mating effort. The arrow highlights the region of additional expenditure required to match the initial payoff for mating effort compared to parental effort (although, once that threshold is reached, additional matings may cost very little). (B) As a result, risk taking is more profitable for mating-effort specialists (usually male), who show lower survivorship than parental-effort specialists.

with sex from the fusion of gametes and with cytoplasmic warfare. Separate sexes, numbering two, are an effective way to get genes repackaged and to resolve some of the conflict between parents' genes. One answer, then, to Professor Higgins's sad cry, "Why can't a woman be more like a man?" is that once there are two sexes, each having different paths to success in reproduction, the strategies that work for each are likely to be very different.

These behavioral specializations, into mating (seeking) versus parental (nurturing) effort, have profound consequences for trends in behavioral differences between the sexes.[17] Mating and parental effort have very different patterns in the ways individuals profit from them ("reproductive payoff curves"; fig. 3.2A). Mating effort, which is typical of males in most species, has what economists call a large "fixed cost"; that is, much effort must be invested to get any return whatsoever, but after some level, great additional gains come from just a little more investment. So specializing in pure mating effort has great impact on a male's life. A male must grow large if terrestrial physical combat is any part of competition; he may have to range far; perhaps he must grow weapons or decorations, like a moose or a peacock; he may have to fight—all these have costs, and many increase risk. A red deer male, even to attempt a first mating, must grow large, involving both energy and an opportunity cost of delayed maturation, for what is invested into fighting for status cannot be put into growth, and if size matters for success, a male who switches too soon from growth to conflict will remain forever small and unsuccessful. He must grow antlers, and he must fight for dominance and control of good feeding grounds. All of this is required just to break into the mating game. After this great initial investment, though, a mating-specialist male's payoff curve rises steeply, for additional matings cost little.

For parental investors (usually the sex we call female—the large-gamete makers), the starting condition is that each offspring costs approximately as much parental effort as any other, though there can be complications (below). Getting a mate is almost never a problem, though selecting a good one may be. What is invested in one offspring can seldom be recycled and reinvested in another; for example, although nests may represent generalizable parental effort (reused for several clutches), feeding is true *parental investment* and is offspring specific—what this baby eats, that one can't. In contrast, what parents do that can be used for a number of offspring without loss is called *parental effort.*[18] In many species, the maximum possible number of successful offspring is likely to be lower for females than for males, but males will vary more in their success, and more males will fail entirely to reproduce.

THE ECOLOGY OF BEING MALE AND FEMALE

The "male-female phenomenon" is pervasive.[19] As we have seen, beginning with simple rules of physics (those jellyfish gametes floating in the sea), consistent differences arise in the costs and benefits of mating versus parental effort. This means that there are likely to be consistent differences between males and females: in behavior, and in how much success varies.

For both sexes in many mammals, five traits seem to contribute to reproductive success: age, body size, dominance rank, early development patterns (high early growth rate), and quality of mates chosen. For females, these typically contribute to getting good nutrition and converting it for offspring. When males take risks in direct mating competition, they are likely to die sooner than females (fig. 3.2B).[20] Thus, for males, lifetime breeding success is most variable in species with direct conflict over mating access, whether by prolonged defense of territories, as in elephant seals, or in competition over single females, as in some butterflies.[21] Male lifetime breeding success tends to be least variable in species in which males compete indirectly, as do men in many human societies.

The roles of intrasexual conflict and resource striving have been thoroughly studied in red deer and elephant seals, and these species reflect the general mammalian picture well.[22] In elephant seals, males compete physically; they grow to be much larger than females, but this means that they suffer the attendant cost of maturing later, and maturing late means more exposure to possible death before any chance of reproducing. Males compete for the control of sandy beaches on which females give birth. More than 80% of all male elephant seals ever born die without reproducing, but a highly successful male may have over ninety offspring in his life. Reproductive success varies greatly among males, whether it is measured seasonally or across lifetimes.

Female elephant seals, too, suffer great variation in success; approximately 60% of females ever born die without giving birth. Most of these are killed as youngsters before they mature; unlike males, almost no adult females fail to reproduce. The most successful females each have about ten offspring in their lifetimes. The reproductive stakes are much greater for males than females because so many males fail and variance in reproductive success is so high. In a species in which males spend only mating effort, and no parental effort, a male who does not compete fiercely gets nothing: so males take many risks and fight intensely. Seal pups are often injured or killed as males fight. Because females return to the same or nearby beaches every year but male territory owners change, the probability that an infant killed is the offspring of the territorial male is difficult to determine.

In red deer, a few males control harems of from one to more than twenty females, or hinds, while most remain "bachelors," lurking nearby with no females. Harem holders do virtually all of the breeding with the hinds in their harems. However, because controlling a harem is so expensive, the tenure of harem holders is typically shorter than the breeding season, and male reproductive success does not vary as much as harem size. Variation in reproductive success among stags is a function of harem size, duration of harem-holding tenure, rutting area, fighting ability, and (less so) life span. Stags fight to get harems and matings. Big, long-lived stags who are good fighters have the most offspring. Because most females have one calf each year, their reproductive variation is small; it depends on female life span and calf mortality.

Thus, for stags, size and ability to gain dominance are crucial; for hinds, keeping oneself and one's calf alive is important. The reproductive return for effort expended by the two sexes is quite different (fig. 3.2A) and follows the general pattern: male red deer reproductive success varies more than female reproductive success, and male competition is more direct and physically riskier than female competition. Both poorly and well-invested hinds can be successful in producing offspring, although their condition does matter subtly. Male calves that are born early in the season (with a long time to grow before their first winter) and at a high birth weight are more successful than those born "late and light." Not surprisingly, the sons of dominant hinds, which are in good condition, are more likely to be born heavy and early.

Sons of dominant hinds have greater reproductive success than daughters of dominant hinds; daughters of lower-status hinds have greater reproductive success than sons of low-dominance hinds. (This pattern has interesting parallels in some human societies, in which the degree of polygyny and wealth are related to relative investment in sons versus daughters; see chapters 6, 7.) Further, dominant hinds produce more sons than do low-status hinds. Dominance in hinds is related to their own birth weight and their weight as adults. Resources still matter for hinds' reproductive success, even though the impact is less than that on stags and may be seen more strongly, after a generation's lag, in the sons' success. Not only genetic endowment, but also social and historical factors, influence what happens to any individual.[23]

Roamers and Homers

The relatively simple physics of "getting there" versus "investing" at the gamete stage means that the most profitable investment strategies of males and females are likely to differ. It shouldn't surprise us that the bearers of little gametes— whose only advantage, remember, was getting gametes into warm, safe places— might be predicted to roam about more. In polygynous species with no male parental care, males do simply spend most of their reproductive effort in searching for mates.

A variety of interesting consequences follow, highlighted by comparing males and females in two different but closely related species of voles: small, blunt-nosed, small-eared meadow mice. *Microtus pennsylvanicus*, the meadow vole, is a typical polygynous rodent. Male *M. pennsylvanicus*, who roam more widely than females as they seek mates, have much better spatial abilities than females, who are risk-averse and stay much closer to home. If you look at their brain structure, the brain parts devoted to spatial ability and memory are larger in male than female *M. pennsylvanicus*. On the other hand, both male and female *M. pinetorum*, a monogamous close relative, have a brain structure and spatial abilities similar to those of female meadow voles.[24] The sex differences arise out of the ecology of reproducing successfully in monogamous (more parental males) versus polygynous (more "searching" males) systems.

Given what we know about the polygynous background of humans, you'd suspect sex differences in human spatial ability as well, so long as men and women have differed in their use of space. Indeed, that is what we see, both in traditional societies in which people have been tested or interviewed and in modern tests of various spatial abilities.[25] Men and women seem to be particularly good at different kinds of spatial tasks: in modern studies men outscore women on mental rotation tasks, while women outscore men on remembering the location of objects. Perhaps men's mental-rotation, map-reading skills are related to their history of hunting (more on the sexual division of labor in chapter 7); women, as gatherers, might profit more from noticing and tracking the location of small things—landmarks—in the environment. In experiments involving looking at pictures or sitting in rooms and being required to recall items they had seen, women outscored men.

The basic trends toward sex differences lead to a central and pervasive life-history tradeoff. How is it most profitable to allocate reproductive effort: as high-risk, high-gain mating effort, or as offspring-specific, true parental investment? There is a real conflict between what an individual does to be successful in each. Although ecological conditions can change the relative benefits of specialization versus combining mating and parental effort and there is really a "family" of curves in figure 3.2A, the "default" condition is a specialization by each sex into either mating or parental effort.

In mammals, these tradeoffs are set into sharp relief, for females, having mammae, are specialized to nurse offspring. This specialization of females for nutritional investment means that males typically can profit by specializing in mating effort rather than parental effort. Within each specialization, however, there is still diversity in how an individual can spend either mating or parental effort.

MATING EFFORT

Mating-specialist males have a very limited number of possible strategies to secure mates: they can try to control females, control or gather resources useful to females, or display for females independent of resources. Females can expend mating effort, but because females produce the larger gametes with higher offspring-specific costs, males are more likely, in most environments, to profit from mating effort than females. This is especially true for mammals (see above). Each of these mating strategies depends on the abundance, predictability, and defensibility of resources and gives rise to a different mating system.[26]

Resource Control

In many species, males that can gather, commandeer, or sequester resources that are useful to females have a distinct reproductive advantage. When such resources are reliably—predictably—abundant *and* economically defensible, males will attempt to control territories and exclude other males. Thus, male elephant seals come to shore before females and compete to control the few sandy patches along the rocky coast, where females must come ashore to give birth; red deer stags control good feeding grounds, excluding other males.[27] In nonhuman species, territoriality is likely only when resources useful to females are both reliable and economically defensible.

Human systems are far more complex. Even considering only "territory," for example, humans have lineage-held lands that are inherited. We also have recognized property rights. That is, we humans have third-party interests and interventions, so that we see not only simple individual or coalitional territory defense, but situations in which intruders are punished by uninvolved third parties (judges, juries) for invading territories accepted by those same third parties as reasonable.

Land and its contents are only one sort of resource. Status (dominance), in other species, and status and wealth in humans are very real resources (chapters 4, 7, 8). Finally, there are non-resource-based mating criteria: "sense of humor," and "considerateness" (chapter 5).[28] Underlying this diversity, it is nonetheless true that men who control resources have a reproductive advantage in most societies.

Harems and Mate Guarding

Harems, which are of great interest to most of the men I know, are fairly common in other mammals, and their form is influenced by ecological conditions. When females tend to remain in groups (usually to avoid predators) but useful resources are not reliable or not economically defensible, males may attempt to control females' movements, and thus the access of other males to females. Unsuccessful males (often addolescent bachelor males) cruise in groups near the harem of a successful male and may try to entice females away.

Men are polygynous in many human societies, having more than one mate (whether a marital partner or not). In these societies, guarding and controlling access to women is a large problem for men. In some famous large societies, harem control is elaborated, with eunuchs to protect the harem and severe punishments for transgressing males. In other species, a male controls a harem as large as he can, alone or (rarely) with a reciprocator male. In human history, there have been caliphs and emperors with literally thousands of concubines, who were maintained with the assistance of men who had been castrated well before they reached an age to make such a choice themselves.[29] Human harems can therefore be larger and their maintenance more complex than those in other species.

By now it should be no surprise that reproductive matters in many societies are treated as absolutely core and that legal measures to discourage sexual transgressions are among the harshest of punishments. Here's an example from the Ashanti about the punishment for a man who sought favors from a king's wife (the woman was beheaded; more about the Ashanti in chapter 12).

> The culprit, through whose cheeks a sepow knife has already been thrust, is taken. The nasal septum is now pierced, and through the aperture is threaded a thorny creeper by which he is led about. Four other sepow knives are now thrust through various parts of his body, care being taken not to press them so deeply as to wound any vital spot. He is now led by the rope creeper to Akyeremade, where the chief of that stool would scrape his leg, facetiously remarking as he did so, "I am scraping perfume for my wives." Next to the house of the chief of Asafo, where his left ear is cut off; thence to Bantama where the Ashanti generalissimo scrapes bare the right shin bone.[30]

Then he was made to dance all day. After dark, his arms were cut off at the elbows, and his legs at the knee. He was ordered to continue dancing, but since he couldn't, his buttock flesh was cut off and he was set on a pile of gunpowder, which was then set alight. Eventually, the chief gave permission to cut off the offender's head. Thus, although my male colleagues who dream of harems always imagine being the harem master, a little consideration to the individual impacts of reproductive variance is in order!

Tending Bonds

Tending bonds are a form of short-term mate guarding: a male identifies a receptive female and remains with her during her fertile period, chasing rival males away in a temporary "consort" relationship. Then he leaves.[31] The pattern we see, of timing and length of the consortship, suggests that the reproductive point is simply for the male to deny other males any access to that female until his sperm have successfully fertilized her.

In nonhuman vertebrates, tending bonds occur in specific ecological conditions. When females can raise offspring successfully alone and there is safe heavy cover (or females are quite widely dispersed), females tend to be solitary; males can seek out females with relatively little harassment from other males. When the terrain is very open, large groups are common, and tending males usually lead the female away from the group for mating during her fertile period (typically a male's dominance status is crucial in his success in this process).

Humans, of course, are extraordinary in that females can maintain continual receptivity, and ovulation is quite difficult to detect. Indeed, some scholars suggest that the evolutionary point of this peculiar situation is to make it difficult for males to get away with such temporary alliances. So, for a variety of reasons, human males may profit from rather steady and constant association with one or more females (longer-term mate guarding) rather than brief "tending bond" periods.[32]

Humans are unusual in another way. So far, the story of mating sounds simple: males seek matings and take risks doing so; females are coy and hard to convince. Not only are there diverse mating systems in humans (more in next chapter), but how much it "pays" males to seek to be polygynous can vary widely. We seem not to be entirely bound by Bateman's principles (that males vary more than females in both mating and reproductive success and that mating and reproductive success are more strongly related in males than females). Transnationally today, these relationships vary across human populations.[33]

Leks

Leks are typically systems in which males defend reproductively useless resources, often with considerable display and at great risk.[34] Leks may look like territorial systems, but nothing of reproductive value is defended. The profitability of such a strategy is not immediately obvious. Generally, if males can control females or resources that females need, they have a stronger bid in the mating game; if neither resources nor females are controllable but appropriate staging areas exist, they may display in leks. By lekking, they simply advertise their "quality"—how well they have survived, what great risks they can take by displaying, how strong their sons might be.

It is important that the males are making unbluffably costly advertisements. When males have nothing to offer but their genes, females assess male ability to perform costly feats, which is a result of both a males' genes and his (e.g., nutritional) history. Males may advertise costly looks (bright colors are expensive to make and maintain and visible to predators) or dangerous calling behavior, depending on the environment. For example, among grouse, prairie chickens (*Tympanuchus cupido*) display visually, whereas forest-dwelling ruffed grouse (*Bonasa umbellus*) display where sight is limited, "drum" on a perch, and perform a short flight in which they make sharp snapping sounds with their wings.

Females visit the lek, assess male quality, and choose a male. Male reproductive success can be extremely skewed, for females typically choose to mate only with those males able to control the central, hardest-to-defend territories. There is an unsolved problem with the maintenance of such systems. Imagine an extreme situation, in which all the females in a generation chose a single male. The sons born to these females vary in their abilities; where does this variation arise? Since all sons carry genes from the same father, even if Dad is heterozygous little variation comes from this source; more comes from differences among genes from their various mothers or Mom's nutritional condition or environmental conditions while growing. It would seem that what females can use to choose (current male ability) is imperfectly related to male genetic quality; the formal arguments make this clear. The question remains, how are such systems maintained?[35] One likely explanation, introduced above, is the parasite version of the Red Queen hypothesis: parasites with particular characteristics exert pressures that mediate selection and polymorphism in the host population. More recently, suggestions have been made that some gene-environment interactions render no single male's strategy best overall, and that the shifting territories of smaller, peripheral males may sometimes gain them a central location temporarily.[36]

Scramble Competition

When neither resources nor females can be controlled economically and no good display grounds exist, males may display to females independent of either real or symbolic resources. This is most likely when females gather in predictable, somewhat centralized places (e.g., mayflies over appropriate water for laying their eggs); then males also congregate and display. This strategy is, in some sense, the weakest from a male's point of view.

And Humans?

Human mating effort can have elements of all of these kinds of competition, but, of course, some are more common than others, and what's effective may differ during a man's lifetime. Boy gangs on corners are a form of "bachelor groups" lurking about, hoping to score. Our variation of "tending bonds," in which males temporarily sequester females away from the attention of other males, is common, usually among males who have little but genes to offer.[37] Perhaps resource-based polygyny is the most common pattern cross-culturally; as we will see in chapter 7, in quite diverse societies males with higher status or wealth manage to have more mates than others.[38] Some strategies are likelier than others at different times in one's life; teenage males often behave in what looks like scramble competition to a biologist, and sometimes I can't help visualizing gang territories—and perhaps discos and soirees among the elite?—as kinds of leks.

PARENTAL EFFORT

After mating, both sexes face the problem of whether and how best to invest in their offspring. In other species, specialization by each sex into either mating or parental effort is most common, but human males, by and large, do invest parentally in one form or another. Parental expenditure or effort has obvious costs to mating effort: what is spent irretrievably on offspring is unavailable for mate-seeking.

As Darwin noted: "The only check to a continued augmentation of fertility in each organism seems to be either the expenditure of more power and the greater risks run by the parents that produce a more numerous progeny, or the contingency of very numerous eggs and young being produced of smaller size, or less vigorous, or subsequently not so well nurtured." [39] He understood the trade-offs well.

Thus, in addition to the evolutionary pressure to optimize investment in getting mates versus raising offspring, organisms have to solve the problem of how best to raise their offspring. Parental investment, you will remember, is the amount of effort invested in any offspring that would otherwise be available for investing in other offspring. From the point of view of the particular offspring, this is a most appropriate concept, although it may be difficult to measure. From the parents' point of view, problems of investment must be considered over more than one clutch or litter—in fact, over a lifetime of parenting. Parental effort is the sum of all parental investment over any defined period.[40] It can include not only offspring-specific true parental investment (like nursing), but also "reusable" effort, such as building a large nest, or digging a den—parental care that can be used for many offspring.

Spending parental effort effectively is not a simple problem. Parental effort can be invested once (*semelparity*, single birth) or more than once (*iteroparity*). Whether "once" or "more than once" is most effective depends on the parent's extrinsic chances of survival; what are its chances, independent of reproducing, of surviving for another attempt? If those chances are very low, or quite uncertain, an individual may do best to devote everything, even dying in the attempt, to reproducing this time.[41] Across species there is great variety. Atlantic salmon (*Salmo salar*) return to their spawning grounds from the ocean, breed, go back to the sea, and return repeatedly. A really successful adult can return many times. Pacific salmon (*Oncorhynchus* sp.), whose chances of living another year are far lower (large marine predators), if they manage to return once, exhaust themselves in breeding.

Perhaps the most curious cases are the species in which one sex lives to reproduce several times, while the other sex of the same species dies after reproducing once. In one octopus, females use themselves up protecting offspring, whereas males mate as many times as they can. In contrast, there is a marsupial mouse in Australia in which the male dies after copulation, while the female is iteroparous. And, of course, there are a number of insects (praying mantises, some crickets) in which the male dies or is killed by the female, and his body is used to feed either the female or the offspring.

Obviously, parental effort can be expended on male or female offspring, and the profits from, and costs of, male versus female offspring can differ, so it is not a trivial life history decision (see chapter 6).[42] Parental effort can be expended as physical biomass or as behavioral investment, but that represents only the first strategic layer. Within effort spent as biomass, there is still the problem of optimizing the size of each offspring versus number of offspring. Parental effort can be spent as behavioral care (feeding, teaching, guarding) or in other ways (e.g., biochemical defense). In sexual organisms, because of the evolution of large eggs and small hydrodynamic sperm, there is an overall trend toward greater parental expenditure by females than males.

VARIANCE IN REPRODUCTIVE SUCCESS: MATING VERSUS PARENTAL STRATEGISTS

There is a clear bias then: the carriers of small, "finder" gametes will profit by behaving in ways that enhance the finding capabilities of the small gametes; the carriers of big, "nurturing" gametes will profit by enhancing nurturance and seeking safety.[43] Males can be parental, however, and in nonhuman species there are important ecological correlates to great male parental effort. One common form of male true parental investment is feeding, as in many songbirds. Another is offspring protection, either with the female (as in geese) or providing care while the mother recovers (e.g., several arctic-breeding sandpipers).[44] In poison-arrow frogs, the female lays the eggs, and the male carries them about until metamorphosis in his vocal pouch; since his skin is full of curare, both he and the eggs are safe from predation.

But male parental care can be a very mixed bag for all involved. Obviously we do not expect expensive male parental care if a male's paternity is uncertain; so mate guarding typically accompanies male parental care. Further, some forms of male parental care may really function as mating effort. Particularly in some primates, for example, a male may be much more likely to care for a juvenile when he may reap matings from the mother by doing so. And the relative worth of guarding versus caring versus mating effort is often frequency dependent and not obvious. The bottom line is that male mating competition may have much stronger influence in shaping male strategies than in parenting payoffs, and even when males appear to be exerting parental effort, they well may be spending mating effort.[45]

The story doesn't end there, however; if that were all there were to it, all species would be polygynous, and mating systems would not vary. In fact, there are further ecological correlates of polygynous, monogamous, and polyandrous mating systems. An important aspect of mating systems for biologists is the relative *variance in reproductive success* (fig. 3.3).[46] In mammals, especially, since females are equipped to feed dependent offspring, there is a bias toward polygyny. In polygynous systems, because only a few males typically ever get a mate, great expenditure and risk taking may be worthwhile. As a result, in polygynous systems more males die than females at most ages (fig. 3.2B). Being a male is usually a high-risk, high-gain strategy. For a female, making a son is thus also typically a costly high-risk, high-gain strategy.[47]

Rarely can males gain sufficient reproductive success from parental care to compensate for the lost reproductive success of forfeited matings, so we see many polygynous systems either with no male parental care or with "generalizable" male parental effort. For instance, male redwinged blackbirds (*Agelaius phoeniceus*) keep watch for predators, protecting several mates and their nests; and males in a number of primate species (including humans) act as mentors and protectors for their offspring (chapter 4). But both groups are polygynous.

Genetic monogamy is defined by biologists as a system in which variance in male and female reproductive success is equal and there are no extra-pair copulations. Such monogamy is typically associated with phenomena such as complex learned behavior, as in many vertebrate predators, or highly competitive environments faced by offspring (so that only superbly invested, costly offspring can succeed). At issue is whether offspring can profit enough from male parental care (of the offspring-specific sort, like feeding) that it "pays" a male genetically to give that care. And whenever a male can either spend paternal investment in generalizable ways (so he is free to seek more matings) or disguise mating effort as paternal effort, he is likely to do so. Thus, many systems are socially monogamous but genetically polygynous.[48] The reproductive efforts of males and females may converge in monogamous systems, compared to, say,

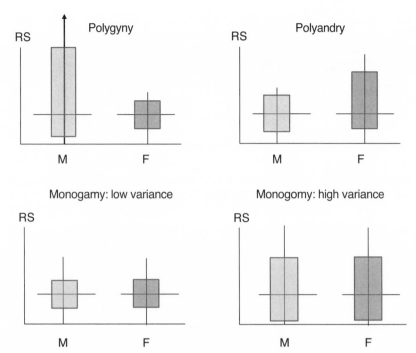

Figure 3.3. The average reproductive success (RS) of all males in a generation must equal the average RS of all females—but the *variance* in RS may differ greatly between the sexes. In polygynous species in which males specialize in mating effort and suffer high failure and mortality rates, the variance in RS typically is greater in males than in females. In polyandrous species, the variance in RS is typically greater for females than for males, but probably never reaches the extremes seen in polygyny because of the cost of the egg. In genetically monogamous systems the variances will be the same, but as you see, the variance may be high or low; the definition simply requires that variances in RS be equal for males and females, whether the variance is high or low.

harem polygynous systems, but those interests are still identical only in restricted circumstances.

Humans exhibit all known mating systems.[49] Although we call many human societies monogamous, this is clearly a biological misnomer; most are socially, but not genetically, monogamous. Most societies with one-spouse-at-a-time rules are genetically polygynous: more men than women fail to marry, and more men than women remarry after death or divorce, producing families in these later unions. The most reproductive men have many more children than the most fertile women. Thus, variance in male reproductive success is higher than variance in females'. Nonetheless, the reality may vary, at least in humans, with population or culture. In analyzing serial social monogamy among the Pimbwe of Tanzania, Monique Borgerhoff Mulder found evidence that it functioned not only as "hidden" polygyny, but also as polyandry, even though females remained the main caregivers. Perhaps polyandry is not so rare as we once believed.[50]

We clearly need to look further at such examples, even though typical polyandrous mating systems are both rare and ecological curiosities. In these systems, females are likely to spend more mating effort than males, while males specialize in parental care. Polyandrous systems are relatively more common among birds compared to mammals, perhaps because in bird species with high parental care, the males, in contrast to most mammals, can feed offspring as well as females. Because egg biomass is true parental investment that cannot be reused for another offspring or for mating, and because there is some size threshold for success in eggs, females in polyandrous systems probably never achieve the number of offspring (and thus the skewed variance) of males in many polygynous systems. But we do see the expected trend: the eggs of monogamous bird species tend to be larger and less numerous than the eggs of closely related polyandrous species.

Defining mating systems by the relative variance in male versus female reproductive success can be useful. For one thing, it allows us to calculate the relative intensity of sexual selection on the two sexes in any system. It is important to remember that it is the *relative* variance in reproductive success of males and females that is important, rather than the absolute level of either variance. Genetically monogamous systems can show high variance (fig. 3.3), for example, if many offspring die—but the variance is still similar for the two sexes.[51]

In the next three chapters, I will look first at how the interaction of ecological constraints and these trends toward differences play out in primates, and then at how resource conflicts of interest develop and persist through human lifetimes.

© The New Yorker Collection 1993 Bruce Eric Kaplan from cartoonbank.com. All Rights Reserved.

4

Sex, Status, and Reproduction among the Apes

Power is the ultimate aphrodisiac.
—Henry Kissinger

I don't know a single head of state who hasn't yielded to some kind of carnal temptation, small or large. That in itself is reason to govern.
—François Mitterand

W E MUST REMEMBER that humans are primates, not deer or seals. Like other primates, we are complicated and diverse; ecological, social, and historical conditions can all contribute to the patterns we observe. Yet there are real regularities to how these forces interact, and sometimes we can make better sense of complex, apparently eclectic happenings by using the selection lens Consider witchcraft trials (more in chapter 10), during which Katherine Harrison was first immune from accusations, then targeted for death as a witch, then became safe once again through a powerful male accuser's turnabout to protector. Self-interest and power differentials clearly had influence. But the basic relationship between resources and reproduction is also complicated by ecological constraints on how striving can occur. Let us explore further the impacts of male and female resource and reproductive value in different environments.

Most primates, like elephant seals, red deer, and other mammals, are polygynous; males specialize in mating effort (though, as we shall see, there can be considerable male parental and alloparental care), and females specialize in parental effort.[1] This has important effects on male-versus-female payoffs for striving and risk taking. In red deer (*Cervus elaphus*) and elephant seals (*Mirounga angustirosis*), male striving for dominance is central to succeeding; strategy matters, but pure size and strength are still very important. Primate species show a diversity of social arrangements, and males' and females' costs and benefits vary with the social arrangements or, more specifically, the breeding system. While strength isn't irrelevant to a male's success in most primates, intelligence and social skills also appear to be important in gaining status and mates.

Here I want to focus on two main questions: How does the (spatial and temporal) distribution of important resources affect male and female reproductive options? How do the resulting mating systems vary?

THE ECOLOGY OF DOMINANCE AND
REPRODUCTIVE SUCCESS IN PRIMATES

Henry Kissinger, quoted at the beginning of this chapter, fairly described the mating market among most primates, including, arguably, humans, though there are many complexities. Even in nonhuman primates, the relationship of dominance to reproductive success is complex. External environmental conditions (e.g., cover, food distribution, predation risk) influence the sort of living arrangements that succeed, and the living arrangements influence male and female strategies, similarities, and differences. The strategies of the two sexes further interact, and for both sexes there can be tradeoffs. Under all the variation, in most species there is a simple bottom line: reliably, higher-status males get more copulations and/or more offspring than other males.[2]

Simple individual dominance in male-male competition doesn't explain what we see. Alternate male strategies include furtive copulations, coalitions between males, affiliative relationships with females, and intrusions by outsider males.[3] Female choice can matter also. Sometimes outsider males join harems during the females' receptive periods and appear to mate without constraint. It may pay males to seek attention and reciprocity from females, even when females are not sexually receptive, for in many primates females can influence male success through mate choice, through "special relationships," and by supporting particular males in social interactions.[4] In some species, such as Japanese (*Macaca fuscata*) and Barbary macaques (*Macaca sylvanicus*), high-ranking males clearly get more matings than others. But the number of copulations is not necessarily the issue; in many primates, females (like human females, but less dramatically so) may mate when they are not actually fertile.[5] Male success comes not simply from maximizing copulations, but from mating with females during their very short periods of actual fecundability; this access appears to be commandeered by the dominant male(s) in most studied species. Interestingly, however, in savannah baboons (*Papio cyanocephalus*), females mate more broadly than simply with the very dominant males; it was almost as if the "queue" for a dominant was too long. It would be interesting to know just what correlates with very narrow, versus wider, female choice.[6]

In most species, the intensity of competition is related to the number of males in a group: the more male competitors are present, the more intensely males must compete. This intensity is further influenced by the payoff: the number of fecundable females in the group. The competition works as we would guess: genetic markers allowing accurate paternity assessment typically (though not always) show that rank and sexual access are positively correlated for male primates.[7]

The relationship between rank and reproduction went unrecognized for years, perhaps because interactions in many primates can be subtle.[8] Consider chimpanzees: status is complicated. There is a formal dominance hierarchy, which is established through aggressive encounters; once the matter is settled, subordinates display submissive greetings to dominants, and dominants show friendly behavior toward subordinates. Acceptance behaviors lower tension in the group, and cooperative relationships exist. "Real" dominance in any particular situation does not always reflect formal rank, and individuals may achieve power in particular situations without challenging the formal hierarchy; primatologists today recognize the power aspects of dominance as "power" (fighting potential, alliance partners) and "social power" (implicit agreement among group members about whether an individual is capable of introducing force into certain social situations). So, for example, female chimps sometimes calmly appropriate food or resting spots from males. Males sometimes negotiate access to estrous females, avoiding overt aggression (but remember that females may be receptive but not fecundable).[9]

Patterns differ among populations of chimpanzees (*Pan troglodytes*) in reproductive matters as well as in many other aspects: for example, in the degree of "possessiveness" by the alpha male, the proportion of matings that appear to be opportunistic, and the role of females in initiating sexual encounters. However, in most studies, "rank hath its privileges" for male chimpanzees.[10] Among bonobos, or pygmy chimpanzees (*Pan paniscus*), females form alliances, and high-ranking females, at least, appear to be able to make choices among males without restriction. Male aggression appears milder and more quickly reconciled than in chimpanzees, possibly because females are gregarious and receptive longer. Male dominance rank still affects mating chances, though, and male-male competition is more intense than was previously thought.[11]

In polygynous New World primates, as among the great apes, most studies find that high-ranking males have an advantage in sexual access: golden lion tamarins (*Leontopithecus rosalia*), some capuchins, red howler monkeys (*Alouatta juara*), Costa Rican squirrel monkeys (*Saimiri oerstedii*), spider monkeys (genus *Ateles*), and more.[12]

What about females? Female primates are physiologically limited in the number of offspring they can produce, and the relationship between females' rank and their reproduction is perhaps less clear than male patterns.[13] Not only growth, but other individual characteristics, such as reproductive rate, offspring survival, and longevity, contribute to female success; very young and very old females tend to have lower reproductive rates. The relative importance of individual factors may vary, but (depending on the species) reproductive longevity and infant survival appear particularly important. Female rank, particularly in cercopithecines, can affect offspring size, growth rate, and age at

maturation, birth rates and interbirth intervals. External factors like food availability, influences on infant survival such as predation and infanticide, and
group size (which in some species is negatively correlated with reproductive
rate) can all be important.[14]

Female striving carries costs in at least some primates: in some populations,
very high status female savannah baboons, who compete to retain their status,
suffer some fertility costs.[15] High-ranking females can show characteristics we
would associate with stress in humans; they suffer more miscarriages and fertility problems than other females. These are unevenly distributed; some high-
ranking females are extremely successful, others have no offspring. If one ex-
cludes the no-offspring females, there is a positive relationship between rank
and lifetime reproduction. In analyzing rank and fertility of female primates,
we have a lot to learn.[16]

Ecological Aspects of Mating Systems

Whenever females can raise offspring successfully alone, the likely outcome is
polygyny, with female specialization in parental effort and male specialization
in mating effort. However, ecological conditions influence solitary-versus-
group living, the kind of group, and the mating system. About half of primate
species live in multifemale groups. Avoiding predation through "selfish herd"
groups and the defense of food resources by multifemale groups are two major
hypotheses for the formation of primate groups.[17]

While females tend to distribute themselves in response to food pressures,
predation, and risk of infanticide, males tend to distribute themselves in response to female distributions. Male-male rivalry can be a deadly business. But
it is also variable. Among baboons, there is a famous example in which three
relatively closely related baboon species appear to live quite differently, depending on the combined ecological pressures of food richness and predation pressure: one lives in multimale groups in an open, dangerous habitat with predators (several males, but no single male, can protect against the predators); the
second lives in a resource-poor but safer habitat in single-male groups; the
third, which moves between both sorts of habitats, switches social structure
with habitat. There is an ecological chain of influences that makes sense of the
variation. A caveat is in order, however: today we know so much more about
interpopulation variation, that such a picture is rarely, if ever, so simple.[18]

The rare socially monogamous single-pair primates tend to live in heavy
cover. There are three hypotheses about the evolution of monogamy in primates. One simply argues that when two-parent care is markedly more effective
than maternal care alone, monogamy will evolve. The second suggests that social monogamy arises when mated males are able to protect their infants from
infanticide by other males. The third suggests that for primates, as for many

other species, ecological factors such as female density and dispersion influence a male's ability to monopolize more than one female.[19]

Among the New World primates, there are additional good examples of the relationships among ecological factors and male and female strategies. When female reproduction is strongly constrained by seasonality, males are more likely to stay in their natal groups, forfeit exclusive mating opportunities, and tolerate infants. Males are more likely to disperse and compete for exclusivity when the reproductive payoffs are higher. Saddleback tamarins (*Saguinus fuscicollis*) show an interesting ecological response. They live in heavy cover, and one might expect monogamous pairing. However, females give birth to twins, who weigh half their mother's weight by the time they are weaned; not surprisingly, mothers do relatively little infant carrying. A commonly observed group is a pair plus an extra male, in a polyandrous trio.[20] Here, male and female constraints obviously interact.

Primates also vary in the degree of male-female size differences; gorilla males are huge compared to females; gibbon males are barely larger than females.[21] Here, too, ecological influences are important. Pair-living, single-male, and multimale group species differ in extraordinary ways. Males in single-male group species, as in gorillas (genus *Gorilla*), face high-risk, high-gain challenges from other males to fight for control of the females; they are physically much larger than females and have very large canines. Males in multimale group species have a somewhat more subtle problem to deal with as well: other males are always about, and sometimes in shifting coalitions. Sneak copulations by other males in the group are, like overt fights, a serious problem. Males in such species are bigger than females (though not so strikingly as in the single-male species) and have large canines, but they also have extremely large testes relative to their body size; here competition between males exists not only between individuals, but also carries down to the level of sperm competition. Males in pair-living species are not much larger than females and have relatively the smallest canines—and the smallest testes. The ecology of group living influences male-male competition—and thus not only behavior but also body size, canine size, and testis size.[22]

SEX, RESOURCES, AND THE ECOLOGY OF HUMAN REPRODUCTION

Differences in the reproductive ecology of the two sexes in humans create opportunities for men and women to use resources quite differently in reproductive competition. These conditions foster important sex differences.[23] We expect to find some general trends in behavioral differences between the sexes, and we do, beginning at birth (and even in utero). Newborn boys cry more than girls, respond less to parental comforting, and require more holding. Newborn girls are more "cuddly" than boys; they respond more strongly to adult faces and to being held. Boys are somewhat more interested than girls in inanimate, nonsocial objects. Boys seem to begin technical problem-solving sooner and wander farther from home earlier. These differences are seen very early and occur across cultures.[24]

While these sex differences have genetic components and proximate hormonal influences, none are "genetically determined" in any straightforward way. Although most sexual traits have genetic influence, very few sex-related differences have a clear single-chromosome-trait correlation; of those that do, most are disabling medical conditions.[25] Here is an exceptional example: babies missing an X chromosome (XO rather than XX; Turner's syndrome) have a 98% probability of dying before birth; those surviving show mental deficiency. Consider such a girl: her single X chromosome comes either from her mother or her father. If her X comes from her father, she will have better social skills than if the X comes from her mother.[26] Now, in normal infants, sons have XY, and the X always comes from the mother, whereas daughters have an X from their mother and an X from their father. It is far too early to make much of this, but there is surely a possibility that the "sugar and spice" versus "snakes and snails" folk wisdom reflects some genetic influence.

A well-known geneticist was quoted as saying that "we already have a genetic marker for violent behavior: it's called the Y chromosome." Indeed, in many species, the Y chromosome is a good "proximate" marker for competitive behaviors, for all the ecological reasons I have reviewed. From Norway rats (*Rattus norvegicus*) to rhesus monkeys (*Macaca mullata*) to humans, young males are more prone to rough-and-tumble play than young females.[27]

Genetic patterns reflect the ultimate evolutionary causes: most differences between normal XX and XY individuals arise not because an X or a Y chromosome "dictates" anything, but because the *ecology of achieving reproductive success* differs for males and females. Even if the Y chromosome alone dictated nothing, its bearers live in environments that influence what strategies will be successful in getting genes passed into offspring. The reproductive return curves of figure 3.2A set many parameters of success for mating-effort specialists (usually male in mammals, XY),who must compete, sometimes violently, and parental specialists (usually female in mammals, XX), who typically reproduce successfully by avoiding risk.

Some ecological conditions favor great male expenditure of offspring-specific effort; in these environments we expect behavioral, and even physical, convergence between the sexes. The less each sex specializes and the more they do the same things, the more alike they will be. We thus expect sex differences to vary among environments, among mating systems, and among cultures as a result of more subtle cultural influences. And indeed, humans are extraordinarily variable, as we will see.

Resource Value and Men's Reproductive Success in Human Societies

Just as the above discussion implies, men in most societies that have been studied use resources—wealth or status—to gain reproductively, typically through polygyny: additional wives, as many as possible. In such polygynous societies, men's ability to marry and to reproduce successfully varies, sometimes enormously, so great expenditure and great risk taking may be profitable. I will elaborate on this point later, but here I want to suggest the general patterns we expect and, in fact, see.

In more than one hundred societies that have been well studied cross-culturally, there are clear formal reproductive rewards for men associated with status: high-ranking men have the right to more wives, and they have significantly more children than others.[28] In many other societies, there are no formal societal rules (such as "men of status X may have two wives," etc.), but wealthy men simply can afford to marry more often than poorer men. Among the Iranian Turkmen, richer men have more wives and more children than poorer men (chapter 7). Among the African Kipsigis, richer men marry younger (higher reproductive value) wives and produce more children than poorer men. On the Pacific island of Ifaluk, men who hold political power have more wives and more children than others. The status–reproductive success pattern holds not only in these societies, but in others as diverse as the Meru of Kenya, the east African pastoralist Mukogodo, the agricultural Hausa, the Trinidadians, and the Micronesian islanders.

Even in societies such as the Yanomamö and Ache of South America and the !Kung of the Kalahari in Southern Africa, in which few physical resources are owned, male striving results in male status, which is effective in marital negotiations. Among the Yanomamö, coalitions of related men are important (chapters 7, 13). So male kin for coalitions represent a resource, and men manipulate kinship terms to maximize their affiliations with powerful men. Further, men can marry only women in lineages that have a particular relationship to their own, so men try to "redefine" their standing in ways that make more women available for mates. Among the Ache, good hunters have more children than other men (chapter 7). In the Kalahari Desert, the !Kung, living in a resource-limited environment, are almost entirely monogamous—but 5% of the men manage to have two wives. Thus in quite varied societies, wealth or status and reproductive success are positively correlated for men. Have we changed today? I explore comparative aspects of this problem in chapters 7, 8, and 15.

Sometimes men's status is not a function of direct fighting, but of resource acquisition and control. As noted above, men do the majority of hunting in hunter-gatherer societies. Who is a better mate? A poor hunter, or an optimal hunter? There is an excellent direct test of "conserving" versus "hunting optimally." The anthropologist Michael Alvard, who worked among the Piro,

Table 4.1. Specific testable predictions arise, given the definitions of "optimal foraging" and "conserving" (see text) that allow us to understand observed human hunting (after Alvard 1994). As in figure 15.6, low impact alone is insufficient to infer conservation ethic: optimal foragers are unlikely to deplete their food resource, especially if they are few, their technology is not highly efficient, and there is no point to taking more than can be eaten or traded at the moment.

Optimal Forager	Conservator
Maximize return rate	Avoid vulnerable species
Take sexes in proportion	Avoid females (polygynous)
Choose adults; ignore young if nonoptimal size	Choose very young and very old
Choose most profitable patches overall	Choose patches in proportion to appropriate (e.g., not endangered) prey abundance
Prey may or may not become depleted	Prey not locally depleted

Amazonian hunter-gatherers, is one of the few scholars to ask specifically and in a testable way about conservation strategies among people in a traditional society.[29] The Piro data allow real testing between these competing hypotheses: that hunters are conservationists ("noble savages"), or optimal foragers, just like other species. What does it mean, "to conserve"? Suppose a Piro hunter encounters a prime reproductive-age female of an endangered species. Do we imagine he will kill her? There is no doubt that hunters in traditional societies know quite well which species are common, which are rare, and which are extremely rare. If a Piro man is an optimal-returns hunter, he will kill her: she is excellent caloric return for his hunting investment. On the other hand, if he is a deliberate conservator, he will pass her by. You probably already suspect that he will kill her, and he does (see table 4.1). In fact, the Piro are extremely good at maximizing their caloric returns—individuals are passed up whose costs would exceed benefits (e.g., very small, usually young); and prime animals are always taken, even if they are an endangered species.[30]

But hunting has additional advantages, including in sexual selection. Men's sharing of meat with others is nonrandom and probably is a form of political favor-trading Further, men in some societies hunt in ways that look clearly suboptimal—for example, passing up small (but useful) game. Why? One strong possibility is that really big, "bonanza" kills (elephants, wildebeest) are really sexy; men who can make big kills gain more matings.[31]

Reproductive Value and Women's Reproductive Success

What about women? "Reproductive value" is the probable number of daughters a female will have in the rest of her lifetime; it is a function of the probability of living from one's current age to any given age, and of the average number of children born to a woman at each age. It has been used to make predictions about migration, contraception, and population growth.[32] It is also useful in understanding trends in marriage age and remarriage rates. Thus, in societies with bridewealth (74% of the 862 societies in the *Ethnographic Atlas*) or some other exchange of goods at marriage, younger women might be expected to command a higher "brideprice" than older women, for they will likely have more children in the course of the marriage, and they do.[33]

Men typically set great store on high reproductive value, though it may not be explicitly identified as such—men may simply note that they think old women aren't very interesting. Under such conditions, men with greater economic resources may be able to command, or be chosen by, women with higher reproductive value in the marriage market.[34] And indeed, in empirical work such as that on the agricultural and pastoral Kipsigis, Monique Borgerhoff Mulder found that the bridewealth required for a woman was directly related to her reproductive value, although negotiations could be influenced by the relationship of the negotiators (e.g., if they were good friends). With the introduction of Western technology and medicine (novel evolutionary events), the differentials have been reduced (chapter 7).

Poor men might choose to court older (lower reproductive value) women who have accumulated their own resources, explicitly trading reproductive value for resource value. In eighteenth- and nineteenth-century Scandinavia, daughters of upper-middle-class men (who would marry richer men) were considered women (marriageable) at eighteen years, whereas daughters of poorer men, who would marry poorer men, were not considered marriageable until years later, in their mid- to late twenties (chapter 8). Meanwhile, richer men provided resources themselves, married younger women, and gained high reproductive value. Similar patterns with men's wealth and women's reproductive value existed in eighteenth-century England.[35]

Resources and status also affect women's reproduction, but, as in other primates, apparently not in the same way or at the same levels as for males. In traditional societies, resources strongly affect women's reproduction when they are limiting (e.g., malnutrition) and result in fewer children, but women can almost never use rich resources to gain the extraordinary reproductive success of highly polygynous males. There is a possible exception: societies in which the descent system allows highly successful women to concentrate resources in their sons who (in contrast to their mothers and sisters) can use those resources to become successful polygynists (chapter 12).[36]

Male resources increase male fertility at the "high end" of reproductive variation—rich resources make the most successful males very successful, typically because they can acquire more than one wife (or in modern times, invest intensely in their fewer children).[37] Women's resources avert failure at the "low end" of the variation—women need enough resources to raise healthy children.

However, through most of our evolutionary history it has not profited women to strive for great amounts, since they typically could not convert such excess resources into reproductive gain.[38] And, as I noted above, the fact that true parental investment is offspring-specific means that women face a conflict between what they can gain from spending effort to get resources versus investing in offspring. For men, since status and resources are so often currencies in mating effort, what's spent on getting resources typically enhances reproductive success; for women (indeed, for female primates in general) such striving may lower their reproduction.[39]

Do Men and Women Compete Differently?

Both males and females undergo considerable intrasexual selection. In all cases, we are talking about competition: what traits, in any particular environment, help an individual survive and reproduce, and its lineage persist and grow, compared to others? This can be a complex and subtle business; for example, maximizing reproductive success in terms of number of offspring may not lead to optimization of lineage persistence and growth, relative to other family lines; it depends on the environment. But all competition, in the Darwinian sense, is focused on survival and reproduction (or, more properly, lineage persistence and relative increase).[40]

Across species and across human societies, males tend to compete more directly and more physically than do females. Remember the examples above: mortality in about a third of adult male red deer arises from injuries sustained in dominance fights. In most mammals, which, as we have seen, tend to have polygynous mating systems with intense male competition, males mature reproductively later than females (so they can grow larger) and, because of intense competition and resultant injuries, die earlier than females. In northern elephant seals, although 82% of males die before reproductive age, and many adult males sire no offspring, the most successful male sires almost ten times as many offspring as the most successful female. On the other hand, any male mammal that has been castrated lives a longer time than its fertile competitors but is unsuccessful at leaving offspring. This pattern holds for humans, although the patterns are seldom so dramatic.[41]

Cross-culturally, in the ninety-three odd-numbered societies of the Standard Cross-Cultural Sample,[42] from about age seven on, boys, but not girls, are taught to be aggressive and to show fortitude, whereas girls are taught to be industrious, responsible, and obedient. Curiously, there was no significant difference in the degree to which either sex was taught to be competitive. Today, just as boys' play is more rough-and-tumble than girls', men tend to compete more physically than women in most conditions: most murders are committed by men (often over access to women), men are more inclined to play physically dangerous sports, and so forth. Women's competition is seldom as openly aggressive or violent as men's.[43]

Both sexes compete in a variety of ways, but girls and women seem to be more subtle in the ways they compete—they appear to avoid potentially risky direct aggression; instead, they turn away at critical moments, shift their voice tone, and spread rumors. It would be hard to assess girls' behavior as less competitive than that of boys; the competition is just conducted differently.[44] Women compete reproductively within families. A number of external factors (e.g., conception risk, using contraception) affect how women's competition plays out.[45]

Differences extend even to games. Girls and young women tend to participate in sports less than do boys and young men. Cross-culturally, using the Human Relations Area Files (which contain more society data than the Standard Cross-Cultural Sample but may contain observations by others than experienced ethnographers) researchers found that in fifty societies, there were more male sports than female sports, and some (hunting, combat sports) were exclusively male, thus reflecting a robust sex difference. In the contemporary United States, women represented 51% of noncompetitive exercisers, 24% of total sports participants, and 20% of team-sports participants. There was little evidence that these trends were reversing. As Robert Deaner and colleagues have noted, through evolutionary time, such competing and team-building men were likelier than women (and other men) to gain status, and thus better access to mates, and perhaps it is not surprising that women's intrasexual competition is far more common in other venues.[46]

Sex, Gender, and Sexual Orientation

Nonheterosexual orientation is something of a puzzle in evolutionary analyses when it precludes ordinary reproduction. Currently no clear answer exists, but a number of scholars have suggested that it may be adaptive in a nonobvious way (e.g., caring for nieces and nephews), or a by-product of heterozygote advantage. In Samoa, there is some evidence that male homosexuals ("fa'afafine") may have the effect of increasing reproductive success of their direct maternal and paternal lineages (though not of aunts and uncles). Results from Western societies related to this hypothesis are inconsistent; perhaps this is related to fertility—low in Western contries, high in Samoa, so that small differences are easier to detect.[47]

Whatever the answer may turn out to be, it is important to remember that these questions of sexual orientation and gender identification—homosexuality, LGBT (and sometimes I), for example—are among those most important for us to separate analysis from attitudes, and avoid the naturalistic fallacy. There is considerable discussion about many categorizations: for example, in L(esbian), G(ay), B(isexual) T(ranssexual) I(ntersex), there is disagreement whether LGB are really allied to T or to I. To complicate matters further, individuals' experiences within each group can vary considerably: these are *major* orientations, not always *exclusive*.

Sexual orientation (attraction to which sex) and gender identification (perception of one's own gender) are separate issues as well. Although both have been highly politicized, especially in recent years, they are, like other sexual aspects of life history, individual traits. Cultural norms can have a tremendous influence on acceptance of an individual's gender identity; many societies are relatively intolerant of "mismatched" physical and psychological features. In contrast, "two-spirit" individuals can hold important social statuses among the Zuni (there originally called "berdaches"), Winnebago, Oglala Lakota, Cheyenne, Omaha, Crow, and more. Among modern nations, Thailand is probably among the most open in this regard: the sex trade offers virtually all combinations to customers.

Homosexual activity is rare but widespread across species, from unbred heifers mounting other females to pairs of female gulls nesting together when males are absent. Across the Standard Cross-Cultural Sample, for the 69 (of 186) societies for which data exist, homosexuality is coded as common in 43.5% (30) societies and as rare in the rest. Data for attitudes toward homosexuality exist for only 40 of the societies; these are coded as "accepted" or "of no concern" in 32.5% (13) societies. In the remainder, attitudes range from "scorn" (15%) to "strong disapproval" (42.5%). Of course, such data are difficult to get and to trust: asking about sex and sexual preference is fraught with difficulty in many societies. Perhaps the most famous examples of male homosexuality are those of the Sacred Band of Thebes and the Spartans.

Cross-culturally today, it appears that there is male homosexuality in all cultures studied; the percentage of homosexuals is low, ranges little, and appears relatively stable over time; social norms do not appear to affect the emergence of homosexual orientation (although, of course, it affects how openly individuals feel comfortable in expressing their orientation). All societies studied produce a full continuum from overtly masculine to overtly feminine homosexuals.[48] Like virtually every other trait studied, the origins of homosexuality appear to involve both genetics and environment. Most of what we know concerns male homosexuality and is correlative. First, identical twins (monozygotic: from a single egg and genetically identical) reared apart are more likely both to be gay ("concordant") than are fraternal twins (who are no more alike than ordinary full siblings), although the precise degree of difference varies among studies. Maternal uncles of gay men are likely to be gay. A large study of all twins in Sweden (not just twins in which one or both were gay) modeled genetic effects; this study found that genetic effects explained 34–39% of the variance, shared environment explained nothing, and individual environment, 61–66% of the variance. (For women, the percentages were 18–19%, 16–17%, and 64–66%). In an old study of incarcerated men and men from the "homosexual underworld," 100% of monozygotic twins were concordant, compared to 15% of dizygotic twins.[49]

Earlier work suggested an influence of a locus in the q28 region of the X chromosome, leading to correlations with homosexual brothers and maternal uncles; this has proved difficult to replicate, and much more research will be needed. For men, there is a correlation between the number of older brothers (each additional older brother increases ego's chance of being homosexual by 33%); the mechanism is postulated to be a maternal immunization response to Y-linked traits (cf. note 15 in chapter 6 on genomic imprinting).The uterine environment is considered extremely important; the idea is that homosexual orientation may arise from an atypical action of androgen. This generates sex differences in brain structure, cognition, gender identity, and orientation. Genetic variation may play a role, but it is unclear just which loci are involved and how they act. There are cases in which the intrauterine environment differs hormonally for twins, possibly affecting sexual orientation through the mother's physiological responses to carrying a male fetus. This is one proposed explanation for the observation that the more older brothers a man has, the more likely he is to be homosexual in orientation.[50]

Of "unusual" opposite-sex preferences, there can also be confusion. Men's preferences for females of high reproductive value have meant that in many societies, and throughout much of history, daughters may have been married off in their early teens. This preference among adult men for young girls is now called "hebephilia" (girls ages 11–14). Today, the American Psychiatric Association and the World Health Organization classifications of hebephilia include most of those ages ("13 and younger" and "pre-pubertal or early puberty"). This evolved preference with deep roots is still common among, for example, hunter-gatherer societies, although sexual activity may be delayed well after marriage, which may function as a mate-guarding strategy (chapters 5, 6).[51]

THE ECOLOGY OF HUMAN MATING SYSTEMS

There are ecological influences on the general trend for men and women, as among other primate males and females, to succeed through different strategies. As in other mammals, there are no clear dominant ecological influences leading to polygyny in humans; rather, it is the "default" strategy. Among other primates, polygyny is associated with a female and her offspring being "economically" independent: females can feed their offspring without male assistance. Such polygyny would often represent a fairly clear sexual specialization into the mating and parental effort curves of figure 3.2A, although male primates in a number of species carry, and sometimes play with, infants.

Human polygyny is a cultural institution involving not just mating but also social rules about marriage. It has great biological impact, and to make matters more confusing, it is defined differently by social scientists and biologists,[52] so it is important to clarify here what I mean. As I noted earlier, behavioral ecologists use the terms "genetically monogamous" versus "socially monoga-mous" (which is typically genetically polygynous)[53] to focus on the impact of sexual selection; it is the *relative variance* in reproductive success between the two sexes that is of interest (fig. 3.3). Societies, such as Western industrial na-tions today, that impose a one-spouse-at-a-time rule are only socially monoga-mous. In most such societies, typically more men remarry than women, and men have higher variance in their fertility than do women (men have children in second and subsequent marriages more than women, but also more men fail to reproduce).[54]

The degree of polygyny in humans has some ecological correlates similar to those of polygyny in nonhuman species.[55] Patterns of parasite risk, rainfall sea-sonality, irrigation, and hunting explain 46% of the observed patterns in human polygyny.[56] The most powerful ecological correlate of the degree of polygyny found so far is perhaps a surprising one: pathogen stress (fig. 4.1). There are good, though perhaps not obvious, ecological reasons. Environmental unpre-dictability may make it difficult to "track" the best phenotypes for an environ-ment; in this case, the most successful parent will be one who produces off-spring with genotypes likely, in turn, to produce new genetic combinations.[57]

Successfully polygynous men, of course, have not only more variable off-spring but also more offspring than monogamous men, so we must look further before claiming that offspring variability might have a functional role. Powerful men will promote polygyny whenever they can, whether pathogen stress is present or not. But monogamy is absent in high-pathogen areas. The degree of polygyny is really a threshold pattern rather than a linear relationship.

A clue that genetic variability in children is important comes from marriage preference: polygynous men in pathogen-laden parts of the world are more likely to marry exogamously (outside their group), especially through capture of women from other societies. Such marital outreach results in enhanced vari-ability in children. Sororal polygyny, in which a man marries sisters (whose children vary less genetically), is rare in areas of pathogen stress. Thus, there is a difference in the *kind* of polygyny if we compare high- and low-pathogen areas: co-wives are more genetically different in high-pathogen areas. From a woman's (or her family's) point of view, being the second wife of a healthy man may be preferable to being the sole wife of a parasitized man; thus, women may prefer polygyny in highly parasitized or pathogen-ridden areas, and men's and women's interests (typically more divergent in polygyny than in monogamy) may converge.[58]

Polygyny is much more common in Africa and South America than, for example, in Europe, so it is important to ask if there is simply some sort of co-variance of pathogens and socially determined patterns of polygyny. We can examine patterns within each region; if the covariation around the world is a side effect of simple geography, then we do not expect pathogen stress and po-lygyny to covary within regions. Within the high-pathogen but socially diverse

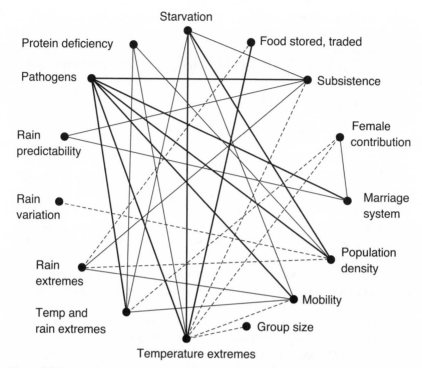

Figure 4.1. Ecological conditions and human social arrangements show a number of associations, summarized here. Bold lines indicate at least one relationship of significance $p < .001$, or more than one relationship of significance $p < .01$. Light solid lines indicate at least one relationship of $p < .01$ or more than one of $p < .05$. Dotted lines indicate at least one relationship of significance $p < .05$. Perhaps the most striking, and possibly underappreciated, result is that exposure to serious pathogens correlates with a number of social phenomena. (See text for further details; also see Thornhill et al. 2010 and Cashdan and Steele 2013 for additional correlations, including stress, collectivism, and xenophobia.) (Modified from Low 1990b.)

tropics and within Africa, Eurasia, South and North America, and the Mediterranean, pathogen stress and polygyny *do* covary. Thus, neither simple geography nor cultural diffusion of polygynous practices within high pathogen regions is likely to be the source of the patterns we see. In the Pacific, there have been no societies in the "high" pathogen stress category, and because the relationship is a threshold relationship, no statistical relationship is apparent there.

Pathogen stress alone accounts for 28% of the variation in the degree of polygyny around the world, independent of geographic region or any other factor. Polygyny, no matter how it is measured, increases with pathogen stress, and the *type* of polygyny (nonsororal versus sororal) is further correlated with pathogen stress. Thus, pathogen stress may be a real environmental uncertainty—a stress—that both renders fewer men acceptable as mates and favors the production of offspring destined to produce radical new parasite resistance in descendant generations.

Pathogen stress may have many social impacts as well. Elizabeth Cashdan, Randy Thornhill, and colleagues have linked pathogen stress to such important behavioral phenomena as individualism and collectivism, assortative sociality, xenophobia, and warfare—conditions in which many others could pose a risk because they carry pathogens not locally common.[59]

Resources and the Kind of Polygyny

As we saw in the last chapter, the ways in which males compete, and females choose, varies ecologically even among polygynous species. A major determinant is whether males can control resources useful to females; if they can, resource-based polygyny will exist. When controllable resources are lacking, males are reduced to trying to control females or to scramble competition (chapter 3).[60] When resources can be accumulated and defended, they are seldom distributed equally among individuals, and when resources are unequally distributed, so, typically, is reproductive success.

In one way, humans are highly unusual polygynists. In other species, as we have seen, typically males compete and females choose. But "third-party" patterns in humans extend to mate choice. In many societies, others (often parents or various male relatives of both bride and groom), not the bride-to-be, make the choice.

When societies lack rules of inheritance, suggesting that there is little to inherit, men typically do not exchange goods for women, but rather, exchange women;[61] when there are no societal rules about wealth or hereditary class stratification, men are similarly more likely to exchange women than goods. But even in such societies, resources are relevant to the pattern of exchange.

When men purchase wives (bridewealth societies), it is clear that younger (higher reproductive value) women are worth a higher brideprice.[62] The currency of choice varies: sometimes women are purchased with cattle, as among the Kanuri people; sometimes with sheep, as among the Yomut Turkmen; sometimes with pigs, as among the Tsembaga-Maring; or sometimes a combination.[63] When large surpluses can be stored and men control important resources, wealthy men can negotiate for more wives than others.[64] If the variance in men's number of wives arises from the choice of men (by women or their families) based on the men's resource control, then men's use of resources in reproductive effort should correlate with the degree of polygyny. Indeed, the use of some payment by men (e.g., bridewealth or bride service), rather than exchange of women or payment of dowry, is strongly associated with the degree of polygyny.[65] When men can accumulate resources, the variability in how many wives they can afford increases; when men cannot, and they exchange women, there is less variance.

In some societies, a bride's family pays a dowry, reflecting an interesting twist on the sexual utility of resources. Dowry is fifty times more common in socially monogamous, stratified societies than in polygynous or nonstratified societies. In stratified monogamous societies, males vary greatly in their status and wealth, and women married to wealthy, high-status men will benefit reproductively. Thus, it may pay fathers of brides to compete, bargaining for wealthier men as mates. In some of these societies, it appears that poorer women's families must pay more dowry than wealthy women's families.[66] Insofar as they are unable to do this, the stratification is intensified.

One example of dowry as female competition is that of modern rural India. Since about 1950, demographic shifts have resulted in a decline in potential grooms for potential brides of marriageable ages—and dowries have risen steadily.[67] By 1990, a dowry was likely to be over 50% of a poorer household's assets. Wives from poor families, able to pay less in dowry, may be less likely to marry; if they marry, they have a high risk of spousal abuse.[68]

Perhaps surprising to those of us in Western societies, in which close relatives are typically forbidden to marry, in many societies first cousins are the preferred marriage partners. Four kinds of first-cousin marriages are possible. A man could marry (1) his father's brother's daughter (FBD), (2) his mother's brother's daughter (MBD), (3) his father's sister's daughter (FZD), or (4) his mother's sister's daughter (MZD). Anthropologists also distinguish "parallel" (FBD, MZD) and "cross" (FZD, MBD) cousin marriage patterns.

These combinations have very different implications for resource control and coalitions, and despite a great deal of complexity, real patterns emerge in the way resources and kin coalitions influence the choice of mates among cousins. FBD strengthens reciprocity and nepotism among male paternal kin; it is associated with patrilocal residence, and men have relatively great social power and resource control in such societies.[69] MBD, which strengthens reciprocity and nepotism among male maternal kin, is associated with matrilocal and avunculocal (two ways of living with the wives' relatives) residence.[70] FZD marriage allows high-status men to keep control of resources, even in matrilocal societies, and is often practiced by such men.[71] MZD could enhance reciprocity and nepotism among female kin; however, perhaps because women so seldom control significant heritable resources, it is virtually unknown.

When men's sources of power are unpredictable and women have sufficient resources to be independent, men cannot always control women. In such societies polygyny or serial social monogamy (including many Western, developed nations with high divorce rates) result in high variance in men's reproductive success. This is the case among the Ache (see chapter 7) and the Cuna Indians.[72] Ache men and women have perhaps ten spouses in a lifetime, the Cuna four or five. Resource distribution, coalitions for mate competition, inbreeding avoid-ance, and nepotism (familial coalitions) are all important in marriage patterns.

Stratification and Striving in Polygynous Societies

The patterns we see in many species, of a few males who are highly successful in reproducing and of many males who fail, suggest that for males, at least, vari-ance in reproductive success may become very high. This is what lies behind the high degree of risk taking in males of many species. In general, the degree of polygyny is thought to be roughly correlated with the intensity of sexual se-lection, and measures of the intensity of sexual selection typically involve cal-culating variance in reproductive success.[73]

Social stratification, however, can complicate our attempts to measure this correlation. All calculations of variance are by definition based on a sample from a single population. Such measures are unambiguous, for example, in comparing all adult males in a population with scramble competition: there will be a mean success and a range of variation in success. However, in many species, all indi-viduals of one sex are not in the "same" population with regard to reproductive competition. In some fig wasps, for example, there are two kinds of males, winged and wingless. They follow very different strategies and have access to different populations of females; wingless males mate with their sisters before the sisters emerge from the fig, while winged males are larger and emerge to seek other fe-males and fight other males for females.[74] The two kinds of males are not really in the same "population" of competitors. In fact, in any such species with clear "al-ternate male strategies," a *reproductive stratification* exists that makes it inappro-priate to combine individuals from different groups or strata.

Stratification can be permanent, as in the above examples, or it can be onto-genetic, as, for example, in species such as bullfrogs, in which males move through age or size categories and have considerably different means and vari-ance in their success, depending on their category. In species in which there are true "conditional" strategies (all males may switch among strategies),[75] stratifi-cation is not an issue except as it interacts with survivorship; but when geno-type or age influence possible strategies, stratification can have a powerful in-fluence. In humans, cultural stratification (e.g., heritable class) can have the same reproductive impact as permanent stratification.[76]

One important ramification of sexual selection theory is predicting how hard males should "strive." For this reason, it is important to incorporate any stratification in analyses, for it influences success of striving, as we saw in chap-ter 3.[77] How hard one should strive and how many risks one should take de-pend greatly on the costs and benefits of striving, and these depend on whether payoffs are constrained by one's stratum.

The intensity of male fights depends on how closely matched the males are in size and other similar attributes of power, and this can create stratification.[78] In bluegill sunfish (*Lepomis macrochirus*) and in ruffs (*Philomachus pugnax*), a European, lek-breeding bird,[79] there are clearly two sorts of males who behave very differently. In sunfish, large males pursuing a territorial strategy exert more effort and take more risks than small males, who become female mimics, lurking until they can insinuate themselves between a big male and a real female in a mating display. Both the mean success and the variation in success are greater for large males than for small ones, and any attempt to understand the intensity of sexual selection would fail if these males were lumped in the same analysis.

To begin thinking about such systems effectively, it might be useful to partition the variance. For striving behavior by an individual in any stratum, what are the costs? What are the opportunities for gain—from striving and from being in a particular stratum? This question of "what produces variation in success?" is not simply a male question; for example, in rhesus monkeys and several other primates, female status has a matrilineally inherited component,[80] suggesting that female reproductive success associated with status can also sometimes be partitioned into an inherited-rank component and a behavioral-striving component.

Our ability to study behavioral dimorphisms in the context of sexual selection can be more precise if we first partition variances in male and female reproductive successes into a nonbehavioral component, including morphological characters and inherited resources, and a second component that would predict behavior.[81] Informally, the biologists Steve Frank and James Crow have suggested a simple method for quantifying the *opportunity for reproductive gain* through striving and risky competitive behavior, in which competitive success has both a stratum-related component independent of competitive behavior and a variable component in which success depends on the intensity of striving.[82] Their method suggests that

1. when opportunities for gain through striving differ among classes or strata, different behavioral patterns are expected: *high-variance strata will contain the most competitive and risk-taking individuals*;
2. within-class variance is most important in strata that are on average most successful; and
3. among species in which status explains a similar proportion of variance, if two species differ in the amount of total variance explained by heritable rank, then the two species are expected to differ correspondingly in the levels of aggression over status.

Using this model, an example can be placed in the wider context of partitioning variance into behavioral and nonbehavioral components. Some interesting insights follow from the general model. Among human societies, variance in male reproductive success can differ greatly.[83] The extent to which resources (status, wealth, etc.) are inherited also varies widely. For a society in which the variance in resource control among males is high and resources or status are heritable (e.g., strong patrilines), resource control can create stratification and, thus, may well influence the utility of striving and achievement—and risk taking.

The heritability of resources should be inversely related to striving behavior. When the heritability of resources is high, then the opportunity for gain within strata is likely to be low and competitive behavior muted. Data on striving behavior are far more difficult to obtain than data showing that resources influence reproductive success, but some patterns are clear. The more polygynous the society (the higher the potential rewards), the more sons are taught to strive using several measures—but only in nonstratified societies, in which a man's striving can make a difference in his ability to marry (chapter 6). When a man's reproductive success is largely set by his heritable resources or social/class position, and unlikely to be changed much by potentially expensive and dangerous striving, parents are unlikely to teach the value of striving.

Women's Gains and Losses in Polygyny

Successfully polygynous men are always reproductively better off than their non-polygynous competitors; that's why it is worth all the cost and all the risk. But the situation is more complicated for women. Above, it appeared that when serious pathogens made some men poor choices, women might prefer polygyny. And sometimes polygynous marriage with a high-status man appears to be preferred by women or their families, even when there are no apparent reproductive benefits.[84]

Women often suffer costs in polygynous systems: in a number of societies, second and subsequent polygynous wives have lower fertility than monogamous wives or than first wives in polygynous households.[85] Children are likely to survive less well in polygynous households, and a major cause of divorce in polygynous societies is conflict among co-wives.[86] But in some systems, women can achieve protection by forming alliances with co-wives, for example.[87]

A variety of proximate factors undoubtedly interact: for example, men may be older when they marry their "later" wives; women who are not considered desirable are likely both to marry late (and thus have low reproductive value) and to be a later wife. Nonetheless, the net result is that within a polygynous society, a woman's (or her family's) choice between an already married man and a not-yet-married man may be complicated.

THE ECOLOGY OF MONOGAMY AND POLYANDRY

Females seldom fully share males' reproductive interests; a rare exception is in genetic monogamy. Males will strive for polygyny when resources are sufficient; when females can be an independent unit with their offspring, polygyny predominates. We expect to see genetic monogamy and polyandry when resources are more limited in harsh and unproductive habitats, when men may do better reproductively by helping to raise a child with true parental investment, rather than continuing their mating effort.[88] We expect social groups in such environments to be small and to have relatively little variation in the resources controlled by individuals.

At a crude level, this pattern holds: in the Standard Cross-Cultural Sample, highly polygynous societies are found in areas of the world with significantly higher plant productivity (a measure of environmental richness) than others; polyandrous societies are found in areas of significantly lower plant productivity; and there is no difference in the plant productivity of "monogamous" and mildly polygynous societies. Here, more than ever, it is important to distinguish between the anthropological and the behavioral ecological definitions of "monogamy."[89] In this case, when the definitions of marriage systems of the societies in the Standard Cross-Cultural Sample come from ethnographies, the term "monogamous" means monogamous to mildly polygynous, and thus it is not surprising that such socially monogamous societies show no difference from mildly polygynous ones. Similarly, reading the ethnographic descriptions of societies deemed "monogamous" leads one to conclude that even when a few men manage (through skill in hunting or getting novel sources of income) to be polygynous when most men remain monogamous, it is in habitats with a poor resource base—those insufficient for many men to manage to gain more than a single wife.

Polyandry is extremely rare;[90] of the 186 societies in the Standard Cross-Cultural Sample, only 3 are reported as polyandrous.[91] Almost all polyandrous systems are fraternal: co-husbands are brothers. Of course, the interests of the two brothers differ, and the impact on fitness will also differ by birth order, sex, and opportunity costs (other available options).[92] Polyandry seems to occur under two circumstances, both related to the conservation and concentration of (scarce) resources. Among the Lepcha of northern India, for example, brothers marry the same woman.[93] The land is extremely poor, and it apparently takes the work of two men to support one woman and her children.

Prince Peter of Greece argued that in Tibet both resources and familial considerations lay at the heart of Tibetan polyandry:

> Taking more than one wife for each son of a house would oblige them to partition the property, something which simply could not be done in a difficult environment such as that of Tibet. . . . Polyandry, then, is the ideal solution to the problem, for if the wife were tempted in the absence of her husband to have sexual relations with someone else, at least by having them with his brother, her offspring will always be of the same family blood.[94]

While Tibetan polyandry is common among wealthier families, among the Kandyans of Sri Lanka, polyandry apparently occurs among poor families. Yet here, too, polyandry appears to be an arrangement by which two brothers join their land and maintain a common family, minimizing the number of potential heirs and raising living standards.[95]

Polyandry thus can result from brother-brother coalitions in order to combat resource scarcity or from attempts to control the distribution of a resource like land, which is immobile and loses its value when too-much divided; it is "a rare but adaptive system for preserving family estates, and hence reliably supporting lineal descendants, across the generations."[96]

The very rarity of polyandry led Ray Hames and Karen Starkweather to sample polyandrous societies not covered in the standard samples (mostly, above, Himalayan and Marquesan). In a survey of fifty-three societies, they found that polyandry was not so rare as we tend to think: it is found worldwide, is most common in relatively egalitarian societies, and may be a response to a high operational sex ratios, and perhaps to men being away for long periods.[97]

5

~~~~~~

Sex, Resources, Appearance, and Mate Choice

I've been called handsome, adventurous, athletic, humble, honest,
intelligent, and messy. Seeking woman, 27–34, with most of those
traits.
—*Ann Arbor Observer*, personal ad (April 1998)

IN SHAKESPEARE'S *Romeo and Juliet*, Juliet's father had promised her in marriage to a man his own age when she was twelve. Because she was in love with young and handsome Romeo, she complained. Her father's answer was:

> An you will not wed, I'll pardon you!
> Graze where you will, you shall not house with me;
> An you be mine, I'll give you to my friend;
> An you be not, hang, beg, starve, die in the streets,
> For, by my soul, I'll ne'er acknowledge thee,
> Nor what is mine shall never do thee good.
> (*Romeo and Juliet*, act 3, scene 5)

Today, in the United States, Juliet would probably sue her father for child abuse. And she'd be likely to win, even though she might not be allowed to marry at the age of twelve. Juliet's dilemma highlights one of the major ways modern Western patterns of mating and marriage differ both from those of most other mammals and from those of people in our evolutionary past: a man and a woman did not, through most of our evolutionary history, simply decide for themselves whom and when they would marry. Most negotiations were done by relatives in the older generation, and often in the service, as we shall see, of family alliances. This pattern is apparently ancient, probably going back to the first human migrations out of Africa.[1]

The second major difference in mating today, compared to past times, concerns how each sex competes for mates (e.g., how they are ornamented, whether they fight). Men and women are not all that different in size or appearance compared to many other primates; no naive biologist would take two human specimens, male and female, and think they were different species, as has happened among some birds. In general, male humans, regardless of the current marriage system, are slightly larger than females, a characteristic that is consistent with our evolutionary history of mild polygyny. This is because much, perhaps most, male-male competition in humans is not a matter of size but of other traits: wealth, power, intelligence, political savvy, and so on: traits that help in complex social competition more than sheer size.

As we saw in the last chapter, females in many other species simply choose their mates, and while the criteria may vary, the choice process looks relatively straightforward. Depending on the ecology of parental care, females choose different traits. They might seek "good genes" in species with no male parental care; this can involve something as straightforward as expensive displays (see below), or something as subtle as choosing an individual whose genetic makeup is different, so offspring will be heterozygous. Females might, as in some grasshoppers, choose males with the best foraging abilities (and in response, male grasshoppers with poor foraging ability may try to force copulations); and females' choices can be modified by their social environment. Or, as in some cockroaches, females may be able to use a male's status and pheromone cues to discriminate against males reared in poor environments. For some species, resource control is so central that females appear simply to choose the resource and get whatever male controls it, as do elephant seal females (choosing a beach on which to give birth) and red-winged blackbird females (*Agelaius phoeniceus*) choosing a rich marsh in which to provision and rear their young.[2]

In contrast, the prospective bride and groom in many (especially traditional) human societies may have little to say in choosing their mates. Or the bride's and groom's preferences may count, but in an informal way that is difficult to document. Because marriage and mating in humans involves others besides a male and a female, a whole set of conflicts of interests may exist. Not only might the man and woman desire quite different qualities in a potential mate, so might their families (see below).

Beauty, Resources, and Mate Choice

Even simple physical differences between the sexes reflect that what is valuable in a wife is likely to differ from what is valuable in a husband; differences and preferences are relatively consistent across quite different societies. Put bluntly, in our evolutionary history, it seems likely that a woman's value was usually her reproductive value and a man's value was his resource value. Cross-culturally today, while everyone values such traits as a sense of humor in both sexes, women still typically seek signals of resource control in potential mates, whereas men seek signals of youth, health, and "beauty."[3] As we saw, assessments of beauty vary across cultures, but typically they reflect health and youth (and a low waist-hip ratio; see below).

Occasionally, at least for periods of time, individuals in very wealthy subgroups may value something that reflects female helplessness, but in general "pale and wan" does not become and remain a widespread preference. Exceptions appear when men and women have conflicts of interest and men hold enough power to resolve those differences in their own favor. Consider women with bound feet in Mandarin China; their fathers and families favored a condition that reduced women's fitness under ordinary selection—except that, because supporting an essentially helpless wife who has bound feet reflects a man's wealth, suitors favored it, and fathers helped enforce it. Female circumcision, as practiced in parts of Africa, probably also represents a male-female conflict of interest: the practice clearly does not increase women's general health, fecundity, or desire, but so long as men demand it and refuse marriage to uncircumcised women (in a closed society), the practice will continue. Exceptions like these not only reflect important cultural variation but, as exceptions, suggest the strength of the general correlation.[4]

A woman's (usually her family's) resources are not irrelevant in marriage choices. In some societies, men with few resources may explicitly trade off reproductive value for resource value (see below)—picking an older, reproductively less valuable woman who controls, in her own right, some resources; perhaps today's pattern of famous actresses "of a certain age" marrying younger men is relevant here. Subtle biases across societies, and fluctuations over time, have usually given an advantage to wealthier families in marrying. When most women worked in fields in western Europe, the standard of beauty was a pale complexion (which only the daughters of the rich could maintain); when we all began to work indoors, the Caucasian standard of beauty became a winter tan, suggesting that one could afford a trip to warm climates.[5]

What Men and Women Want

Freud asked: what do women want? Resources, as Shakespeare knew,

> Dumb jewels in their silent kind
> More than quick words do move a woman's mind.
> (*Two Gentlemen of Verona*)

Anita Loos was pithier, if not as poetic: "Kissing your hand may make you feel very, very good, but a diamond and sapphire bracelet lasts forever." In our evolutionary past, women whose mates provided resources for them and their children did better than others; so did the women's families. Women tend to favor men with intelligence (which may be a proxy for resource-garnering ability), strength, "masculine" faces (e.g., strong jaws), energy, and vigor—all, again likely proxies for assessing a man's competitive ability. For example, for men, social status and dominance have fitness advantages in the Tsimane.[6] In physically dangerous environments, women appear to seek aggressive and formidable men. This appears to be true for traditional societies as well; if we examine the ethnographies, comments people in traditional societies make are largely related to direct measures of fitness rather than to "purely cultural" or conditional cultural interaction traits.[7]

When males invest parentally, as in humans, males as well as females may profit reproductively from exercising choice in mates.[8] If both sexes can exercise some choice, and if men and women have been reproductively successful through different strategies, they are likely to look for quite different things in their mates.

Of course, ideas about beauty or desirable traits in a mate will surely be influenced by cultural norms. Even in relatively simple species like guppies, females not only choose (costly) male signals and displays that reflect good condition, no parasites, high energy, and so forth, but young females copy the choices of older females. There is evidence that in modern humans, both men and women show mate copying.[9]

To complicate matters, we humans manipulate all sorts of signals: hair, eye, and skin color, body shape; what possible reproductive value could, for example, blue hair and nose rings reflect, unless perhaps (1) a cultural identity, or (2) an ability to lead trends in fashion? Cross-culturally, cultural augmentation of sexual signals or ornaments is virtually universal, favored for the same reasons that selection favors physical ornaments and displays in other species. Remember the old adage, "If you've got it, flaunt it." Males and females profit by signaling, or flaunting, different attributes. Humans are actually rather paradoxical with regard to sexual selection and sexual displays: most scholars agree that human evolutionary history, like that of most other primates, is polygynous: 83% of (primarily traditional) societies for which we have information are polygynous.[10] All of these patterns suggest that in humans, males should be the "ornamented" sex, yet most people talk about women's adornments. But ornaments can be what either sex advertises.

We can break relatively stable preferences down into signals that reflect health (shiny hair, clear skin) or youth (no wrinkles or sags) and current reproductive stage (for women, waist-hip ratio, color of nipples); signals that suggest other reproductively important attributes like wealth; signals that reflect social awareness (stylishness, which may be purely culturally defined); signals of belonging to a certain group. Cross-culturally among traditional societies, the things people describe as attractive in the other sex turns up all of these categories.

Signals of women's youth and health, directly related to selective advantage, are universally described as desirable: clear, unwrinkled skin, firm breasts, lustrous hair. Some traits reflect an interplay of cultural practices and selection; for example, among the Malaysian Iban people, short hair is undesirable: they cut a sick person's hair very short, so short hair in this case reflects recent illness. Some of these "interplay" traits may reflect directly favored traits; in a number of societies, a man's tattoos reflect his status, and, of course, undergoing the tattooing further reflects pain tolerance and fortitude. But we should be cautious; I don't think we have good data to tell us just what all the functions are. Finally, some preferences may simply be "purely cultural": in some societies, relative hairlessness is a sign of beauty in a woman; in others, robust and luxurious hair is desired.[11]

Women's preferences may be complicated by proximate physiological cues that appear to have some selective import. Across the menstrual cycle, women's assessments of what male traits are desirable appear to change. During their fertile phase (late luteal and ovulatory), women may prefer men with symmetrical and masculine features, and may even prefer the scent of such men; there may be some interaction between these phenomena, because when photographs are used there is some variation in results: the masculinity preference is not so strong. Women's preferences for masculine faces and for men's scents associated with testosterone and cortisol levels change across their menstrual cycle, though their preference for intelligence is stable. In our evolutionarily novel environment, women use hormonal contraceptives; these might influence the choice of partner and, if contraceptive use changes, partner dynamics. The prediction is that if current contraceptive use is congruent with initial use (e.g., when a couple met), women should be happier with their partners than if contraceptive use changes—in other words, that women's sexual satisfaction is associated with changes in hormonal contraceptive use.[12]

An important widespread physical preference by men—one that would not occur to most of us but which is intriguing in light of this argument—is for a particular relationship between a woman's waist and hip size. Across all sorts of cultures with quite different specific ideas about beauty, both men and women see as most attractive a female waist-hip ratio of about 7:10 or 8:10. This is true whether the preference is for rather generous, Reubenesque proportions, or for slender Angelina Jolie builds. Why? Women of reproductive age, unless they are pregnant, tend to put fat on their hips, breasts, and buttocks. Older women (of lower reproductive value) and pregnant women (not currently fecundable) thicken at the waist, giving a higher waist-hip ratio. The relative size of waist versus hips gives important reproductive cues. A relatively narrow waist means "I'm female, I'm young, and I'm not pregnant." The waist-hip ratio reflects many complex relationships—but they all boil down to: Is she fertile? Is she fecundable? Even more subtle cues may matter, because today, women who delay having their first child until after age 30 are likely to have few children—and thus a low reproductive value.[13]

Hips, breasts, and buttocks are physical signals that communicate age, whether there have been prior births, and even, in the case of buttocks, one's ability to metabolize scarce food efficiently. At least some of these physical signals can be deceptive, even without deliberate manipulation. Consider how we put fat on

Figure 5.1. Steatopygy, the preferential deposition of fat on the buttocks, is associated with harsh and unproductive environments. It probably represents an honest signal of nutritional competence: "even in this harsh environment I can not only maintain myself, but store fat." (Photo courtesy of the Image Archives, Denver Museum of Nature and Science)

our bodies. Little children tend to have fat on their faces, fingers, and toes, presumably for protection against temperature extremes. All other age and sex categories distribute fat relatively evenly over the body—with one exception. Reproductive-aged women, unlike all other age-sex categories, deposit fat preferentially on the breasts, hips, and buttocks, contributing to that approximately

7:10 waist-hip ratio men prefer. If a woman's hips are broad because she stores fat there, rather than because her pelvic structure is wide, a man gets potentially inaccurate or confusing information (wide pelvic girdles provide a head start on the easy birth of large-headed infants; fat doesn't). It isn't that fat is "bad," but it denotes energy reserves rather than a structurally wide pelvis. Similarly, if a woman's breasts are large not because they comprise mammary tissue for milk production but because fat is stored there, a male gets information not about a woman's lactational capability, but about her stored energy. This seems an irrelevant issue today, when food supplements and medical attention are readily available, but it may well have been an issue in mate choice in our evolutionary history. Even today, insufficient mammary tissue means a woman may have difficulty nursing effectively.[14]

In extreme, food-limited environments, obvious fatty deposits on women's buttocks signal the ability to gain sufficient nutrition on a limited diet—a subtle reflection of maternal quality, which is important in a male-parental species. Thus, it is no surprise that extreme, harsh environments are the context for both steatopygy (the condition in which fat is obviously concentrated on the buttocks; fig. 5.1) and a cultural preference for extremely fat women. Darwin gave a secondhand report of one example of steatopygy as a sexual preference trait. Among the !Kung (then called Hottentot), Darwin's informant reported, a truly sexy woman was one who was unable to rise from level ground because of the weight of fat on her buttocks. Fat on the buttocks is probably a "true" signal of ability to store fat on any particular diet. In contrast, fat deposits on the breasts and hips are likely to be confusing and even deceptive signals; at the least, such fat is likely to be confused with mammary tissue and wide pelvises, traits contributing to two very different aspects of maternal quality.[15]

INFLUENCES ON MATE CHOICE
IN MODERN WESTERN SOCIETIES

In Western societies today, a man's mate choice is still likely to focus on a woman's health, her reproductive value (which means her youth), and her current reproductive status (fertile, not pregnant)—but other traits and environmental conditions can matter, as we are finding with increased technological ability.[16] In many societies, the ideal is a healthy young virgin. In an interesting twist, a man's preference might depend on whether he was seeking a short-term mate (in which case, we would expect him to prefer a woman at peak ability to conceive, perhaps in her twenties) or a longterm mate (in which case we would expect him to prefer a woman with peak reproductive value, about age 17–19).[17]

Even today, men's resource value matters. Women choosing mates are interested in men's resource control. Among French Canadians, men (but not women) of higher social status have more sexual partners than others, suggesting that status is important for men in the mating game. Further, women's (but not men's) number of partners decreased linearly with age, suggesting that women's reproductive potential is important for them. In a cross-cultural study of thirty-seven modern cultures, while some particulars about mate preference vary across the cultures, there are some consistent preferences that are obviously related to evolved sex differences in mate quality. Women rank men's ability to get resources high, and men rank women's youth and health high; both sexes rank social abilities like "sense of humor" high as well.[18]

We don't usually think (after we graduate from high school, anyway) much about physical cues in mate choice, but certain physical cues may matter. The anthropologists Doug Jones and Kim Hill, in a modern cross-cultural study of facial attractiveness in contemporary societies, found that selectively relevant traits consistently rank high and that men prefer neotenic (childlike) faces in women in a number of cultures—reflecting, of course, youth and high reproductive value. Both sexes prefer symmetry in both face and body, possibly reflecting health. I have argued, as have others, that physical sex differences beyond size reflect our polygynous background—that breasts, hips, and buttocks have served as sexual signals when females compete for the attention and investment of powerful, parental, resource-investing males.[19]

Both individual and cultural conditions, as well as whether a long-term or a short-term mate is sought, can influence the relative importance of different traits in mate choice. In an experiment using facial morphing techniques, stressed men preferred women who were dissimilar from themselves. Women's established level of preference for masculine faces, and their memory of details of earlier episodes, affected their choices of men's faces; in an experiment that showed a face and an object, women with established preferences for masculine faces remembered the objects (microscope, fish) shown with more masculine faces. When the cost of making a mistake about a partner's commitment would be high (i.e., for young fertile women), women are more skeptical than older, postmenopausal women, who may not worry so much about commitment. Both sexes prefer selflessness—but when directed to themselves or associates, rather than random selflessness. And women using oral contraception showed lower sexual satisfaction and partner attraction than women not using oral contraception; however, these same women ranked their partners higher on paternal provision, their relationships lasted longer, and they were less likely to separate.[20]

A history of unfaithfulness or of having been recently "dumped" by a partner can affect how individuals play the mating game. For example, women interested in finding a long-term partner are less likely to agree to go to a concert with another woman known as a cheater than to go with a nonflirtatious woman. And if a partner, male or female, discovers that his/her current partner has been dumped by the previous partner, men and women react differently. Neither men nor women were attracted to the idea of sharing past history, but when a man found that his current partner had dumped her last partner, his ratings of her fell. In contrast, when a woman found the same information about her current partner, her ranking of him was likely to increase![21]

Cultural influences can affect individual preferences, too. For example, women in healthy nations show less strong preferences for very masculine faces than women in nations in which poor health is dangerous to survival. Certain religious inculcation (e.g., stressing monogamy, high fertility, and high parental care) caused men, but not women, to act less impulsively than comparable control individuals.[22]

Some men's preferences—healthy, young, not pregnant—are virtually universal. In Thailand, for example, men, regardless of whether they are rich or working class, living in the city or countryside, tend to prefer young virgins (although virginity is not so profound a preference as in some societies); they insist that while they themselves may have extramarital intercourse, their wives may not—their wives are to make a good home, stay there, and be faithful.[23] Women tend to accept that men will be sexually active outside marriage, and their first concern is that any such activity does not divert financial resources from the home and children; a good husband, whatever else he does, provides for his family faithfully.

An interesting attitude shift is occurring in Thailand that is related to the ecology of HIV transmission. Most women still prefer, when their husband has other women, that he visit commercial sex workers: the cost is modest, and the transaction is complete with the payment. (Traditionally, wives objected less to commercial sex workers than to minor wives.) But as knowledge about HIV has become prevalent, some women are beginning to prefer that their husband have a steady mistress, or even a minor wife; although such women represent a greater threat to a wife's resources, they represent a smaller disease risk.

Signals of Desirability and Their Manipulation

Status signals have a cost and will be maintained only if they benefit the bearer. Some signals make actual confrontation less likely, saving calories and avoiding risk. Other signals serve as sexual attractants and are the source of much physical dimorphism. In nonhumans, the sex competing for mates is the sex that gives such signals. Males in polygynous species are usually big, colorful, and likely to have weapons, because those are the males who win status and are chosen by females (see above). Females in the few polyandrous species are usually a little bigger, and sometimes a bit more brightly colored, than males. Because much of the selection on displays is preference by the choosing sex rather than relative survival enhancement, sexual selection on expensive—and possibly dangerous—displays can "run away" beyond the point of usefulness and can decrease survivorship.[24]

Thus, males in polygynous species are likely to grow antlers, or large horns, or bright feathers, or long decorative tail feathers—all costly, and usually risky, displays that may do nothing more than advertise a male's ability to take these risks: the Handicap Principle. The message is: I am so fit that I can support this expensive handicap, which would kill a lesser individual.[25] And when females prefer these costly displays, they work. For example, female European swallows prefer to mate with longer-tailed males: these males more often get mates, and get them sooner, than other males. The success of longer-tailed males is high—but these tails carry a cost in terms of survivorship; long-tailed males die sooner than other males.[26]

Most nonhuman examples, including those given above, principally involve male physical (energetic) resources, even when, as in bower birds (for whom the criteria are the number and color of decorations on the bower), the display is not simply a physical part of the displayer's body.[27]

In contrast, humans invent, augment, and change signals; and females do a great deal of signaling. Because of our polygynous history, men's and women's cultural augmentation of sexual signals should give information about different characteristics. Men are likely to signal wealth and power status, and members of cultural subgroups with limited real resources seem likely to concentrate those resources in highly visible signals. Sociological studies of wealth and status signals among contemporary poor groups, for example, find the "ghetto Cadillac" phenomenon common.[28] Because humans show male parental investment, a woman's reproductive value becomes important; thus, women should signal reproductive value, traits that reflect youth and health (no wrinkles, a 7:10 waist-hip ratio, shiny hair, clear skin). Today, billion-dollar industries exist to help women signal just such traits: makeup and cosmetic surgery, for example, are designed to communicate youth, health, and sexual interest—the products and processes are aimed at making the skin tauter and less wrinkled. Though a few men indulge themselves this way, most clients are women, who get facelifts and undergo liposuction to obtain an attractive waist-hip ratio. In the nineteenth century, women put belladonna in their eyes, dilating their pupils, ordinarily a strong signal of sexual interest. Women use rouge to make the cheeks rosy (indicating they have either been exercising or are sexually aroused) and lipstick to mimic the dark, blood-engorged state of the lips during sexual excitement. The specifics change across time and societies, but the desired result does not.

Women, like females of other species, can signal interest and availability behaviorally. Patterns of eye contact in flirting appear to be virtually universal and invariant in widely differing societies.[29] The fact that women frequently wear signals of "unavailability" (e.g., wedding rings, styles of clothing or hair worn only by married women) and that in some cultures they undergo treatment that may decrease their general health and vigor (foot binding, clitoridectomy; see above) are suggestive. Men and women's interests often conflict, and women are at least sometimes manipulated by men, for example, by proclaiming unavailability in return for parental investment or by undergoing foot binding to get a wealthy mate. Such ornaments of unavailability should be more common in societies in which women are more dependent on men for resources.

There are intriguing cross-cultural trends. Men's wealth or power status is shown by ornaments in 87 of 138 (mostly traditional) societies studied.[30] Only 4 societies distinguish men's marital status by ornament, and 2 of these are genetically or ecologically monogamous,[31] living in very poor environments where men have trouble becoming successfully polygynous (chapter 4). In contrast, women's marital status is signaled in all but 3 societies.

One function of marriage in most societies is that of building alliances between families. In such societies, women might be expected to signal family wealth, although not necessarily any separate wealth of their own. Indeed, women show wealth by ornaments in 49 of 138 societies. Is it their own wealth, or their family's (father's before marriage, husband's after)? Although men usually have greater resource control than women, there are societies in which women control significant resources or wield considerable influence over resource distribution—but the societies in which women have power and influence are *not* those in which women wear ornaments of status and power. Put bluntly, the advertisement of women's status is less likely to be effective when directed at close kin, members of the household or of the small band with whom one interacts daily—*they* can't be fooled.

Across most traditional cultures, women's signals of status largely reflect their husband's or male kin's wealth or standing, which is consistent with the prediction. Such signals also may represent a conflict of reproductive interest between the man and woman, since male resources are used to acquire mates, and signals of "excess" resources, even if worn by a man's wife, can constitute his sexual advertisement of mating effort.[32] These patterns are consistent with the observation that males seek resources as mating effort, competing against other males to whom they are variously, though often not at all, related and sometimes interacting with individuals they know less well; females, on the other hand, seek resources as a form of parental effort, working at or near home with sisters or co-wives.

Female ornaments of power show one significant relationship cross-culturally: women's ability to hold political posts. However, the relationship is not positive but negative: societies in which women can hold political posts are societies in which women do *not* wear ornaments of power or status.[33] The question then remains: when women do operate independently in the extrafamilial, community sphere, why do they not signal position and power in the same way, and to the same extent, as men?

Cultural specifics obviously can influence the general pattern. For example, ecologically monogamous societies show distinct patterns compared to other societies. It's true, for example, that women signal marital status far more than

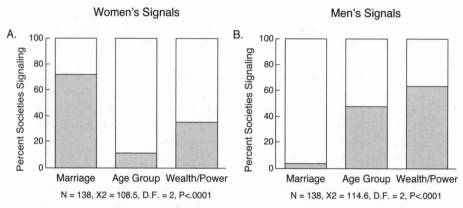

Figure 5.2. Across cultures, women (A) and men (B) signal different information by their dress and ornament.

men. The societies in which men do signal marital status (and thus "unavailability") are ecologically (genetically) monogamous or large nation-states (fig. 5.2). Ecological constraints mean that men can't profit from polygyny, anyway, leading to genetic monogamy. Men in 20% of genetically monogamous societies signal marital status; men in 1.5% of other societies do so. Pubertal and/or age-group status is not discernible for either men or women in ecologically monogamous societies (fig. 5.2), while women in 12% and men in 67% of other societies signal this. Finally, men signal wealth and power in 67% of socially monogamous societies—those in which such signals might be potent advertisements of their ability to take on additional mates; men in only 10% of genetically monogamous societies do so. Women in genetically monogamous societies do not signal wealth or power; women in 38% of other societies do so, though this is not a woman's own wealth or power but rather a reflection of her male relatives' status. When women signal the wealth of men, there is potential for a great conflict of interest.

Who Can Choose?

What is common, approved, and thought ethical varies widely across human cultures, temporally and spatially. In other species, mate choice is a relatively straightforward affair: typically, males display and females choose. And in Western industrial nations today, once again individuals are relatively free to choose their mates. But in human history this has not always been—and perhaps seldom was—the case.

Juliet's father, above, was seeking a powerful ally, and Juliet's ability to exercise mate choice was nil. Across human societies, this is far from unusual; in most societies, the reproductive interests of more than the two who mate can matter greatly. In traditional societies, the potential bride had greater say than the would-be groom in marriage negotiations in only 3.7% (3/81) of societies.[34] Grooms had greater—or sole—say in 39.5% (32/81). In most of these societies, the older generation had considerable power in these decisions. Among the Kipsigis of Kenya, for example, two men decide the brideprice for a young woman; a younger woman, of higher reproductive value, commands a higher brideprice, though a special friendship between the men might lead to a discount (chapter 7).[35] In some societies, such as the Arunta of central Australia, a couple is betrothed before the female is even born. A boy grows up prohibited from talking to the girl his age who is his designated mother-in-law. Even though older men make the formal decisions, a young woman may have either great or no influence. For example, among the Kipsigis, though men set the brideprice, women do exercise some choice.

Wealth, marriage "market forces," and ecology all influence marriage patterns. Juliet's father was wealthy; otherwise he could not have contemplated marrying her off at age 12. Recall from chapter 4 that dowry is common in societies that are stratified and monogamous; wealthy, parentally investing men are at a premium, and families of marriageable women compete for them. In such societies, wealthy families are able to marry their daughters off earlier, and brides are chosen not only on their own characteristics (youth, beauty) but also on their family's (wealth, father's schooling and occupation). Poorer families, because they not only must pay dowry but are harder pressed to replace the daughter's labor, gain if they can delay a daughter's marriage, so daughters in wealthy families marry earlier and are likely to have higher fertility than poorer women.[36] In bridewealth societies like the Kipsigis, wealth makes a difference in a woman's age at marriage—but in these societies, the economic forces mean that wealthy men (or sons in wealthy families) can afford brides who are younger (of higher reproductive value). In both, family wealth contributes to family fertility and growth.

6

~~~~~~

Sex, Resources, and Human Lifetimes

And one man in his time plays many parts,

His acts being seven ages.
—Shakespeare, *As You Like It*, Act 2, Scene 7

A GREAT PHILOSOPHER never married, and, in a possibly apocryphal story I read as a child, on his deathbed he called for all of his works to be set upon his lap. When the works had been brought, he sighed, "All of this is less than the weight of one grandchild." (And, to complete the story tidily, he promptly died.) His insight is an important one: what are resources for, anyway, if not to build our families? As we move through the stages of our lives, our struggle for resources never ceases; in fact, our very lifetimes are shaped by the struggle. And males and females follow different life paths and struggle differently.

Our individual reproductive costs and benefits depend not only on the ecological conditions outlined in the last chapter, but also on our age, our sex, our condition, our conspecific competitors, our resources. Biologists, who are perhaps more sanguine than other scholars about humans as biological creatures, argue that the same rules that apply to other primates apply to our own costs and benefits at different ages and stages. But along with the generalities of "being a primate" and the ecological variation we encounter, we have some traits that are particularly human, and these help set the stage for the things we can do in our lives and for the patterns of sex differences we see.

Though we are a primate, we are in some ways rather an unusual one.[1] To examine how our patterns relate to those of other primates, we use life history theory, which links behavior, natural selection, and historical accident and phylogenetic constraint to explore variation in maturation, birth, death, and behavior patterns. This approach lies at the heart of understanding diversity, precisely because it deals with natural selection, adaptation, and constraint.[2] It is a subset of natural selection theory and shares the same logic: it argues that the characteristics we see represent trade-offs in allocation of effort (energy and risk) between survival and current reproduction; between current versus future reproduction; and, within current reproduction, among offspring of different sex, size, and number. As in any constant-sum game, an organism's effort spent in one endeavor cannot be spent in another. A few life history traits are central to any analysis: size at birth; growth pattern; age and size at maturity; allocation of reproductive effort; age schedules of birth and death; number and sex ratio of offspring. The tradeoffs among these traits lead to a variety of patterns in, for example, mating, parental care, and senescence.

We humans have what life history theorists call a "slow" life history, shaped by what strategies succeeded in our past;[3] we have our first babies at a later age than one would predict from our size (as do our close relatives, chimpanzees and bonobos, though not gorillas; fig. 6.1). Though our gestation length is on average just what one would predict, our newborns are larger than expected; we have costly babies. Maternal access to resources, energy exchanges, and intrasexual selection are key to understanding these issues.[4]

Our life expectancy at birth is much longer than we would predict from our size—in contrast to our closest relatives (fig. 6.1). Newborn boys are larger and are carried longer in utero than girls, on average, and thus, as in most mammals, more expensive. The overall time from conception to weaning is shorter than we would predict; we wean our babies early and small (fig. 6.1). How is this possible? One important contribution, which we will explore later, is that of "cooperative breeding"—while mothers in most of our closely related species nurse their offspring alone, in humans, others besides the mother help feed and care for both mother and infant.

Our babies remain helpless, or altricial, for much longer than other primates while their parents protect and care for them.[5] Human babies roll over for the first time at about the age young chimpanzees are already frolicking in their group, climbing happily up, down, over, and about their mothers, consorts, and playmates.

In newborn primates, brain and body size are correlated. However, human infants have large brains for a primate of their size.[6] After birth, human babies' brains continue to grow rapidly for another year, unlike other primates, in which brain growth slows significantly soon after birth. This pattern may represent a trade-off between human brain size and our upright posture, for our posture creates great difficulties in giving birth to large-brained offspring.[7]

Because human infants are large at birth and grow rapidly, if they fit the "usual" primate pattern, they would mature early. But we are "slow"; we have our first offspring only slightly later than predicted (see above; fig. 6.1).[8] Further, young women in many societies are "subfecund" for several years—they are less fertile and suffer greater infant loss if they conceive than women in their late teens and twenties, suggesting that the resource demands of pregnancy conflict with a young woman's own continued growth. Even in societies in which a girl is married and sexually active before puberty and in which there is no evidence of deliberate fertility control, she is unlikely to have her first child before her mid- to late teens.[9] This is a conflict: other things being equal, early

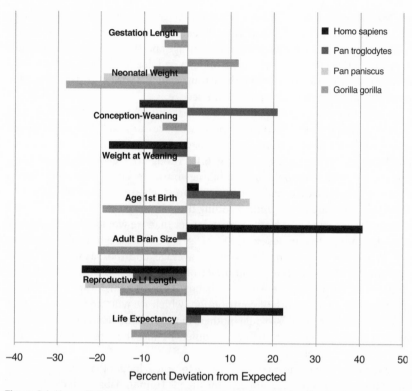

Figure 6.1. Human lifetimes are unusual in a number of regards, as are the other hominins. The earliest comparative work was that of Harvey et al. (1987), but both phylogeny and data have improved since then. These data represent only wild populations and are from sources in Rowe 1996 and Lindenfors 2002. See also van Schaik and Isler 2012. The natural log of each variable was orthogonally regressed against the natural log of adult female size (because size drives much of the pace of life history). Residuals were standardized. Human age at first birth is slightly older than predicted. Infants are heavy at birth after an "expected" gestation length, but they are weaned early and light. Life expectancy and adult brain size are much greater than expected, but the human female reproductive life span is much shorter than expected, reflecting menopause (modified from Low 2013a).

reproduction is more profitable than waiting. It's not surprising, then, that human menarche is earlier when there is better nutrition,[10] or that in traditional and historical societies in which there is differential access to resources, women paired with wealthy men began reproducing earlier than others (chapter 8). Male reproduction, as in many other mammals, is typically delayed even more, primarily because of the social forces of male-male competition. So both human females and males begin to have children later than one would expect from their size.

We make up for our late reproductive start, however: we are proportionately longer lived than, for example, chimpanzees, and our interbirth intervals are short. So we can increase rapidly once we start to reproduce. The average interbirth interval for surviving offspring among hunter-gatherers ranges from 30 to 45 months, which is similar to interbirth intervals in nineteenth-century Sweden (27–37 months) and nineteenth-century Germany (27–44 months). Chimpanzees, in contrast, even though they are smaller and should therefore have shorter interbirth intervals, average 60 months between births. We thus mature late but produce children rapidly once we begin.[11]

Finally, women stop reproducing in most human societies by their fifties, because they no longer ovulate (menopause), even though they live for many additional years; among our close relatives (fig. 6.1) a shorter-than-predicted reproductive life is also apparent. Most other female mammals spend perhaps 10% of life as "postreproductive"; women spend about 30%. This may represent not a shortening of reproductive capacity but rather a lengthening of life, although it is also true for the other hominins (fig. 6.1).[12] In sum, humans are typical in some simple life history patterns and unusual in others; we are extreme, but not unique, in a variety of ways.

These life history facts are not unconnected, amusing trivia about humans: they influence what we can and cannot do. Many of our peculiarities are linked to our complex intelligence, our social competition and cooperation, our intense resource gathering, and our practices of inheritance. We have extensively developed reciprocity, including indirect reciprocity, and often we have complex sets of group-imposed rules, mores, and laws. Here, too, it may be useful to start by treating humans as complex but not qualitatively different mammals, at least until the simple models fail to predict what we see. Our complexity means that the frequencies of different strategies can be influenced both by natural and by cultural selection (chapters 8–15),[13] giving us much about human lifetimes, and sex differences in those lifetimes, to examine.

STARTING OUT: RESOURCE STRIVING IN THE WOMB

Typically, perhaps sentimentally, we tend to view pregnancy as a time of maternal support and care for the growing embryo. But genetic conflicts over resources start here. An infant in utero is only half like its mother (this relationship is exact, for the child gets half its genes from its father). A child is only genetically on average half like any other full sibling (half siblings are only a quarter like each other on average). Thus the stage is set for conflicts of interest, both with mother and any siblings sharing the womb, and later, of course, with father and other siblings.[14]

And conflict there is. The biologist Robert Trivers first pointed out that any particular offspring gains if it can get more maternal care than is optimum for the mother. Pregnancy, far from a romantic interlude, more closely resembles an arms race, one manipulated by genes in the fetus with interests different from those of the mother. As the biologist David Haig showed, fetal genes from Dad that increase Mom's transfer of nutrients to the fetus will be favored. Will such a transfer harm Mom? No matter from the fetus's point of view, so long as the mother is healthy enough to continue investing in the fetus. In fact, if the mother's ability to produce other (competing) siblings is reduced, so much the better. On the other hand, the mother's genes, both in herself and in her fetus, that limit such transfer when it is detrimental for the mother—for example, when she is young and still growing physically herself—will be favored.[15]

As early as the implantation of the zygote on the uterine wall, trophoblast cells (fetally derived) invade the lining of the mother's womb and remodel certain arteries so that they cannot constrict to shut down blood flow to the fetus. This means several things. A mother cannot control the blood or nutrient flow to the fetus without affecting herself as well, and the placenta can now release hormones directly into the mother's blood stream. Some of these manipulations are countered by maternal strategies. Fetus and mother are truly combatants in an arms race. David Haig has cogently pointed out that a number of unpleasant accompaniments of pregnancy (as well as serious medical conditions like preeclampsia) are better explained as maternal-fetal conflict than by any competing theories.[16]

Sex Differences in Development

Over and above the issues of maternal-fetal conflict, sex differences also begin in the womb. Of course, the XX versus XY genetic differences have impact, but we are still learning how the genetics interacts with the hormonal environment in the womb, for example. There is growing evidence that early exposure to testosterone affects not only the development of infant genitalia and later brain structural differences between the sexes, but also the development of sex-specific traits, including behaviors. Interestingly, it is not estrogen that promotes female-typical development, but rather it is low exposure to testosterone. The impact is dose-dependent: the higher the exposure to testosterone, the more male-typical both brain structure and behavior become. Girl-typical and boy-typical play appear to be influenced by prenatal testosterone concentrations—and these differences show up in other primates, such as vervet monkeys (*Chlorocebus*) playing with human toys! Prenatal testosterone exposure also appears to be related to later gender identity and sexual orientation (chapter 4) and to personality (especially empathy [negative] and tendency to physical aggression [positive]). The sexual differentiation of the brain may be instrumental not only in gender identity and sexual orientation but also in patterns of some neuropsychiatric disorders.[17]

The brain continues to differentiate sexually in a variety of ways through childhood and adolescence, and some of these changes are correlated with behavior differences. In terms of physical activity and motor skills, boys tend to have the edge (sometimes this is a disadvantage in school, where they are expected to sit still for periods and thus may become discipline problems). After puberty, boys gain advantages through height and muscular strength, and as noted in chapter 4, are more likely to play competitive team sports.[18]

Boys and girls perform similarly on tests of general intelligence, in part because questions favoring one or the other sex have been avoided. In general, the data suggest that boys do better on some kinds of visual-spatial tests (e.g., imaginary manipulation of one-dimensionally represented figures) and girls tend to have larger vocabularies; curiously, boys' performances show more variance. Overall, however, the differences are small, and the results sometimes are in conflict. There is some evidence that boys' and girls' performances have been converging in recent years.[19]

Many behavioral differences are not clearly related to anatomy or physiology but are more likely the result of socialization; compared to boys, girls care more about how they are doing in school; they underestimate their abilities and care more about going to college. Boys and girls interact with their peers differently, both in terms of group size and in the degree of rough-and-tumble play.[20]

WHAT'S A MOTHER TO DO? OPTIMIZING MATERNAL EFFORT AMONG OFFSPRING

Switch viewpoints for a moment, back to the mother's perspective: it is clear that maternal investment in one child usually comes at the expense of investment in others or a mother's investment in herself.[21] Robert Trivers' phrase "parent-offspring conflict" highlights an important issue: offspring profit by getting as much, free, from Mom as possible—but Mom, unless she has only one offspring in her lifetime, must apportion her effort among *all* her offspring and herself. Large sibships can be particularly costly for later-born children.[22] Closely spaced pregnancies, when nutrition or other factors are limiting, may lower a woman's lifetime reproduction. Among the African Efe, for example, women's ovarian function and resulting birth schedules show a seasonal pattern that correlates only with food availability—a clear reproductive response to changing ecological conditions. Even subtle maternal responses (e.g., adjustment of blood flow to the uterus) during pregnancy fit a life history model of lifetime reproductive optimization.

There may be further conflicts for women between maternal investment and the effort required to acquire resources (whether gathering food or finding child care in order to work): what a woman invests in work typically cannot be invested in child care, though the intensity of conflict probably varies cross-culturally (chapters 8, 15).[23] Ecological factors constrain these conflicts. For example, among the !Kung of southern Africa, women have interbirth intervals of about four years. Predators are prevalent, and !Kung women who depend on bush foods may carry their children, at least occasionally, for up to six years.

But women must carry the foods they gather, and they can only carry so much. Women's "backloads" (weight of child plus foraged material) are good predictors of interbirth intervals and mortality patterns. !Kung women living in compounds, who are not dependent on bush foods, have children close together. For bush-living women, the number of successful descendants is maximized not by maximizing the rate of births, but by responding to the conflict between the production of a new child and the cost of such production on the survivorship of other children. Women who live in compounds do not face these conflicts.[24] The issue of trading off number versus quantity of offspring seems remote in wealthy Western developed nations today, but it may in fact still be related to patterns of resource consumption and fertility (chapters 8, 15).[25]

Mothers, more than fathers, face conflicts of getting versus allocating resources.[26] Among the South American Ache, Hiwi, and Ye'kwana, nursing women can forage less than others.[27] In some societies, other children, usually siblings, help with child care, and the availability of such children can have an impact on a mother's lifetime fertility. On the Pacific island of Ifaluk, for example, a woman's lifetime fertility is correlated with the sex of her first two children: women whose first two children are girls have greater lifetime fertility than others. Daughters assist in child care on Ifaluk, so mothers whose first children are daughters defray some costs and can have more children.[28]

In other societies, resources to hire wet nurses might help minimize the conflict. A dramatic example is given by the anthropologist Sarah Hrdy, who found that in eighteenth-century Paris, interbirth interval, fertility, and infant mortality all varied with the mother's status.[29] The richest women had very short interbirth intervals, very high fertility, and low infant mortality (although chronic anemia and prolapsed uteri were costly); a linear relationship between the cost of the wet nurse and infant survivorship meant that the richest women, who could afford the best wet nurses, fared best in terms of fertility. But hiring a wet nurse had unintended consequences.[30] Among the bourgeois, complexities created more variation in pattern. Poor women had long interbirth intervals, low fertility, and high infant mortality; and the wet nurses fared worst of all, with very long interbirth intervals, very low fertility, and very high infant mortality.

Conflicts of Interest: Abortion, Infanticide, Abandonment, Neglect

Parents seldom kill their children. After all, even though a parent's interests are not identical to a child's, an infant's death means the loss of considerable parental investment. Indeed, infanticide in most species is typically committed by reproductive competitors rather than by parents; for example, in lions, langurs, and gorillas, when a male takes over a harem he is likely to kill all infants under a certain age. The mother then becomes sexually receptive, and the male profits both by eliminating an offspring with a competitor's genes and by gaining a mating. In humans, also, stepparents (whose reproductive interests do not coincide with the child's) are more likely to abuse or neglect children than genetic parents, and, regardless of the old fairy tales, stepfathers are more likely to commit infanticide than stepmothers. In light of these facts, patterns of parental investment in families of adopted children (in which both parents likely participated in the decisions) versus stepfamilies is interesting. Stepchildren are more often abused, neglected, and even murdered than genetic children. Adopted children, perhaps because they have more need and parents' interests are convergent, may sometimes get more investment than genetic children.[31]

Human parents, like parents in other species, do sometimes commit infanticide and abort and abandon their infants. When is infanticide not pathological, but adaptive? Once again, we must remember the trade-offs: if each infant requires great investment, parents must apportion their effort, and parental investment biases, even to the extent of infanticide, can be reproductively profitable: for example, if the mother is alone and without family or resources to help with care, or if the child is unlikely to be successful.[32]

Cross-culturally, deformed or seriously ill newborns are killed most often, and there is evidence that some cultural conceptions of "ill omens" leading to infanticide are real reflections of low newborn quality. Similarly, mothers are more likely to commit infanticide when external circumstances reduce their chances of successful investment; interbirth periods that are too short, twins, lack of an investing mate or stable pair-bond—all increase the likelihood of infanticide or neglect.[33]

Historical studies of child abandonment also reflect such considerations: a mother's ability to invest in the child (including her own health, familial resources, economic conditions), and the child's health, legitimacy, and sex. In France, Spain, and Russia, infanticide was related to economic factors and a mother's abilities.[34] Similarly, although he failed to discern any pattern, historian John Boswell's overview of child abandonment reveals that 46% (29/63) of cases he examined were, despite great variation in time, country, and other circumstances (as well as absence of data), related to maternal ability to invest. When resource allocation problems (16/63; 25.5%) and offspring quality (4/63; 6.3%) were considered, selective reasons were apparent in 49/63 cases, or 77%.[35]

Abortion, too, appears more common in circumstances in which the birth of a child is likely to reduce the mother's lifetime reproductive success. As women age, and their reproductive value declines (their future reproductive opportunities wane), they are less likely to seek abortion. Even attitudes toward abortion in our society are related to the proportion of women in any group who are "at risk" of an unwanted pregnancy.[36]

Of course, none of these behaviors or attitudes about them are set or in any way "determined"; they can be influenced not only by an individual's own condition but also by the attitudes of those around them. For example, in the United States today, a woman is likely to favor abortion if she is still fertile and thus potentially vulnerable to an unwanted pregnancy. Opinions on abortion also become political as individuals and party leaders influence each other. From 1972 to 1994 in the United States, Democratic and Republican party positions on abortion have changed gradually (as reflected by House and Senate votes); as this has occurred, individual voters have switched party alliances to align with their own attitudes about abortion. And attitudes and behavior can change over time; for example, the Guttmacher Institute recently reported (May 2014) that teen pregnancy, birth, and abortion rates have reached a historic low.[37]

Invest or Desert?

Let us put these empirical data into context. The trade-offs of using resources for oneself (e.g., for growth or maintenance) versus reproducing are relatively clear in other species. Humans, too, appear to be influenced by resource availability in their reproductive behavior, whether consciously or not, and those trade-offs are very different for men and women.[38] Figure 6.2 shows a combination of reproductive trade-off curves for three people: a man, a woman, and their child. For each parent, there are three families of trade-off curves (A, B, and C) representing different possible trade-offs for the effect of allocating resources to oneself for future mating success. In addition, there are three curves for the effect of a parent's self-investment on an offspring's mating success (offspring A, B, and C in fig. 6.2); that is, too-selfish parents, in this model, reduce their children's chances of surviving and marrying.

When the "return from child" curve is higher than the "return from new mating" curves for *both* parents at a given resource level, both partners profit from being parental, and both should cooperate in rearing a child (fig. 6.2: shaded area in superimposed curves). Thus, in figure 6.2 (#1), when parental curves are concave (monotonic increasing) and the child curve is convex (monotonic decreasing), both parents profit most by investing in the child, and cooperative rearing is likely. When the payoffs for self-investment are higher for both parents than for investing in the child, the child is in danger (fig. 6.2, #2).

When parental curves are identical, no conflict of interest exists between the parents, whether they are likely to keep or desert the child. But conflicts are likely in some age and resource combinations. Consider, for example, a married couple, both forty-two years old, in which the man is rich and powerful and the woman becomes pregnant (fig. 6.2, #3). The woman will profit by investing in this child. For the husband in this example, using half of the available resources for his own new mating benefits him more than investing in this child. This

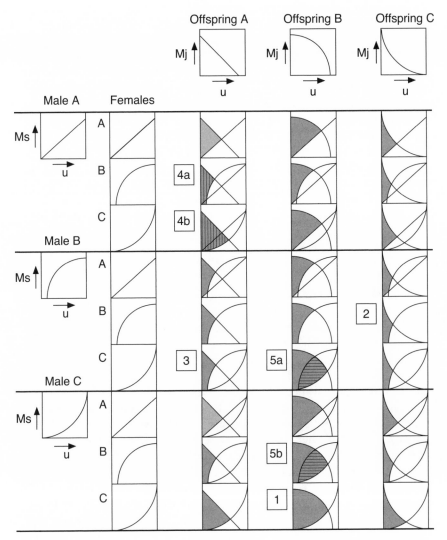

Figure 6.2. Fathers, mothers, and children face different tradeoffs, here superimposed. For each parent, there are three families of trade-off curves (A, B, and C) for the effect of allocating resources (u; x-axis) to oneself on own future mating success (Ms). In addition, there are three curves for the effect of investment by a parent in itself (u; x-axis) on an offspring's mating success (Mj for offspring A, B, C). The shaded area represents the difference in gain between investment in offspring and investment in self.. That difference is zero where the curves cross. Examples 1 and 2 focus on how parental confluences of interest can vary (shaded area); examples 4a and 4b show how female interests might change with age (vertical hatching); examples 5a and 5b show male and female confluence (shaded) and conflict (horizontal hatching) of interest over further investment in the child vary with age. (Modified from E. Hill and Low 1991.)

situation has implications for male allocation of effort between mating and parental care; whichever is more fitness-limiting will be important in the decisions.[39]

Because women's reproductive value peaks at the age of first reproduction and declines thereafter, age affects the shape of women's return curves more than men's.[40] Holding the male and child curves identical, consider the trade-offs for an older versus a younger woman: #4a represents the conflicts for a younger woman, #4b for an older woman. The vertical hatching represents the area in which it pays the female to continue to invest. This area is greater for the older woman; she gains more from investing. Because the trade-off curves are shaped so differently, there will always be a greater benefit for the older woman to stay and invest in this child—thus suggesting that women are less likely to abort even an unintended pregnancy when they are older. Another influential factor is resource availability for females: women with considerable resources can enhance their own chances of mating again without significantly altering the child's chances for success; women with fewer resources have a more significant conflict of interests.

Sex Differences in Reproductive Lifetimes

When women's and men's lives are compared, some reproductive patterns must, of course, be similar. The average number of children can never be higher for women than for men, for example. Other phenomena—like age at marriage, variance in reproductive success, and the rate and impact of remarriage—can, and do, differ strikingly between men and women.

I have argued that men and women, like other male and female mammals, seek resources toward somewhat different reproductive ends. Much of men's striving, like that of other primate males, centers around the mating aspects of reproductive effort.[41] Male-male coalitions are associated with competition for status or resources and with resources that can be more effectively obtained and protected by groups of males than lone males: heritable land of lasting value, some large game. These conflicts may become warfare (chapters 13, 14). While coalitions are likely to originate among brothers and be more common and stronger when related men live together, more or less fluid coalitions of this sort arise among men of various relatedness and among nonrelatives in many societies. Men's coalitions typically are broad-reaching and fluid, involve both more risk and higher reproductive stakes than women's coalitions, exert considerable power, and control significant resources (chapter 11).[42]

The principal difference between these expenditure patterns is that mating effort and non-offspring-specific parental effort, unlike true parental investment, may have a high fixed cost: much must be spent before any success is realized, and later successes cost little compared to the cost of achieving the first success. Thus, many males will fail to reproduce, but a successful male may have many times more offspring than the most successful female, just as in the elephant seal and red deer examples in chapter 3.

This fact has profound implications not only for male risk taking and survivorship, but also for *parents* in polygynous societies: parental expenditure may not be optimized by equal investment in the two sexes (see below) but rather by biasing investment toward the sex in which success is more strongly correlated with investment and varies more—that is, the sex for which the maximum possible payoff is higher for investment. Often but not always, in polygynous mammals this investment biased toward males (for a conspicuous exception in humans, the Mukogodo, see chapter 7).[43] Successful individual responses vary in different ecological and social conditions.

Finally, these individual costs and benefits can lead to different population patterns, depending on the composition of individuals, and environmental richness, evenness or patchiness, and predictability. That is, population growth rates arise from individual reproductive striving, ecological richness, and predictability—and these vary around the world (chapter 15).

Differential Investment in Sons and Daughters

In a way, we have come full circle, from fetal strategies for resources in the womb, to issues of differential parental investment in children. Some cases (see above) of differential investment or withholding are issues of individual quality of child or parental ability. But some broad, general patterns of differential investment— by sex of offspring—arise from the differences in the return curves shown in chapter 3. The importance of resource value for men versus reproductive value for women means that in many societies it may be harder for a man to get a wife than for a woman to get a husband (though a really successful man can have many more children than the most successful woman). Age affects men's and women's reproductive possibilities differently.[44] If the reproduction of sons is more variable than that of daughters in a particular environment, and especially if wealth or status also matters more to men's success than women's, investment (including inheritance, where relevant) is likely to be biased toward sons. Whenever the two sexes succeed by acting differently, interesting consequences follow: parents who invest in and train their sons and daughters differently will prosper reproductively, and the evidence is that parents do. Investment can involve the energy and other physiological costs of carrying a child, differential investment in training and education, and differential inheritance practices.[45]

Mating Market Forces: Fisher's Insights

As long ago as 1928, Sir Ronald Fisher, a British mathematician who identified and clarified several important problems while dabbling in biology (his work centered on statistics), noted that the population sex ratio should influence the value of relative investment by parents in offspring of the two sexes. He noted:

> The sex ratio shall so adjust itself, under the influence of natural selection, that the total expenditure incurred in respect of children of each sex, shall be equal; for if this were not so and the total expenditure incurred in producing males, for instance, were less than the total expenditure incurred in producing females, then since the total reproductive value of the males is equal to that of the females, it would follow that those parents, the innate tendencies of which caused them to produce males in excess, would, for the same expenditure, produce a greater amount of reproductive value; and in consequence would be the progenitors of a larger fraction of future generations than would parents with a congenital bias toward the production of females.[46]

Fisher's point is not universally true, as we will see in a moment, but it sets the stage. He focused on the population-wide impact of the relative average contribution of males and females. Since, in a sexual species, every child has a mother and a father, if one sex is rare in the mating market it becomes more valuable. The contribution to the next generation of *all* the males is exactly equal to that of *all* the females; so if there are only half as many males as females, each male is, on average, twice as valuable.

There are several assumptions here that, if not made explicit, will lead us astray: this model assumes outbreeding and does not consider sexual selection at all except as a function of "numbers available." But it does suggest why, in outbreeding populations, the sex ratio at birth across the population will tend to equilibrate at about 1:1 (50% male, 50% female).[47] It also suggests that extreme biases in population sex ratios might affect social systems. This has rarely been important for human societies; there have been short disruptions of population sex ratios (e.g., male deaths in war), followed by shifts in the birth sex ratio toward more males, but in most populations, birth sex ratios do indeed equilibrate at about 1:1.[48] Social and cultural influences, however, might easily interact with natural selection to result in unusual sex ratios (chapter 10).

Sex Biases in the Womb: Trivers–Willard Effects

In many polygynous mammals, including humans, because only the best-invested males have much chance of reproducing, mothers invest more in each successful son than in each daughter. Sons, compared to daughters, are carried longer in utero, they are larger at birth, they nurse longer and more frequently, and they are weaned later.[49] Robert Trivers (see parent-offspring conflict, above) and Dan Willard argued that in polygynous species under such conditions, females in good nutritional condition should be likelier to bear sons than daughters. A more broadly applicable statement might be: when variance in reproductive success of one sex exceeds that of the other sex, and when parental investment can influence the reproductive success of one sex more than the other (as in elephant seals and baboons), there should be a correlation between parental condition and investment in that sex.[50]

Trivers and Willard assumed that a mother's physiological condition (resources available to rear a successful offspring) would decline with age. In most nonhuman species, and in many preindustrial societies and developing countries, this may be true.[51] But whenever the nutritional condition of mothers does *not* decline with age, a *male* bias in sex ratio might be found in older mothers. If a female's condition is good, as she nears the end of her reproduction it may pay to invest ever more heavily, with a greater potential reproductive profit if successful; for example, older female gorillas tend to have sons. In some nutritionally rich societies, such as nineteenth-century Sweden, mothers over age 35 showed a sex-ratio bias toward sons, and mothers under 25 toward daughters. It is interesting, in light of these physiological patterns, that parents can shift their production of sons versus daughters under different circumstances.[52]

Sex Biases in Inheritance and Survival

Perhaps no other species exhibits the degree of resource transfer that can take place through inheritance in human families; parents invest, often differentially, in their children even after death. Inheritance is typically biased by legitimacy, birth order, and sex.[53] Why? Remember that, in many societies, relatively few males can be highly successful, and resources are central to men's success. Parents can maximize the survivorship and reproductive chances of their sons and daughters—their net reproductive profit—in such societies by allocating resources unevenly.

In societies with heritable goods, cultural factors, the size of a family, and the sex of siblings may influence men's and women's ability to marry somewhat differently. Within traditional polygynous marital systems, inheritance is strikingly male biased, precisely the pattern predicted if male reproductive success varies more than female reproductive success, and male success is influenced by resource control.[54] In other situations, for example, the hypergynous Indian societies (see below), and the Mukogodo of Kenya (next chapter), poor parents may find that daughters are more marriageable (and thus reproductively more valuable) than sons.

Even within sex, resources are not necessarily allocated evenly. In many societies, earlier-born sons tend to inherit the greatest proportion of the resources; they literally have fewer chances to die before putting those resources to reproductive use. Among fifteenth- and sixteenth-century Portuguese nobles, the proportion of ever-married men and women decreased with birth order, as did fertility for married individuals. Among the Kenyan Kipsigis (also next chapter), men's reproductive success declines with the number of brothers and increases with the number of sisters—and parents show reduced parental investment in sons with many brothers and heightened investment in sons with many sisters. Among the Gabbra pastoralists of Kenya, too, investment is biased: men with many older brothers have lowered reproduction.[55]

Sex-biased patterns of investment are common even where more equal distribution is stipulated by law. For example, in nineteenth-century Sweden (chapter 8), firstborn sons tended to inherit land, later sons and daughters made do with other goods. Women's lifetime reproduction decreased as their number of siblings increased. For men, only the number of brothers mattered, suggesting that (1) brothers represent resource competitors for men, and (2) as total sibship size increased for women, they were more likely to be drawn into caring for their siblings (regardless of sex), at some cost to their own reproduction.[56]

In nineteenth-century Germany, the overall sex ratio of children born was almost exactly even. A fair proportion of children died in their first year of life, and the pattern of these deaths (many due to parental neglect) were status related: farmers' daughters tended to die, whereas in other classes, sons were more likely to die. For farmers, daughters were likely to be considered less desirable than sons; for other classes, the reverse appeared to be true. Similar sex biases show up in the early (pre-1860) history of the United States and many older agricultural societies (though not in the nineteenth-century Swedish data: chapter 8). There is thus possible evidence of uneven parental investment tied to the perceived value of each sex for parents in different social classes.[57]

A variety of other Trivers–Willard effects may appear in contemporary American society. Interbirth interval, birth weight, and the proportion of children nursed are related to income and the presence of an adult male in the household. As income increases, so does interbirth interval and the percentage breastfed—for sons, but not for daughters. Daughters receive relatively more from low-investment mothers, and sons get more from high-investment mothers. In Tennessee, sons in higher-status families fare better than others. Among polygynous Mormons, sex ratio and parental status covary as predicted by Trivers and Willard.[58]

Not only does the sex of the recipient matter in terms of how much and what kind of help is received, but so may the sex of the giver. In many societies, not only parents but also grandparents, aunts, and uncles give to younger relatives. Because of the uncertainty of paternity in mammals (giving rise, for example, to the mate-guarding tactics described in chapter 3, and raising reproductive conflicts of interests), we might expect some sex differences in how the two sexes give, and are given, bequests. In the United States, women tend to spread their bequests more widely than men, leaving more but smaller bequests. Men are more likely to leave bequests to their widows, with instructions about dispersal to the couple's children. Perhaps this is because men are more likely to remarry and have more children (creating a reproductive conflict of interest), so that women are more likely to leave bequests directly to their children. Bias toward sons is not only associated with polygyny but also with larger families and agricultural holdings; this pattern becomes rarer as land becomes less important and as families become smaller. In a study of American college students, aunts invested more in their nieces and nephews than uncles did, and matrilateral aunts (with the more certain degree of relatedness) gave more than patrilateral aunts.[59]

In some societies, a sex bias in abortion and infanticide exists; this represents a conundrum if it becomes widespread and persistent, for the rare sex comes to be more valuable in any marriage or mating market (above).[60] Recent experience under China's one-child policy and the sex preferences in India demonstrate this dilemma (see chapter 10). Many, possibly most, cases of long-term sex-biased infanticide may simply be maladaptive, but there are some examples that suggest evolved parental strategies. Among the Inuit Eskimos, for example, the female-biased infanticide pattern suggests that parents may be trying to match their number of sons to local prevailing sex ratios, keeping daughters when the local sex-ratio is male biased and killing them when females are overabundant.

In hypergynous societies, women may marry "up" and men "down," but the reverse is not allowed. Thus, daughters are valuable to lower-class families but costly to upper-class families. There appears to be no single across-society sex bias; but infanticide is female-biased in high-status families, and son preference is less strong in low-status families. Male-biased infanticide is very rare. Thus the infanticide patterns above are consistent with other fitness-striving in hypergynous societies.[61]

Training Boys versus Girls

By now, it is an obvious prediction that parents are likely to raise children of the two sexes differently, not only for ecological but also for resulting physiological and physical reasons; and it should be clear that these differences should be exaggerated or minimized, depending on ecological conditions and the nature of the society. Cross-cultural research based on children's behavior in natural circumstances suggests that the sex or gender differences we observe in American and English children are not limited to Anglo-Saxon cultures, and that there are patterns to both differences and similarities cross-culturally in how boys are treated compared to girls.[62]

The ecology of mating versus parental returns (chapter 3, and especially fig. 3.2A) and makes predictions about raising boys and girls differently across cultures. Because the intensity of sexual selection differs between the sexes in polygynous systems, it seems likely that male and female humans, like males and females of other polygynous species, have maximized their reproductive success through different behaviors throughout their evolutionary history. Cross-culturally, sons are more strongly trained than daughters in behaviors useful in open competition (e.g., fortitude), while daughters are more strongly trained in such values as sexual restraint, obedience, and responsibility—traits widely sought by men in their wives.[63]

The above predicts nothing more than universal sex differences in training.[64] However, further thought suggests that the more polygynous the society (so long as it is not stratified), the more should boys be taught to strive. Variance in reproductive success increases for men as the degree of polygyny increases, and very few men may be extremely successful while many men fail, just as in other polygynous species. In such situations, the rewards may be great if boys are trained to strive.[65] But because there is some association between stratification and polygyny (stratified societies tend to be polygynous), and the impacts of polygyny and stratification are opposed, it is important to separate these conditions analytically. In stratified societies, men's potential reproductive rewards for striving are constrained. Whether the stratification is by wealth or is hereditary, the more stratified the society (the less striving and reproductive payoffs covary), the less boys are taught to strive openly (cf. chapter 4).

Cross-culturally, the more women actually control important resources or exercise power, the less daughters are taught to be submissive. For example, the more women are able to inherit property, the less daughters are taught to be obedient. The more formal power women have within the kin group, the more daughters are taught to be aggressive and the less they are trained to be industrious. In societies in which women can hold political office, daughters are more strongly inculcated in achievement and striving than in societies in which women cannot hold office, although the difference is only marginally significant. The more authority women have over children older than four, the less daughters are taught to be obedient.

In sum, across cultures, sons and daughters are trained differently in ways that relate to the evolutionary history of each sex's reproductive success. Despite the strong inference that childhood inculcation is a parental response to ecological and social pressures affecting reproductive success, I know of no direct data on the relative reproductive success, within any society, of parents training their sons and daughters differently within the society. There are now excellent examples of gene-culture coevolutionary modeling.[66] Empirical data do exist for a number of societies showing that male reproductive success is related to resource control and status (see below), and thus that boys who learn to be successful in obtaining resources and/or status grow up to be reproductively successful men.[67] The positive relationship seems to hold generally for a wide range of societies. It is therefore not surprising that across all societies, boys are trained to strive.

In current American school situations, too, there is evidence that boys and girls are treated differently in ways that are consistent with our evolutionary history, but with effects we might neither expect nor desire. For some time, psychologists and educators have noticed that in school, girls are more likely than boys to attribute their failures to low ability and to respond to failure with decreased effort and performance—a "learned helplessness" (related to development issues above).[68] Yet, on average, girls receive more praise, less criticism, and higher elementary school grades than boys and are viewed by teachers and other adults more favorably. Why doesn't the positive reinforcement result in stronger girls' performances? In one study, the kind of positive feedback teachers gave children differed by sex.[69] Girls were praised for nonintellectual items, like being neat; boys for intellectual cleverness. Eighty-eight percent of negative feedback to girls concerned academic issues (compared to 54% for boys). Nonacademic issues were the focus of 46% of negative feedback for boys (versus 12% for girls). Is it possible that without meaning to, we have been perpetuating a sex difference? (Girls: be reliable, obedient, and neat; boys: be clever and striving.)

Sex Differences in Senescence

Among primates, humans are relatively long-lived, compared to what one would predict from our size (fig. 6.1).[70] It is still not clear why we are so long-lived. Although it is tempting to claim so, it is probably not (at least entirely) because we are so smart.[71] Nor are medical advances the answer; most evidence suggests that people in numerous traditional societies had long lives.[72] The current weight of evidence suggests that while reducing accidental death, for example, has changed human life expectancy slightly, the existing maximum lifespan is not a product of medical advance but of human evolution. One important inference from this is that there is little chance that medical science will be able to change human lifespan significantly.

In humans, as in other primates and most mammals (and most polygynous species in which males fight), males tend to die at younger ages than females. Older explanations argued for "male vulnerability" because males are the "heterogametic sex"—they have an "unprotected" Y chromosome: any deleterious gene on the unprotected portion would cause harm.[73] But such suggestions do not explain the fact that, for example, males in many bird species (in which *females* are heterogametic) still die sooner than females. In fact, the breeding system and the different payoffs for risk taking by males and females are what matter. In polygynous species, especially those with limited or no male parental care, males evolve to be mating specialists, with the attendant high fixed costs, risks of failure, and high payoff for success (chapter 3). In such systems, males who do not take risks will die without offspring, but risk-taking males will have higher death rates.

Old humans do die eventually, but first, like old soldiers, they begin to fade away. This raises the issues of why organisms senesce and why they do so at such varying rates. Older explanations like "wearing out" and "toxin accumulation" would not predict rate differences; all species should senesce at similar rates.

And that is simply not true. Older group-level "adaptive" explanations (clearing the way for new individuals) simply do not make sense; cheaters who stayed alive and well would win. Arguments that senescence is selectively irrelevant do not explain why senescence actually begins at the age of first reproduction, not late in life.[74]

The biologists George Williams and W. D. Hamilton first pointed out that, other things being equal, longer life should be favored by natural selection.[75] Senescence arises in part because of pleiotropic effects (a single gene can have multiple effects). Natural selection favors genes with positive early effects. Because reproductive value[76] declines from the age of first reproduction, at some age selection cannot distinguish between simple early good effects versus early good effects accompanied by later (pleiotropic) deleterious effects, for these later costs affect fewer individuals in any population (many have already died), and they affect a smaller proportion of the reproductive lives of the remaining individuals. In this way, deleterious genetic effects late in life accumulate; and, as we age, if it isn't one complaint, it's another. Senescence, then, is not an evolved phenomenon but a cost, a byproduct of selection favoring early positive traits— we senesce because we're stuck with the process.

GRANDMOTHERS AND OTHERS: COOPERATIVE BREEDING

Humans, unlike individuals in many other species, are typically embedded in a family and a network of friends, all of whom may have an influence on women's fertility, and even on early female reproductive senescence. Certainly fathers can help, and in some societies their absence can have quite dramatic consequences. Children may help defray costs.[77] Additional helpers may have been essential to couples in the past, and may even be so today.[78] Among the Ache of Paraguay, each breeding pair obtains help from about 1.3 nonreproductive adults. Young married men and unmarried men of all ages produce the vast majority of "excess" food. This is important because a reproducing Ache couple can expect food deficits. Among the Aka in the Central African Republic, mothers' energy expenditure can be reduced by helpers. Here, grandmothers are most effective, showing an almost 1:1 substitution of energy production. Fathers are more likely to assist in direct child care, but the impact is a net reduction of total direct care by all caregivers. Among the Tsimane of Bolivia, fathers engage in direct care, especially playing and comforting, suggesting their concern for child well-being (although there is no evidence that mothers view caretaking fathers more highly). There are clearly multiple pathways for helping mothers with dependent children, and the effects of different helpers can be quite distinct.[79]

Kin can have other impacts on mothers as well. In rural Gambia, patrilineal kin (especially the husband's mother) are associated with increased fertility, while matrilineal kin (especially the maternal grandmother and unmarried older sisters) affected children's survivorship. The difference in impact is not surprising, given that the burden of motherhood falls squarely on the mother and her maternal kin. In Thailand, living virilocally (one form of living with the husband's kin) increases total fertility by shortening the time to first birth and later interbirth intervals. Living uxorilocally (one form of living with the wife's kin) also shortens the time to first birth, and marginally improves child survivorship, but has no other effects—so the resulting number of living children does not differ from those of women living neolocally. Some of the variation may be explained by a particular subset of relatedness: where the X chromosome comes from. Girls' X chromosomes can come either from their mother or father, whereas boys' X chromosome always comes from the maternal line. It appears that X chromosome inheritance can be related to child mortality in the presence of a grandmother.[80]

In modern Britain, too, kin still have an impact on fertility. After a woman's first birth, her probability of having a second birth is higher if she has relatives who provide child care and if she has a relatively large number of emotionally close (and often seen) relatives.[81] In a study of nineteen modern high-income countries, kin availability was correlated with fertility, but the nature of correlations depended on the nature of kin relatedness: having a living mother increased fertility (largely through earlier first births); coresidence with parents, however, correlated with decreased fertility. Modern methods also make a difference. In modern rural Ethiopia, women who have relatively high fertility are likely to adopt contraception; there is little evidence for broad social transmission of the idea, although the most predictive forces were individual sociodemographic characteristics (parity, education) and religious norms.[82]

The relative importance of maternal grandmothers may be related to human senescence, which is unusual in ways related to cooperative breeding, as reflected by the patterns of figure 6.1. All physiological systems senesce, and reproductive function is no exception. Yet human female reproductive function, at least today, decays decades earlier than other systems in either sex, including male reproductive function. Most human physiological systems decay at a slow steady rate from about age 30, and function at age 65 is about 60% to 70% of maximum. Among many mammals, female reproductive function decays at about the same rate as other systems, and even very old females retain some fertility. Such a female's life expectancy at her first reproduction is usually not notably longer than her expected reproductive life.[83]

In most other mammals, old females might spend 10% of their lives after their last birth; in contrast, human females live perhaps a third of their years after menopause, and human females lose reproductive function dramatically after age 30.[84] Women in natural fertility (non-contracepting) societies show top fertility between ages 20 and 30; fertility then declines to zero (menopause) typically between ages 45 and 50.[85] In the developed nations today, the average age at menopause is about 50.5 years.[86] Such a difference in the rate of senescence among systems—heart-lung, for example, versus reproductive—is quite striking.

Only in a very few other species do females live a considerable proportion of their lives either in a condition of very low fecundity, or after something similar to menopause—total reproductive senescence.[87]

The anthropologist Kristin Hawkes and her colleagues note that species with a reduction of fecundity, such as elephants (*Loxodonta africana*), horses (*Equus ferus caballus*), odontocyte whales (suborder Odontoceti), bonobos (*Pan paniscus*), and humans, are species in which offspring may depend on their mother for some time. Thus, they argued, it may pay older females to shift from production of additional offspring to continued high-level care of existing offspring and grandchildren. In many traditional societies, grandmothers continue to contribute to the well-being of their families, assisting their daughters and grandchildren. Postmenopausal grandmothers can often be as efficient as younger women in getting resources. And today, some parents support their children across the life course.[88]

Despite the attractiveness of this hypothesis, there are really few data and results conflict. The first tests of the grandmother hypothesis were conducted by the anthropologists Kim Hill and Magdalena Hurtado, working with the Ache of Paraguay (chapter 7). They found positive effects of grandmothers' help; men and women with a living mother do experience slightly higher fertility than others, and children with a living grandmother do survive slightly better than others. These effects are small in the Ache, and women who remained fertile longer had higher reproductive success than others. Recent simulation models similarly find that, probably because of the weak selection on relatively few women toward the end of life, grandmothering benefits are not likely to be large enough for selective benefit.[89] Empirically, the results vary greatly. Because the value of a grandmother may vary with age-specific fertility and mortality patterns, a grandmother's value to a mother's fertility may vary across societies. Studies in the Gambia have found that maternal grandmothers improve the survivorship of their grandchildren and that the presence of a woman's in-laws increase her fertility. In contrast, no such effects existed in Malawi. Similarly, conflicting results arise from cross-cultural reviews. In different societies, various relatives may or may not affect women's fertility and children's survival. Here, as Darwin might say, is a problem for future analysis.[90]

Craig Packer and colleagues tested the grandmother hypothesis with field data from baboons and lions—two results countered the grandmother hypothesis: (1) old-reproducing females had no higher mortality costs of reproduction than younger females; and (2) grandmother help did not improve the fitness of either grandchildren or reproductive-age offspring.[91]

There are alternative hypotheses about the evolution of menopause. The evolutionary anthropologist Hillard Kaplan and colleagues introduced a two-sex learning- and skills-based hypothesis that proposes these phenomena determine economic productivity in traditional societies. Using data from the Tsimane, they found that men, as well as women, become reproductively inactive early, and that both older men and older women contribute to the benefits of their younger relatives (such as grandchildren). Again, we would expect cross-cultural variation. Ruth Mace and Alexandra Alvergne suggest that the combination of patrilocality/virilocality (ways that young couples live with the husband's kin: patri-husband's father, viri-husband's uncle) affects the degree of relatedness between young and older women (the grandmother hypothesis suggests that grandmothers should direct help toward their genetic grandchildren, but these living arrangements, the most common among traditional societies, mean that older women are not likely to be related to younger women). Mace and Alvergne found that the effects of competition on reproductive success in overlapping generations extend to women who do not reside in the same household—a step toward understanding help in patrilocal and virilocal living arrangements.[92]

The evolution of menopause, and the varied effects of different helpers, remain interesting and unsolved problems.

7

Sex and Resource Ecology in Traditional Cultures

The fact that a man bears an excellent reputation among men, is no proof that he may not be the worst possible companion for a woman.

He sat for more than an hour, trying to analyze his feelings. When a woman does that, ten to one she is in love. When a man does it, ten to one he is not.

—Ella Wheeler Wilcox, *Men, Women, and Emotions* (1893)

MEN AND WOMEN seek and use resources for reproductive success, but they can differ as much as peacocks and peahens in how they seek resources, what kinds of resources they seek, and how they use those resources. The reproductive ecology of the two sexes in humans, as in other mammals, creates opportunities for quite different uses of resources in reproductive competition by males and females and for different strategies (e.g., coalitions) to get them. In earlier chapters, I explored how our background of mammalian sex differences interacts with ecological conditions to yield different mating and resource systems. Here I want to explore how the two sexes are likely to use resources for reproduction within different systems.

Sexual Divisions of Labor

It's an odd thing, perhaps, that many of us are not bothered by discussions of male-female differences in other species but find it unsettling to ask about sex differences in ourselves. Although men and women differ relatively little physically, humans are one of the most sexually dimorphic of all the primates in behavior— men and women *do* different things.[1] Except for mating activities, which consume little time, and nursing, male and female chimpanzees or gorillas spend their days in far more similar activities than do men and women. Human behavioral sex differences are not an artifact of Western culture or recent history: across societies, around the world, men's and women's days are spent differently.[2]

The anthropologist George Peter Murdock, with Catherine Provost, looked at men's and women's resource activities cross-culturally.[3] They found a number of activities to be exclusively or predominantly male around the world: hunting large aquatic or terrestrial fauna, ore smelting, mineworking and quarrying, metalworking, lumbering, woodworking, fowling, boatbuilding, stoneworking, manufacture of musical instruments, and bonesetting and other surgery. There was surprisingly little variation around the world: in only 7 of 1,215 cases (society × activity) was an activity reported as predominantly or exclusively female. These cases are truly scattered and often something of a misnomer. For example, "mining" is coded as a female activity among the Fur of Sudan; what Fur women actually do is collect dust containing iron ore for sale to smiths.

Around the world, the activities done principally by women included fuel gathering, drink preparation, gathering and preparation of plant foods, dairy production, spinning, laundering, cooking, and fetching water. There were no technological activities that widely were exclusively female; in fact, men participated, sometimes equally, in many of these "mostly female" activities. When simpler technology is replaced by more complex machines in daily activities across cultures (e.g., the plow in horticulture), men become involved. And when an occupational specialty begins to involve commoditization, profit, and a larger market—even if women otherwise are the principals—men tend to take over. Consider male potters among the Aztecs, Babylonians, Romans, Hebrews, and Ganda; male weavers among the Burusho and Punjabi; and male mat makers among the Aztecs, Babylonians, and Javanese. As Murdock and Provost noted, "Even the most feminine tasks . . . cooking and the preparation of vegetal foods, tend to be assumed by specialized male bakers, chefs, and millers in the more complex civilizations of Europe and Asia"—and our own.

I suspect divisions of labor relate to our mammalian heritage: to women's constraints of pregnancy, child feeding and care, and to the differential return curves of largely female parental effort versus the high-risk, high-gain opportunism of typically male mating effort (fig. 3.2). Women are more likely to do activities that require daily attention, do not require long absences from home, are not life-threatening, allow multitasking, and can be easily resumed after an interruption.[4] In most societies throughout much of our evolutionary history, small children were likely to die if their mother died. Under most ecological conditions, the patterns we see were simply efficient and made ecological sense.[5]

Our mammalian background is reflected even in simple daily patterns of sex differences. For example, men and women share food very differently. Consider the Ache. The resources brought in by men tend to be more unpredictable in payoff than those of women and are eaten more by nonfamily members. Men's foraging occasionally appears to be, at the same time, inefficient: men could more than double their caloric rate of return by taking some foods (e.g., palm starch) that they ignore. Instead, they hunt meat and collect honey, both of which are foods that tend to be shared. When and why foragers share is a fascinating question that Michael Gurven and colleagues have explored successfully.[6]

What are the reasons for this behavior? One obvious hypothesis is risk reduction: helping you now, when I have meat, might oblige you to reciprocate later when I fail and you are successful. But reciprocity would mean that sharers would discriminate, giving to those who give back and excluding "free riders" who take much but return little or nothing, and this does not seem to be the case. Two other hypotheses seem more likely. First, when returns (as from hunting) are sometimes too large for one's own family to consume and storage is impossible, giving and "tolerated theft" may be less costly in many ways than the alternatives. Second, "showoff" men who are more successful hunters receive more attention from group members and often fare better reproductively; more women are willing to mate with them. I suspect that such patterns will turn out to be common.[7]

Sex and Control of Resources

Not only do resources contribute differently to the reproductive success of men and women, but women's access to resources varies greatly cross-culturally. When do women control significant resources? There is no single meaningful measure of women's power or resource control cross-culturally, though there are some patterns. Across cultures, the more women contribute to subsistence, the more control they have over various resources.[8] In our own society as well, there is evidence that an increase in professional working women has resulted in greater economic independence for them.

Women's ability to get and control resources also differs between monogamous and polygynous societies. Within polygynous societies, as the degree of polygyny (percentage of polygynous marriages) increases, women's ability to control the fruits of men's labor decreases and their ability to control the fruits of their own labor increases—women can function as more independent economic units, just as in other primates. Both inheritance and resource control are affected: the greater the degree of polygyny, the less likely are women to be able to inherit property, for sons in polygynous societies are better able than daughters to turn resources into grandchildren (chapter 6). Women are most likely to be able to inherit property in matrilocal societies, when they live among their own kin, and least likely to inherit property in patrilocal societies, living among their husband's kin.

Men, Women, and Resources in Traditional and Historical Cultures

The arguments so far suggest that any relationships between resources and reproduction in humans will be influenced by ecological considerations. Males with greater resource control, for example, are likelier to show higher fertility than poorer but otherwise comparable individuals—a major reason why inheritance is largely male biased in polygynous societies. In traditional societies, this bias typically results from greater polygyny by higher-status or wealthier men.

Men and women are likely to use resources to reproduce somewhat differently. The most fecund woman can have fewer children than the most fertile man, simply because this is a polygynous society, and parental investment imposes different constraints than mating effort. How, then, do resources affect the reproduction of men compared to that of women, and how are the patterns of families affected? Several studies on cultures that vary in many regards (especially in some crucial resource-control issues) can give us insight both about patterns that are common and patterns that vary.

Yanomamö

The Yanomamö, one of the best-studied of all traditional societies, live in small villages in tropical rain-forest regions of southern Venezuela and adjacent regions in Brazil. Their daily activities revolve around collecting wild foods, getting firewood and water, and cultivating their gardens; about three hours daily are spent in these subsistence activities. The anthropologist Napoleon Chagnon worked among the Yanomamö (principally among the resource-rich lowland groups) for more than twenty years, and most of our information comes from his work.[9] The Yanomamö still wage intervillage warfare, and at least one-fourth of all adult men die violently. Warfare has reproductive causes and impacts, as successful warriors gain status—and thereby reproductive success (chapter 13). Men raid other villages, and attack smaller groups in transit, to steal women (more in chapter 13). The frequency of war is ecologically influenced.

Chagnon noted that "social life is organized around the same principles utilized by all tribesmen: kinship relations, descent from ancestors, marriage exchanges between kinship/descent groups, and the transient charisma of distinguished headmen who attempt to keep order within the village."[10] The Yanomamö do not accumulate the sort of heritable resources that would serve as bridewealth. Social dynamics center on the exchange of marriageable girls; marriages are arranged by the male kin of prospective mates and are often highly political.

The Yanomamö, who are slash-and-burn agriculturalists with little in the way of accumulated resources, are one of the societies held up sometimes as "egalitarian," in which everyone behaves as though the interests of all were identical. As Chagnon has shown eloquently and repeatedly, nothing could be further from the truth, at least for reproductive matters. The Yanomamö are polygynous, and men's abilities to marry and have children vary far more than women's.[11] Wealth is indeed largely absent, but in the lowlands, which are ecologically the most productive, a man's demonstrated skill as a warrior and a large male kin group make a great reproductive difference for him. The variance arises largely because skilled warriors, called *unokai*, are able to marry earlier and more often than other men (more on this in chapter 13).

What of women's costs and benefits in this polygynous society? In some societies, polygyny, despite its likely costs, could have real resource benefits for women—if men vary in wealth or the women are able to form alliances.[12] But Yanomamö men do not vary in wealth, so a polygynous Yanomamö woman must share the resources of her husband with co-wives, even if he has nothing more than a monogamous man. Furthermore, Yanomamö men have a large say in arranging marriages, as is true in many societies. Polygynous households tend to have smaller gardens for their family size than monogamous households, but economically, the only difference between polygynous and monogamous households is that polygynous households receive more food from others. A high-status Yanomamö man therefore, though he does not directly provide wealth, may indirectly create some benefits for his wives.[13]

Ache

The Ache live in eastern Paraguay, in the southwestern part of the eastern Brazilian highlands. This area is generally higher and drier than the rest of the Amazon basin. An excellent study of Ache demography, ecology, and life history is that of the evolutionary anthropologists Kim Hill and Magdalena Hurtado.[14] All Ache groups in recent times have been hunter-gatherers. Apparently, for the last four hundred years they have engaged only in hostile interactions with outsiders and have not traded, visited, or intermarried with the nearby Guarani populations. The Ache live in small bands of 15 to 70 individuals, moving throughout the forest. Bands comprise closely related kin and some long-term friends. Large sibling groups of both sexes tend to remain together along with additional kin. Only four Ache groups existed in the second half of the twentieth century before they made permanent contact with outsiders.

Daily life centers around hunting and gathering; men spend almost 50 hours per week hunting white-lipped and collared peccaries, tapir, deer, pacas, agoutis, armadillos, capuchin monkeys, capybara, and coatis; they collect honey, which accounts for 87% of the calories in the Ache diet. Men often hunt in ways that look inefficient from an optimal foraging perspective, but in fact such men seem to be pursuing a high-risk, high-gain showoff strategy that may often fail but can produce big, flashy hunting successes—and, with success, more sexual access to women. Women spend about 2 hours a day gathering; they collect fruits and insect larvae, as well as extract the fiber from palm trees.[15] They also carry the family's children, pets, and possessions, which constrains their ability to forage. Men may travel with the women's group, but more often they set off in small groups to search for game, spending about 7 hours per day hunting.

In the late afternoon, families gather and prepare food (fig. 7.1). Hunters rarely eat from their own kills, and much food is shared, leading early observers to argue

Figure 7.1. In the afternoon, after food-gathering activities, Ache men (here) and women rest and prepare food. (Photo by Kim Hill.)

that the society was completely egalitarian. While meat is apparently shared evenly under most circumstances, honey and gathered items are not. Further, when a man dies, his young dependent children are far more likely to die than if he had lived. While reciprocal sharing of meat is ordinary, when a man has died he can no longer reciprocate, so meat is no longer shared with the widow and children.

Both men and women may have multiple mates over time; during young adulthood they may switch spouses frequently. This pattern seems to have changed little after contact with Europeans. After marriage, residence is typically matrilocal. Many children have multiple recognized fathers. Reproductive success is difficult to measure under such circumstances, but despite the fact that the Ache have little in the way of heritable wealth (which in so many societies correlates with reproductive success for men), status matters: the best hunters have the greatest reproductive success.[16]

The issue of proven success yielding reproductive success is widely accepted, but the data (as for the chimpanzees in chapter 4) are usually correlational and scattered. The evolutionary anthropologist Eric Alden Smith has pulled together both qualitative and quantitative data on five widely dispersed societies: the Ache, Hadza, !Kung, Lamalera, and Merriam. The data indicate that better hunters have higher age-specific fertility than other men. He examined five hypotheses: (1) direct provisioning of hunters' wives and children, (2) dyadic reciprocity among hunters, (3) indirect reciprocity, (4) costly signaling, and (5) phenotypic correlation. No hypothesis could be definitively rejected, although extensive and apparently unconditional sharing of meat with other hunters somewhat weakens the first three possibilities (it would be hard to tell if they were engaging in long-term politics, I think). Costly signaling (think handicap principle) had some support in both other species and humans. Among the Merriam, for example, exceptional fishermen, whose catches varied less than turtle hunters', did not have higher reproductive success than the costly signaling turtle hunters (even more interesting, fishermen who also hunted turtles did have higher reproductive success than men who only fished).[17]

Kipsigis

The Kipsigis are agropastoralists and agriculturalists living in the Rift Valley Province of Kenya. Since the 1930s, they have become more and more involved in commercial agriculture, selling maize.[18] Women do the agricultural and domestic work while everyone, even the children, shares the duties of animal husbandry. The Kipsigis are a polygynous bridewealth society, where resources are given by the groom or groom's family to the bride's family. Half the women are married by age 16, half the men by 23. The father of a prospective groom initiates negotiations, making offers of cows, goats, sheep, or, since 1960, cash. Women (or their families) choose men according to an "ideal free distribution" of wealth: the number of wives a man has is related to his wealth. The average bridewealth is significant: for a man of average wealth, it comprises a third of his cows, half his goats, and two months' salary. The father of the groom alone is responsible for providing the bridewealth for his son's first marriage. Payment of bridewealth gives a man rights to all of his wife's children (legitimate or not) and to her labor. Once, a woman's labor value was significant; this is less true now.[19]

In contrast to the Ache, Kipsigis marriage is quite stable and divorce is almost unknown.[20] If a mistreated wife deserts her husband, the bridewealth is not returned. Though many things can influence the exact amount of marriage payments, high bridewealth is typically paid for younger women (of higher reproductive value) and plump (healthy, well-nourished) women. Older women and women who have given birth before marriage command lower bridewealth. The importance of wealth differences between the groom's and bride's families has varied over time. Once, better-educated and wealthier men had higher bridewealth demands made on them. This has changed, because women's parents now seek men with good economic prospects. Another sign of changing times: women with secondary education command high brideprices and tend to marry better-educated men.

Mate choice operates in both directions: men's wealth (more precisely, land ownership) matters. Men are preferred who are better educated and who can offer more acres to prospective brides or who (after controlling for land per wife) have had fewer wives and a successful paternity record. Women's reproductive fate after marriage suggests that the things men and women choose are biologically significant; both fertility and offspring survival matter. Women who are married to poorer men and have co-wives have long (34.2 months) interbirth intervals and relatively higher infant mortality rates than women married to wealthier men.[21] Social and biological factors interact, yielding a rich and complex system, but here, too, traditionally, wealthier men have more wives and more children than others.

Mukogodo

The Mukogodo, in central Kenya, were foragers and beekeepers until early in this century (fig. 7.2). They were, in the evolutionary anthropologist Lee Cronk's words, "neither wealthy, powerful, nor prestigious" and poor compared to several neighboring groups such as the Maasai, the Samburu, and the Mumonyot.

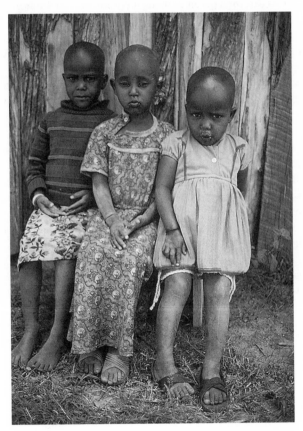

Figure 7.2. The Mukogodo, neighbors of the Maasai, stand at the bottom of a regional hierarchy of wealth and status. Their sons find it difficult to accumulate enough bridewealth to marry non-Mukogodo women, but their daughters, if beautiful, can marry "up" and bring bridewealth back to their natal families. (Photo by Lee Cronk.)

Today the Mukogodo sit at the bottom of a regional socioeconomic hierarchy and are somewhat stigmatized by their neighbors. And this fact has had curious social and health consequences.[22] As in many pastoral societies, the groups in this region use livestock for bridewealth. Because the Mukogodo are so poor, Mukogodo men have little chance of raising the bridewealth required to marry a woman from a neighboring group. It also means that when Mukogodo daughters are married to men of neighboring groups, the Mukogodo family acquires cattle, sheep, and goats from bridewealth.[23] Daughters are more valuable than sons in important ways.

In chapter 6 we explored the general phenomenon of parental biases in the sex of their children: the Trivers–Willard effect. In many societies, parents prefer sons—they give them better care, more inheritance—because sons under many conditions can turn parental investment into grandchildren more effectively than can daughters. Not so among the Mukogodo. Mukogodo parents appear to respond to the different costs and benefits of sons and daughters in several ways—but among the Mukogodo, daughters are worth more in reproductive terms.

In contrast to the general mammalian pattern, Mukogodo mothers nurse their daughters longer than their sons; Mukogodo caregivers (mothers and others) stay closer to girls than to boys and hold them more. Parents take their daughters more frequently than their sons to the dispensary and clinic for treatment and enroll their daughters more often in the local Catholic mission's traveling baby clinic. Non-Mukogodo parents do not show these biases. And when Lee Cronk measured the birth sex ratio of Mukogodo sons to daughters, it was highly unusual: not—as is typical—about even, with a few more boys than girls, but with more than twice as many daughters. Furthermore, probably because of the biased care, girls show better growth than boys: better height for age, weight for age, weight for height.

A Mukogodo son grows up and marries, usually a Mukogodo woman. A daughter may marry a Mukogodo man, but if she can marry a wealthier neighbor, he has to pay bridewealth, and she will bring wealth to her family. Interestingly, the Mukogodo, living among societies that have a strong male preference, claim, if you ask them, to prefer sons to daughters—but they treat their daughters better. Parents among the Mukogodo, as those in a number of societies, appear to favor the more profitable sex in their children.

Turkmen

The Yomut Turkmen are clearly a resource-based society (fig. 7.3). As the evolutionary anthropologist William Irons has noted, a large part of the daily activities of the Turkmen is devoted to economic production, and they clearly seek to maximize wealth.[24] They have long been immersed in a market economy far larger than their tribal boundaries. Production for trade is important,

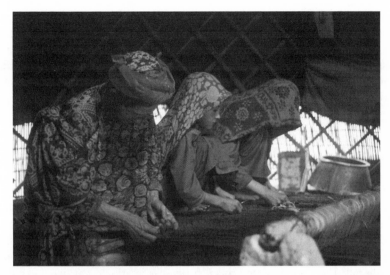

Figure 7.3. The Yomut Turkmen are active in a market economy and clearly seek to maximize wealth. Wealthy men have significantly more children than poor men. Here, women weave a rug for sale. (Photo by William Irons.)

as are accumulation of agricultural land and livestock and savings in the form of money, jewelry, and the like. Men's work appears more important in wealth building than women's work, and extended families appear to be built around closely related male relatives who remain together; that is, it seems to pay off in inclusive fitness for related men to stay together and build wealth.

Marriages are arranged by parents. Women are typically married at about age 15. Sons of wealthy men also marry at about age 15; sons of poorer men marry later, because the Turkmen are a bridewealth society and bridewealth is high. A man of median wealth will pay two to four years' income for a virgin bride. In concrete terms, this is ten camels for the bride's father and one camel for her mother. The bridewealth can also be paid in cash or other livestock. One camel equals two horses, one really good race horse, two cows, ten sheep, or ten goats.

The price of a bride appears to be far more fixed and nonnegotiable than among, for example, the Kipsigis. There are no bridewealth "discounts" for friendship or hardship, just later marriage. Because the Turkmen are polygynous and most men want to marry, and the sex ratio in the adult population is about 110:100 (male biased), most widows do not remarry. It's a seller's market: a man who wishes to argue down the bridewealth is in a poor position to do so. If a widower seeks a virgin bride, he must pay double bridewealth. If a married man wishes to take a second wife, he is supposed to pay triple the brideprice, though in practice it works out to about 2.5 times the bride price. Thus, polygyny is possible for only about the richest 5% of men.

As a result of these conditions, age-specific fertility rates are higher for richer men and women than for poorer men and women (significant only for men). Because the Turkmen are polygynous and bridewealth based, richer men have significantly more children than poorer men. Thus, as Irons noted, the Turkmen, in striving for wealth and social success, are striving for the proximate variables that result in greater reproductive success.[25]

Qing China (1644–1911)

Early in the Qing period in China, in 1652, the nobility established the Office of the Imperial Lineage to register births, marriages, and deaths and to maintain the genealogy of the Qing noble lineage. Membership in the lineage was patrilineal, and members were entitled to a wide range of state perks, including a subsistence allowance, titles of nobility, and other state stipends.[26] To ensure complete and accurate registration, there were overlapping sets of household and vital-events registrations, much as in Sweden (next chapter). These records, commonly called the Jade Register, cover thirteen generations over two and a half centuries. Thus the demography of the Qing nobility is remarkably well known.

Marriage in the lineage was relatively early and virtually universal. The mean age of marriage for both men and women was the early twenties; 99% of women were married by age 30, and 99% of men by age 40. Yet the Qing's marital age-specific fertility, especially at younger ages, was quite low compared to contemporaneous European societies.[27] The Qing lineage was large, and there was complex variation over time, but in general the pattern was as follows: men who were of high nobility, and whose wealth and rewards were therefore greater, were more likely to be polygynous than the lower nobility—and polygynous men had more children than monogamous men.[28]

Both high nobility and polygyny were strongly associated with higher fertility, but the strength of these associations varied over time. Differences in status-related fertility were relatively small between 1750 and 1780 (a period of comparative wealth, when lower nobles could do reasonably well). Differences were greatest during the depressed period of 1780 to 1820, when the government cut subsidies to the lower nobility. As imperial lineage rewards and subsidies declined, polygynous men married fewer wives, and the average polygynous fertility plunged from ten to five children.[29] Status, resources, and fertility were clearly linked for noble men in Qing China.

Resources were important to ordinary men in Qing China as well.[30] In the common rural extended-family households, marriage and fertility differed not only with type of household but also with individual status within the household. Among "senior" relatives, household heads married earlier and in higher proportions than anyone except uncles (who, in this system, were even more senior in their own right).[31] Among junior relatives, sons tended to marry earlier and in larger proportions than other men, such as brothers' and cousins' sons. The higher a man's position, the more likely, and earlier, was marriage. Fertility, like marriage, correlated with a man's position: higher-status men had more sons than others (and fewer daughters, since female infanticide was acute among the elite in rural areas).

Another kind of hierarchy, somewhat more related to a man's own effort and capabilities, existed during this period: the banner hierarchy. Thus we can separate the effects on a man's marriage and fertility of his household status (i.e., his position rather than his individual capability) from his occupation (e.g., soldier, artisan, official, commoner). Demographic behavior of commoners was more strongly related to banner position (their own efforts) than household position, even controlling for possible effects of overlap between the hierarchies. Men who achieved a banner position were more likely to marry and had higher fertility than other commoners. In sum, outside the nobility, marriage was not universal for men, and access to marriage and resulting fertility were a function of resource control, partly due to exogenous factors, partly due to their own efforts.[32]

8

〜〜〜〜〜〜〜〜〜〜〜〜〜〜〜〜〜〜〜〜〜〜〜〜〜

Sex, Resources, and Fertility In Transition

Life history theory deals directly with natural selection, fitness, adaptation, and constraint.

Demography, the key to life history theory, allows us to calculate the strength of selection on life history traits for many conditions.
—Stephen Stearns, *The Evolution of Life Histories* (1992)

A CENTRAL QUESTION of this book is: *exactly what*, in any species or society, contributes to greater or lesser lifetime reproductive success? As we have seen, in most mammals and in the majority of traditional human societies for which data exist, status, power, or resource control enhance lifetime reproductive success, especially for men. Men's reproductive variation in traditional societies arises mostly through differential polygyny—higher-status men can marry earlier and more often than other men, and often they can marry younger women, who are of higher reproductive value.

Does this pattern have any relevance today? Two phenomena make it likely that today we will not see the huge differentials that existed among some traditional societies or in reproductively despotic societies (e.g., those in which a caliph might have a thousand concubines). First, modern nations have for some centuries been (socially) monogamous. Second, in western Europe and North America, total population increased but family size fell dramatically during the nineteenth century—the "demographic transition."[1] Today, societies are large, heterogeneous, and mobile, while family sizes are small; associations between status or resource control and the number of children vary: in some cases (typically rural) they remain strong; in others they have become weak or have even disappeared (e.g., in much of the Western world; fig. 15.2).

To discover whether resources, power, and sex still correlate, an obvious step is to examine patterns during the demographic transition. We may think that our societies today have little connection with our past, but it is important to explore the history and patterns of resources and reproduction.[2]

Nineteenth-century Germany and Sweden provide opportunities to explore the impact of resources on male and female lifetimes—survival, reproduction, and migration. In Sweden, there exist extensive and overlapping civil registers from the 1600s. The completeness of the demographic information is remarkable. This is a population for which excellent records exist throughout the historic demographic transition from large to small family sizes. We are lucky that the Swedish government funded a demographic database designed to track individuals and families over time and to make use of the extraordinary historical records as well as modern data. The records are extremely accurate and include longitudinal (lifetime and across generations) data for all individuals across two to four generations, depending on the parish: sex, date of birth, date of death, age at marriage, best occupation, date of record loss, type of record loss (death, emigration), dates of birth of all children, and comparable data for those children.[3] We can thereby follow the fates of families over time. Let us ask, then, the same questions of Swedish and German families as we asked of families in the traditional societies of chapter 7.

Because of record completeness, some new questions are important. Did higher-status men have more children than others? What bequeathed status? Did children of higher-status parents survive better than others? Did higher-versus lower-status children leave the parish at different rates? As adults, were richer compared to poorer individuals differentially likely to leave the parish? As adults, if they stayed, were they differentially likely to marry? Did the age-specific fertility of richer women differ compared to poorer women? We must ask some questions differently from those we posed for traditional societies: in both Germany and Sweden, for example, social monogamy was the rule—but we can ask if wealthier men, if their wives died, remarried more than other men, and whether those second (or third, or fourth) marriages resulted in richer men having more children than other men.

Germany in the Eighteenth and Nineteenth Centuries

The evolutionary anthropologist Eckart Voland used family reconstitution data to follow families through a number of generations in several parishes in Krummhörn, a northwestern coastal area in Germany. Unlike much of Europe (including Sweden), there was no notable change in population density from the seventeenth to the nineteenth century.

The population was mostly Protestant, and subsistence was dominated by agriculture and dairy farming. Voland followed fitness (as a result of infant and child mortality, marriage and family formation, celibate individuals, etc.) within the Krummhörn region. He used three categories of land ownership: landless, smallholders (who had 1–75 *Grassen*; 1 Grassen ≈ 0.37 ha); and farmers, who had more than 75 Grassen.

Land ownership was the single most important determinant of reproductive success: farmers had greater long-term reproductive fitness than others,

beginning by ten years after marriage, and wealthier farmers had more reproductive success than smallholders. Fifty-eight percent of wealthier farmers
married compared to 41.3% of smallholders and 39.7% of nonlandowners. In
contrast to Sweden, this population practiced ultimogeniture: the youngest
adult son inherited the undivided lands. Differences in reproductive success in
the first generation resulted from differences in three components: total marital fertility, offspring viability, and men's mating success (e.g., through serial
social monogamy).[4] In the long term, farmers were more fit than smallholders,
who were more fit than nonlandowners, although the rank order varied in
particular contributions to that fitness. It is worth remembering, as many researchers have pointed out, that sheer numbers are not always the key to lineage success and persistence.[5]

Who left their natal area? Competition within the family was important.
Among the landless, daughters were likely to leave to find a mate; sons may
have had more work options whether they migrated or not. There were reproductive costs to migration: daughters who migrated married about a year later
than others.

Among farmers, competition for land resources was important (perhaps as
a step to mating opportunities, as in Sweden; see below). Dispersal neither
guaranteed new resources nor guaranteed that loss was inevitable. Dispersal
became an optional choice, and there was no single optimal strategy.[6]

Nineteenth-Century Sweden

There are many similarities, but also important differences, between the German data and those from nineteenth-century Sweden. I used two Swedish data
sets: one follows the lives of men married for the first time between 1824 and
1840, their children, and grandchildren in four geographically separated and
economically diverse parishes (Tuna, Locknevi, Gullholmen, and Nedertorneå);
and the second examines the impact of land ownership in 63 villages in a northern five-parish region, Skellefteå, for which I read land-tax records (the records
ceased in 1896 for privacy reasons).

The four parishes differ in economic and ecological conditions (described
below), and people's lives varied greatly, even within a parish (fig. 8.1), though
there are some important commonalities. Land-tax records exist in this dataset
only for one of the four parishes, Tuna.[7] I will review variation, but I am seeking
the most general possible answer: did the resource-reproduction connection
exist, and if so, did it persist or disappear in nineteenth-century Sweden?[8]

Nineteenth-century Sweden was largely agricultural, with emerging proto-
industrialization. Early geographically scattered market activity involved trans-
forming raw materials into "made" commodities, but a large part of the labor
force worked part time or at home; the family could function as an economic

Figure 8.1. In nineteenth-century Sweden, rules were egalitarian; nonetheless, the lives of migrant farm workers and wealthy families differed considerably. (Photos courtesy of the Skellefteå Historical Museum, Sweden.)

enterprise. Proto-industrialization tends to arise in developing regions that have both an underemployed, land-poor population and urban markets—like Europe in the nineteenth century and much of the developing world today.

People married late in Sweden, and more than half of adults never married. Most women married for the first time in their early to mid-twenties (although their father's wealth affected the age), and men in their late twenties (though this, as we will see, varied among parishes); at marriage the new couple typically set up their own independent household.[9]

These parishes are economically more diverse than those in Krummhörn, Germany, so it is not surprising that both economic and family patterns differed among the parishes. Unless we were asking the behavioral ecologist's questions about resources and family patterns, it would be easy to fail to see a "bottom line." Yet I think there is a clear conclusion underneath all the diversity: whenever wealth—especially land—or resource differentials existed, resources and reproductive success were positively correlated, just as in Germany.

How Economies Varied in Four Parishes

The predictability and richness of resources make a huge difference to how ecological phenomena affect survival and fertility. So, too, does inequality. In these four parishes, Gullholmen, an island principally of fishers, was dependent on highly variable catches; this resulted in late marriage and a variety of strategies to even out resources over the years.

In Locknevi, a single, very limited valley of good agricultural land meant that as the population expanded, those who were able to leave did so, and the poor were left to try to farm. The population grew and then stagnated as opportunities waned, largely because of the limited amount of good farmland. Toward mid-century, the wealthy landowners divided up and sold their large estates, moving out of the parish. As a result, landholdings became progressively smaller, and though more land was brought into agriculture, much was of marginal quality. Thus in Locknevi parish during the period of this study, resources shifted from being relatively uneven with some very large holdings, to being more even but quite limited.

Tuna parish was largely a farming parish, though forest and mining industries were also present in the early 1800s. Many men worked in the local iron foundry as well as farming. Industrialization increased rapidly beginning in 1850. The iron foundry closed in 1879 but reopened in the mid-1880s; in response, people moved out of the parish, then back in. The mixed economy (forestry, ironwork, mixed-crop agriculture) of Tuna was more diverse than Locknevi's, and perhaps as a result (and the availability of nonmarket alternatives such as hunting and fishing), population measures did not vary with economic fluctuations in Tuna as they did in Locknevi.

Nedertorneå, in the far north, was a farming parish, although the land was generally of poor quality and the growing season short, so people hunted and fished to supplement their resources. As a result, crop fluctuations had relatively little impact in Nedertorneå, much as in Tuna. Let us examine in more detail how the resource-reproduction connection varied in each parish.

GULLHOLMEN

Gullholmen is an island parish with almost no very wealthy or very poor people during the nineteenth century. Gullholmen's small population rose steadily, always less than 1,000 individuals—but because the island was small, the population density was the highest of any parish discussed here.

Only 12% of men living to maturity migrated from the parish. There was little variability in the status men held, and men of different statuses survived equally well—although men of different occupations were differentially likely to marry, and married men had more children than unmarried men. Men married at about age 27. In most parishes, only the wife's age at marriage mattered to the couple's fertility, but in Gullholmen, a man's age at marriage was important, too.

Women married very late in Gullholmen (26+ years), and 58% of adult women failed to marry at all while in the parish. Women who married earlier had more children than those who married later; they almost never remarried, and remarriage did not affect their fertility. In Gullholmen, a woman's chances of marrying were independent of her father's status, and neither her father's nor her husband's occupation was related to her age at marriage. In contrast to other parishes, a woman's fertility was only marginally related to her husband's occupation. Nonetheless, the lifetime family size of married individuals was the highest of the four parishes. Family size decreased somewhat over the study period, from about 5.5 ± 2.7 children in generation 1 to about 4.2 ± 2.4 in generations 2–4. Thus marital fertility was reduced by the 1860s in Gullholmen, but survivorship improved, and the number of children surviving to age 10 did not decline; in fact the number of children surviving to age 10 was the highest of the four parishes although fertility was late.[10]

Because of the uncertainty of fish catches, people married late but had large families; perhaps both of these patterns are related to getting established and becoming able to handle unpredictable resource fluctuations. Umeå University historian Anders Brändström told me that in the records and contemporary letters from Gullholmen, a frequent strategy was to use some money in excellent fish-catch years to buy "luxuries" like silver candlesticks, which could be sold easily in poor years.

LOCKNEVI

In Locknevi parish, at the start of nineteenth century there were a few very large landholdings that employed agricultural day workers. A small ironworks in the southwestern part of the parish provided supplemental income for some until the 1880s. Most people's economic lives were uncertain: not only did work depend on crops, but people's purchasing power, as reflected by the number of days' work that were required to purchase a "market basket," varied considerably over time. Family patterns in Locknevi varied with economic times: marriage and fertility rates fluctuated with crop prices.[11]

In contrast to Gullholmen, where people married late, Locknevi folk married early but delayed having children (fig. 8.2). A man's best occupation influenced his chances of marrying: 74% of agricultural workers and servants living their entire life in the parish failed to marry compared to 20% of lower-middle-class

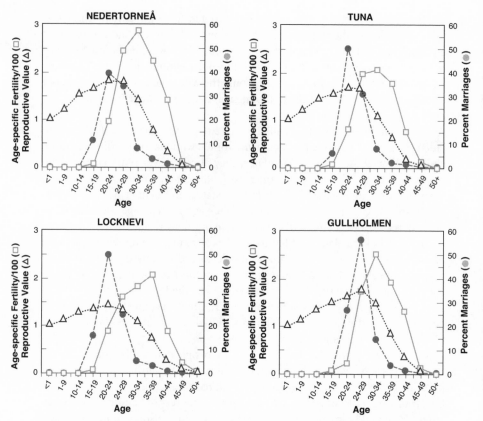

Figure 8.2. Economic conditions and opportunities varied across the parishes, and so did women's reproductive patterns. Δ = Fisher's reproductive value; ● = Percent marriages ; □ = age-specific fertility. Women in Nedertornå, Locknevi, and Tuna parishes married in their early twenties; in Nedertorneå (the most northern parish, with poor soil and short growing seasons) marital fertility was much delayed, and infant mortality was high, depressing early reproductive value. Women in Gullholmen, a fishing parish, married in their late twenties and delayed fertility, but had the highest marital fertility of any of the four parishes. These differences mean that Fisher's reproductive value also differs in pattern across the parishes. See text for further explanation. (Modified from Low and Clarke 1991.)

men. The age at which a man first married was not related to his lifetime fertility. Remarriage after a wife's death was common in Locknevi, and it influenced men's fertility: men who remarried had more children and more children surviving to age ten than men married only once. About a quarter of wealthy upper-middle-class men remarried, more than other groups.

In generation 1, when landholdings were uneven, a man's occupational status predicted the number of his children who would survive to age 10; wealthy men did best. But when the wealthiest landowners sold off their manors and left the parish, fertility became low and the remaining men showed few fertility or status differences.[12]

Sixty-one percent of women failed to marry while in the parish, and those who married delayed their fertility remarkably, perhaps because of increasing resource constrictions.[13] Women who married earlier were somewhat more likely to have more children than those who married later.

A woman's age at marriage was not correlated with either her father's or husband's occupation. Further, women's lifetime production of children, and number of her children alive at age 10, were not related to father's or husband's occupation. Perhaps because most men's resources were so limited, they did not directly influence women's reproductive patterns. Men's patterns were connected to wealth, but women's patterns (perhaps because of the high remarriage rates in the early years of the study) were not.

TUNA

The population of Tuna Parish, in Medelpad, rose from approximately 1,200 in the early nineteenth century to about 3,300 in the late part of the century. Occupation provided no clue about family formation and reproduction. Land ownership, however, did. Landowning men ($n = 102$), who were earlier in the birth order than their brothers, were likelier to marry than other men ($p < 0.00001$): 95% of landowners married, compared to 35% of nonlandowners. Landowners married women about 2.5 years younger than other men and had about 1 to 1.5 more children. In sum, landowners had larger families, measured as children born ($p = 0.0002$) or children alive at age 10 ($p = 0.001$), no matter what the times, and their families were not only larger but also less variable in size than those of nonlandowners. Even within families, the reproductive lives of brothers—landowning and nonlandowning—differed in this way ($p = 0.0003$). Owning land appears to have provided a buffer against hard times, over and above the nonmarket alternatives.[14]

NEDERTORNEÅ

The population of Nedertorneå, in the far north, rose steadily during the nineteenth century.[15] Here, a man's occupation mattered to his lifetime reproduction: 36% of male migrant workers and servants married, compared to 57% of those in the upper middle class, 44% in the lower middle class, 57% of bönder ("farmers," but there are complications with the household records, and the tax records are more reliable: see below), and 44% of torpare (cottars, roughly "tenant farmers"). Thus, a relatively stable association with land had a positive impact.

Historical particulars mattered also. First, there were very rich and very poor people (in contrast, e.g., to Gullholmen); this distribution occurred because early in the nineteenth century the central Swedish government moved upper-level civil servants to the outpost town of Haparanda, to shore up the local economy. Second, infant survivorship was low in part because parents fed their infants often-spoiled cow's milk, rather than breast-feeding them. In the 1840s, a doctor came to Haparanda and began a campaign to reinstitute breast feeding. Because he worked mainly with upper-middle-class families and in the town of Haparanda rather than in the surrounding countryside, there is great variation in interbirth intervals, survivorship, and thus fertility and family size, tied to location, class, and time. Fertility and survivorship differed both with class and town/country residence.[16]

Sixty-four percent of women failed to marry while in the parish—but Nedertorneå had the highest peak of age-specific fertility in the four parishes (fig. 8.2). Women's likelihood of marriage was related to their father's occupational status: daughters of upper-middle-class men and farmers were most likely to marry.

Overview of the Four Parishes

Despite the fact that the four parishes differed in size, population levels and growth patterns, and economic bases and stability, and despite the fact that demographics responded to changes in the environment and resources, there are general principles.[17]

Traditions and ecology in northern and southern Sweden differed in significant ways. In many areas in southern Sweden, single major crops, such as rye and corn, dominated the economy. Bad harvests created real hardships, which were reflected in prices and purchasing power. In the north, barley was an important grain crop, but agriculture was more mixed and was consistently supplemented by fishing and forestry. Thus, in the north, failure of any particular crop was likely to have less impact on people's lives. Harvest and market price information alone are insufficient reflections of conditions in such areas. In areas with single crops, famines and high food prices predict, for example, theft rates, but in parishes in which agriculture was more varied, or in the northern "forest" areas like Nedertorneå, where a less powerful market force was in effect, this was not true.[18] All of these differences are important, particularly in the examination of historical and parish-specific patterns; but they cloud answers to the very basic question we started with: did the resource-reproduction correlation persist or disappear? We need more recent statistical analyses of the parishes to reach a truly general answer.

Although the particulars of reproduction varied in time and among parishes, broad comparisons are possible without ignoring this variation among parishes and through time. Let us classify each individual's wealth as "richer" (owned land and/or had an occupational status of upper middle class, lower middle class, or bönder) versus "poorer" (occupational status of cottar or proletariat and no land ownership record). We can then compare each adult individual's lifetime reproduction to the median for (1) all individuals reaching

"maturity" (defined here as 23 years) in any decade in any parish, or (2) all in-)
dividuals marrying in each decade in each parish.

Thus we compare any individual's lifetime reproduction to the median for his or her parish and decade of maturation or marriage, as appropriate. These data form a picture of life prospects for people born in different conditions and relate to the general problem of resources, family decisions, and demographic transitions, taking regional differences and historical particulars into account but still yielding a general picture.[19]

Female Life Paths in the Four Parishes

Figure 8.3 gives an overall summary of the lives of all individuals in the four-parish study.[20] Despite the rich variation discussed above, there were clear dominant patterns. Daughters of poor fathers (fig. 8.3A) were more likely to die or migrate from their parish of birth before the age of 15 than were daughters of rich men.[21] Women were also more likely to leave as adults if their fathers were poor. Overall, daughters of richer men were more likely to remain in the parish and marry than others; however, of the women who stayed in their parish of birth to reproductive age, daughters of both rich and poor men were equally likely to marry (fig. 8.3A).

The likely remaining lifepaths for married women and unmarried women differed, depending on their father's wealth. Of women who married, virtually all women (98%) born to poor fathers married men who were poor, while 77% of women born to rich fathers married rich men. We compared married women's fertility to that of other women in the same parish of birth who married in the same decade. Daughters of wealthier men and wives of wealthier husbands (mostly the same women; marriage tended to be within class) had higher age-specific fertility at all ages than poorer married women (fig. 8.4).

The fertility of unmarried women was compared to that of all adult females in the same parish of birth who reached age 23 in the same decade. Women who were born to poor fathers and never married were far more likely (86%) to have fewer than the median number of children born to all adult women—most had none. Never-married daughters of rich men still had a 48% chance of having greater than the median number of children, suggesting that illegitimate births occurred at higher rates among the rich. The result of these pathways meant that, at all ages, "richer" women followed a different life path compared to "poorer" women (fig. 8.3A).

Male Life Paths in the Four Parishes

Sons of poor fathers were 7% more likely to migrate before age 15 than were sons of the rich men (fig. 8.3B). As adults, sons of rich men were very likely (91%) to become wealthy themselves, while sons of poor men were likely to stay

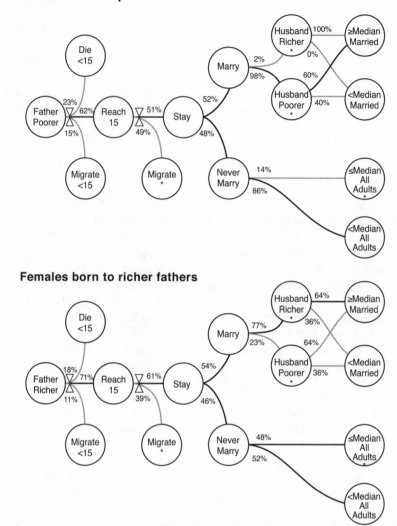

. Females born to poorer fathers

Females born to richer fathers

Figure 8.3. Some paths (heavy lines) were more likely than others for individuals born to richer or poorer fathers in nineteenth-century Sweden. Reproductive comparisons are relative to all adults who reached age 23 in the same decade in the same parish for unmarried individuals, and relative to all individuals marrying for the first time in the same parish during the same decade for married individuals. Strong significant within-sex differences are highlighted by asterisks. Extrinsic factors (indicated by valve symbols) could matter, for example, in the probability of outmigrating or marrying. These greatest-likelihood pathways simply track, for all individuals born in the sample, the percentage at each comparison point who follow one or another path. This is a visual representation, and the numbers diminish at each juncture, so the percentages will not always suggest the results of the statistical analysis (e.g., a statistical difference may be great, while the percentage is small, or vice versa, because numbers are large or small). **(A)** A daughter born to a poorer father was more likely than her richer cohorts to leave the parish after age 15; if she stayed, she was about equally likely to marry. She was overwhelmingly likely to marry a poorer husband. If she did not marry, there was an 86% chance she had fewer than the median number of children for all adults. A daughter born to a richer father

B. Males born to poorer fathers

Males born to richer fathers

had a higher chance of remaining in the parish and an equal chance of marrying. If she married, there was a 77% chance that her husband was richer. If she did not marry, there was a 48% chance she would have greater than or equal to the median number of children for all adults. **(B)** Sons born to poorer fathers were likelier to leave the parish before age 15, compared to sons of richer fathers (15% versus 8%); for sons of poorer men who stayed, there was an 89% chance they would be poor and a 40% chance they would migrate out as adults. Men who stayed were likely never to marry (57%) and of those men, 97% had fewer than the median number of children compared to all adults. Sons born to richer fathers, once they reached 15, had an excellent chance of becoming richer themselves (91%). These men were more likely to stay as adults (69%). Their chance of marrying was about 48%. Those who married tended to have the median or greater number of children compared to other married individuals (59%); those who did not marry were likely to have fewer than the median number of children compared to all adults (55%). (Modified from Low and Clarke 1991.)

poor (89%). As adults, poor sons of rich men were 6% more likely to migrate than poor sons of poor men.[22]

Of men who stayed in their birth parish, poor sons of poor fathers were most likely to remain unmarried (57%); 97% of such poor unmarried men had fewer children than other men (average <1). Rich sons of rich fathers had an approximately equal chance of marrying or not (48% versus 52%) and, once married, had a 59% chance of having relatively large families. Rich sons of rich fathers who did not marry were, like poor sons of poor fathers, likely to have fewer than the median number of children. Although 97% of poor sons had this fate, only 55% of rich sons did.

Two rare but interesting paths occurred. Most dramatically, the few men (11% of the whole sample) who were able to attain wealth, even though born to a poor father, had an 84% chance of marrying and, once married, a 66% chance of having at least the median number of children. Although we have only birth, marriage, tax, and death records (no memoirs), it looks like personal effort and ability mattered for these men.[23] On the other hand, poor sons of rich fathers did not fare as badly as the poor sons of poor fathers. Poor sons of the rich still had a slightly greater probability of marrying (54% versus 46%) and, once married, had a 74% chance of having at least the median number of children. Perhaps these men benefited from their relationship within a wealthy family in ways not reflected by their own occupational status.

Despite all ecological and large-scale economic variations among the parishes, a man's lifetime reproduction still varied with his, and his family's, wealth. Sons born to poor fathers were most likely to remain poor themselves, never to marry, and to have no children. Sons of rich men were most likely to become rich themselves. Rich or poor, if a man married, he was likely to have at least the median number of children.

Wealth mattered throughout life, and "richer" men, like "richer" women (and wealthy and high-status men in a variety of other societies), had higher age-specific fertility at all ages (see fig. 8.4 for women's data). The most dramatic pattern of children's lives with father's wealth is that sons of poorer men, even in egalitarian nineteenth-century Sweden, had almost no chance of becoming wealthy and having more than the median number of children. As figure 8.3 showed, the father's wealth is certainly not all that mattered: some percentage of both sons and daughters of wealthier men failed to marry and have children. But statistically, wealth and status still mattered, even in late-marrying, socially monogamous, demographic-transition Sweden.[24]

The Skellefteå Region and the Importance of Land

The five-parish region of Skellefteå is in the north of Sweden; in the nineteenth century, there were few real class divisions and less proto-industry than in the south. Skellefteå was and is largely forested but has relatively rich farming lands

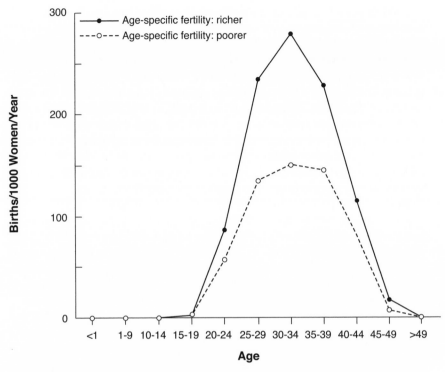

Figure 8.4. In nineteenth-century Sweden, wealthier women had higher age-specific fertility at all ages than poorer women. They began to marry earlier; however, women married to wealthy and poor men had the same interbirth intervals and age at last birth. (Modified from Low and Clarke 1992.)

near the Bay of Bothnia. Farming and forestry were the major occupations throughout the nineteenth century, but hunting, fishing, and tar production also formed part of the economic base. Market penetration was less than in southern and mid-Sweden;[25] despite periods of crop failure, the area was capable of supporting its population. At the beginning of the nineteenth century, Skellefteå's population was largely along the bay and fertile river valleys, but by the end of the century, the interior was fully occupied by settlements and privately and commercially owned forests. For people's lives, the varied supplies and the variety of crops with little market were very important.

Land was the primary resource; it practically constituted a class difference. Unlike in Germany, it was taxed (*mantal*) across Sweden from the 1600s not by area per se, but on the amount of land deemed to be required to maintain a man and his family.[26]

Land ownership is conflated with assigned status in many records. Here I concentrate on men in the land-tax records, which are more reliable regarding ownership than the household ledgers.[27]

The Skellefteå sample focused on approximately 4,000 men age 18 and older, who lived in 63 villages along the Skellefteå River, which flows to the Bay of Bothnia, and for which I examined *mantalslängder* (land-tax records) for the years 1830, 1840, 1850, 1860, 1870, 1879 (records were missing for 1880), and 1890. During the study period, new land came into cultivation and the number of landowners increased. In this sample, landowners comprised a total of 1,008 men with approximately 4,000 entries in the tax records; there were records for 2,997 nonlandowning men. For these villages, all men were linked to fathers, spouses, and children over generations, as for the four-parish sample.

In the 63 villages, the number of landowners of record increased steadily from 288 in 1830 to 516 in 1890; This is largely due to population growth, because the percentage of residents who were landowners decreased about 5% over the period (between 1830 and 1890: from 24.6% to 19.3%). The average tax on land held by landowners declined from 183.75 mantal in 1830 to 105.81 mantal in 1890. Land got divided, and the mean mantal decreased while variation increased, particularly after 1860 (the coefficient of variation ranged from 0.58 in 1830 to 0.83 by 1890). This is because amounts owned by the smaller landowners (represented by the 5%, 25%, 50%, 75% quantiles) all decreased. The largest landholdings (95% quantile) did not decrease at all. In fact, much of the increase in the coefficient of variation arises from increases in the largest landholdings. Inequality became greater over the century.

In northern Sweden, many sons did not inherit land. Sons of landowners were more likely to get land, but sons of wealthy nonlandowners were likelier to get land than sons of poorer nonlandowners. The amount of land owned by landowning sons correlated only loosely with the amount of land owned by their fathers and varied across the decades.

However, overall, sons of landowners were on average 2.5 times likelier to get land than sons of nonlandowning men, regardless of how much land the father owned (fig 8.5A), though the particular likelihood varied across decades. There was no significant relationship between the amount of land a father owned and the amount his son(s) might own when both landowning and nonlandowning sons

Figure 8.5. (A) Sons of landowners were likelier, in every decade studied, to own land as adults than were sons of nonlandowners. Overall, sons of landowners were 2.51 times as likely as sons of nonlandowners to own land as adults. Because the average age at which a man got land in this sample was 38 and records ended in 1890, I did not analyze decades of birth after the 1850s. **(B)** Whether their fathers owned land or not, landowning men in Skellefteå had more children than did their nonlandowning brothers, as was also true in Tuna parish (see Low 1990c).

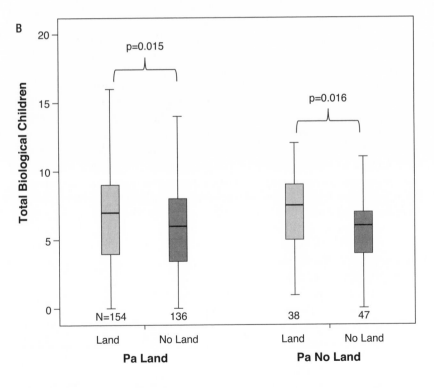

are included. As in Tuna parish, if we compare the reproductive success of men who owned land with their nonlandowning brothers, there is a significant difference (fig. 8.5B). Nonlandowning sons of landed fathers show more reproductive variance than comparable sons of nonlanded men, perhaps suggesting that landowners could help their nonlandowning sons in ways other than giving land.

All these complexities meant that in a broad set of studies on inheritance and inequality across many societies of different economic bases (all using the same statistical approaches),[28] the Swedish data showed up as more equitable than they actually were. Wealth, particularly land, *did* matter to reproductive success; fathers *did* favor certain (mostly first-born) sons, and the vagaries of being born into a wealthy landowning family versus a poor nonlandowning family did matter. These results actually support the main findings of the larger project.

Sex, Resources, and Fertility in Transition

The "bottom line"? Even in socially monogamous, late-marrying, egalitarian Sweden, and even during the demographic transition, wealth augmented fertility. Indeed, wealth, viewed simply as "richer" (upper middle class, lower middle class, land owners) and "poorer" (nonlandowners who were cottars, migrant workers, or household servants), influenced the life patterns of both men and women (fig. 8.3). From decade to decade, as fertility shifted with external conditions, richer families showed less variation in their reproductive success— they not only reproduced more, they did so consistently, and they were less at the mercy of external ecological, economic, and social influences.[29]

I find it striking that, in a Western, late-marrying, monogamous, relatively egalitarian society, wealth differentials still promoted fertility differentials. When resources became constricted, family reproductive differentials shrank; but whenever possible, individuals (especially men) converted resources into children.[30] Thus, family wealth and status made some life paths far more likely than others.

Who left? In general, as in the Krummhörn case, those who would have fewer options if they remained at home left, though that could vary across landowners and nonlandowners. Sons and daughters of poorer families, as in the four-parish study, were likely to seek work elsewhere (there was almost a Poisson distribution of the distance travelled). Oldest sons tended to stay; they were likeliest to inherit the land and to have the mating opportunities such inheritance presented. Younger sons might stay or go. As in Krummhörn, there was no single optimal strategy.

Fertility trends with local, reversible patterns, as in Sweden, seem more likely than a singular demographic transition with a singular cause. The primary components of population change (fertility, mortality, and migration) respond to ecological conditions at the individual family level in ways that are predictable and familiar to students of nonhuman populations, as I explain more fully in chapter 15.

The interplay of ecology, fertility, and lineage persistence is clear. Why then do the patterns not show up in the standard aggregate data for nineteenth-century Sweden? First, it is a question seldom asked. Second, if we fail to measure the appropriate resources, or if we look simply at aggregate measures rather than lineages or only for conscious decisions as mediators, we may fail to uncover the pattern. Even with the intensive recordkeeping in nineteenth-century Sweden, more poor people (with lower fertility) were "lost" to the records than were wealthy land and business owners.[31]

FERTILITY TRANSITIONS: WHAT, IF ANYTHING, DO THEY MEAN?

When resources are moderately low, or relatively ineffective in lowering children's mortality or enhancing their competitive success, fertility will be high. Such conditions obtain in many traditional societies, and probably in a good many pre- and proto-industrial societies as well. In Sweden, with one of the earliest and most advanced social and medical welfare systems known, a man's wealth also had no effect on children's survival, although it did affect his children's likely wealth and fertility. If wealth can make a difference to children's survival and success, then higher per capita investment directly increases children's chances of surviving and becoming successful.[32] Land ownership, wealth, and fertility remained related in the Swedish demographic transition, even while fertility fell. *Fertility shifts were local and reversible and responded to local conditions* (to local crop failures, for example, not to a global effect of, say, industrialization).[33]

I'll say more about this in chapter 15. For now, following this logic, across human societies *complexities in either the ecological or social environment that result in increased effectiveness of parental investment should result in more investment, even at the expense of fertility itself—as Darwin noted.*[34] Thus, it is not surprising that wealth differentials promote fertility differentials, even in Western societies that are socially monogamous and attempt to be egalitarian.[35]

The extent to which we can predict fertility shifts as a result of economic conditions or purchasing power will depend on a number of factors, but especially on how much parental investment assists individual children. Technological advances may require more education or better training to enter the labor market, and thus more investment to produce each competitive child. And if even only a few families increase investment to enhance competitiveness, the stakes are raised for all families in the population.[36] The relative costs and benefits of children themselves are also influential: resources make a difference in marriage, income, and migration opportunities for better-educated children compared to others.[37] Even the richest family's wealth could be dissipated through continued, even, investment in large numbers of children. Unequal investment, even in the face of legal mandates for equal inheritance, is unsurprisingly common.[38] When increased investment in individual children enhances their ability to survive, marry, and reproduce, net lineage success can be enhanced by shifting more resources into investment in children (education, savings, health insurance, resource gifts, etc.). Unless there is a net increase in total resources, the allocation of available resources must be into fewer children.[39]

9

Nice Guys Can Win—In Social Species, Anyway

The real great man is the man who makes every man feel great.
—G. K. Chesterton

Simply stated, an individual who maximizes his friendships and minimizes his antagonisms will have an evolutionary advantage, and selection should favor those characters that promote the optimization of personal relationships. I imagine that this evolutionary factor has increased man's capacity for altruism and compassion and has tempered his ethically less acceptable heritage of sexual and predatory aggressiveness.
—George Williams

W<small>E STARTED</small> with very simple and general hypotheses about how resources affect reproductive success and why men and women typically have quite different resource strategies. But complexity has crept in. By the time we reached the demographic transition societies of the last chapter, forces such as markets, governmental rules, and societal mores clearly influenced men's and women's costs and benefits in resource, mating, and marriage decisions. Cultural transmission and evolution matter; simple hypotheses about selfish genes, favoring themselves and their copies in kin, or even simple reciprocation, can't explain what we see. The world may be nasty, but it is much nicer than selfish genes in solitary animals would make it. As the biologist Richard Dawkins put it, "nice guys finish first" under some conditions.

How do simple systems of kinship and reciprocity generate the complex sociocultural patterns we see? Why do we help others, often others who cannot repay us? Do we do so for selfish or selfless reasons? Do men and women differ? In this and the next chapter, I explore these questions before returning to issues of sex differences in coalitions, competition, and even warfare over resources.

A<small>RE</small> W<small>E</small> L<small>EMMINGS</small>? A C<small>AUTIONARY</small> T<small>ALE</small>

If only we were as selfless as the lemmings! Lemmings, small mousy rodents, breed like crazy and frequently face population crises even more severe than the one some people think we humans are facing now. However, some lemmings, so the story goes, sacrifice themselves willingly in the face of these traumas: they leave their home and soon fling themselves over cliffs, dying so that others may live, thus averting a population disaster. One of the most cogent (and funniest) comments on this story is a Gary Larson cartoon: a group of lemmings is rushing downhill into water to drown; way in the back is a lemming wearing an inner tube.[1] There's a problem here: the facts are quite different from the myth, and the inner-tubed lemming is the one that will survive.

How did this mythical story arise? When conditions get crowded, adrenaline levels rise, and prime adults drive away the newly reproductive adolescents. Ecologist Dennis Chitty, in his book *Why Lemmings Commit Suicide* (1996), referred to the myth of the selfless lemming as "beautiful hypotheses, ugly facts" and raised an interesting question: Why do we get so upset at suggestions that self-interest might be rearing its ugly head? In fact, why do we think self-interest is "ugly"?[2] Yet genetic self-interest, as we have seen in earlier chapters, lies at the heart of most behaviors in other species.

Humans are like lemmings, and all other living organisms, in some basic ways. They, and we, must solve a set of ecological problems in some of which, individual and group "interests" are likely to diverge.[3] Group living introduces subtlety into the evolutionary rules. We cooperate daily, in many ways. The rule "genetically profitable behavior is likely to persist and spread" must be fleshed out considerably before it can explain much that is interesting, including much of our cooperation and sex differences in that cooperation.

WHEN AND WHY DO WE COOPERATE MOST EASILY?

One step in fleshing out our understanding involves honing our definitions. Some interactions have a true genetic cost to the "doer" whereas others that *appear* costly are actually profitable to our genes (table 2.1, chapter 2)—much of our helping behavior arises precisely from this fact. Yet our very language makes this distinction difficult to remember. An unfavorite (of mine) *New York Times* crossword had the clue "was altruistic"; the answer was simply the word "shared." From our perspective, that is insufficient.

This common usage creates problems both in theory and in everyday life. It implies that costly and cheap sharing will be equally common with friends and strangers. It muddles behaviors with very different impacts as "altruism." For many of us, this seems like an old, solved issue—but I assure you that it is not, in fields from wildlife management to some parts of demography.

The distinction between *apparently* (phenotypically) costly versus *genetically* costly behaviors lies at the root of much misunderstanding. If we do not distinguish between actions that benefit our genes through reciprocity and kin selection, on the one hand, and true (genetic) altruism on the other, we will not be able to make sense of many human behaviors. Because the functionally important distinction is the effect of behavior on genes, *I will restrict the meaning of "altruism"* here[4] and will call behavior that incurs a net genetic cost "genetically altruistic" (chapter 2, table 2.1). Saving strangers at the cost of your own life is genetically altruistic when there are no sufficiently compensating rewards for your remaining kin. Those who become celibate in a way that doesn't help their family,[5] and those who give costly help to nonrelatives repeatedly without reciprocation, are also genetically altruistic.

We expect these latter behaviors, termed "genetically costly (altruistic), phenotypically (apparently) costly (altruistic)" (bottom line, table 2.1), to be rare, and they are. Most of the examples we can think of sound a little silly. Because they benefit reproductive competitors *at a reproductive cost to the doer,* they cannot evolve through ordinary natural selection and are always vulnerable to competition from genetically selfish behaviors.[6] This fact, of course, is more easily seen in simpler situations. In chapters 4 through 8, the male and female strategies of nonhuman primates, and of men and women in traditional and transition societies, appear to make ecological sense.

Hamiltonian kin selection—and reciprocity, mutualism, and their "relatives"—should, in contrast, be common because, while they *look* costly, they help their causative genes; they are termed "genetically selfish, phenotypically altruistic" (second line, table 2.1). Thus, many "nice" behaviors—behaviors that look costly—actually help ourselves or our relatives enough to be genetically profitable: nursing one's children costs calories but is important in becoming a successful mammalian mother; helping my neighbor of twenty years when a storm damages his house means he is likely to help me if and when I need him. We are such a social species that to ignore such interactions, to define "self interest" as narrow economic interest, is misleading (though many behavioral economic games can help us understand the larger problems).

Individuals in many species do phenotypically costly things, and sometimes we do not see reciprocation, so indiscriminate cooperation is something of a dilemma for biologists.[7] Lion females nurse one another's young; but because they are relatives, we are watching inclusive fitness maximization. Mother Theresa was quite a different case; she had no children, devoted her life to helping nonrelatives, and did not accumulate benefits to give any relatives. She was a genetic altruist. Analytically, the problem is to distinguish true genetic altruism (predicted to be rare) both from genetically profitable though phenotypically costly behaviors like helping friends and relatives, and from simple poor observation or inference (as in the lemming case). Often, we *assume* a "helping" behavior to be costly when we have, in fact, no idea whether it is. It is important that both our definitions and our observations are correct.[8]

Matters get more complicated as groups—societies—get larger, more fluid, and more heterogeneous. The complexity of group living in highly social intelligent primates like humans raises new possibilities. How can we generate societies of Nice Guys? If selection worked to favor the good of the group, rather than genes, true genetic altruism would be common, as common as nepotism and reciprocity. As I explain below, there is, in fact, no evidence that any organism has evolved through ordinary natural selection to assist (at least if assistance is expensive) unrelated individuals without some form of payback. But that doesn't mean there must be a dearth of Nice Guys; we do a whole array of nice things for one another. At the most basic level, the help of Nice Guys sorts into four main classes: (1) helping kin; (2) reciprocity and related phenomena; (3) a set of behaviors such as "by-product mutualism," as in the cleaner-fish example below, or when one animal in a group spots a predator and flees in its own self-interest but, as a by-product, helps its neighbors (pseudo-reciprocity); [9] and (4) cultural transmission of norms for cooperation: cultural evolution.[10] All are influenced by our prosociality.

In a nepotistic or familial interaction, one individual pays a cost and benefits an individual who also shares genes; there need be no apparent (phenotypic) return because the returns are genetic, as my son so often reminds me (and as his son will soon be reminding him!). The most general statement of Hamilton's rule (chapter 2) is that there is a cost-benefit trade-off to all behaviors, depending on the extent to which they help or hurt others and how closely those others are related to us, as W. D. Hamilton and Jon Seger (see chapter 2) explained.

Clearly that is a small part of the story. Cooperating with nonrelatives can also affect one's genetic success and generate "nice" behaviors. In the simplest sorts of mutualism, the presence or interaction of two individuals is profitable to each—the mutualism is just a by-product of individual profit-seeking. Mutualism occurs between cleaner fish and their "client" fish, for example; cleaner fish, usually boldly marked, set up "stations" in obvious places (e.g., an isolated clump of coral reef). Client fish come and assume a striking head-down "ready" position; the cleaner approaches and picks parasites off the client's body, sometimes even off its teeth and gums. The client is rid of parasites, and the cleaner gets a meal. It is even possible for mutualistic interactions to evolve from originally parasitic ones—for win-win to arise from win-lose.[11]

SIMPLE STRATEGIES FOR WINNING GAMES

The cleaner fish-client mutualism is a fairly simple kind of cooperation, by-product mutualism. But cooperation is sometimes complex and indirect. One way to gain insight into when to cooperate with others is to simplify the problem into strategic games, in which we can examine individual costs and benefits. If we can understand the dynamics of structured games, perhaps we gain insight into less well-defined interactions in a cultural context.

Prisoner's Dilemma

Robert Axelrod and W. D. Hamilton explored a classic example of such a game, the Prisoner's Dilemma.[12] It is usually introduced this way: two prisoners sit in different interrogation rooms in the police station; they are suspected of committing a robbery together, though the police do not really have enough proof to convict them. If each holds fast and does not confess (i.e., cooperates with the other suspect), the police can only arrest them on a minor gun-possession charge: a year in prison (Reward, R). But the police offer a deal independently to each prisoner: rat on your accomplice, and we'll let you go, while giving your accomplice a harsh sentence (perhaps ten years; this offer is called called Temptation, T, for the player who gives away the confederate and goes free, and Sucker, S, for the betrayed player). *However*, if both prisoners tell on their confederate, they each get a five-year sentence (Penalty, P). The general Prisoner's Dilemma has the payoff sequence: $T > R > P > S$; let's define these as $0 > -1 > -5 > -10$ (the particular numbers don't matter, only the sequence). No matter what the other player does, the obvious nasty solution is that it pays each player to defect. The perverse result is that rational players defect, getting five years in prison each, when, had they only cooperated, they could have gotten away free—but the large payoff is possible only if both players were to able to trust what the opponent will do, and so to cooperate.

Without any mechanism to instill trust, it pays any player to defect. And this is the key in this game. Cooperate-cooperate is highly unlikely in a single interaction of this game.[13] Of course, many interactions are not one-shot deals but, rather, involve repeated encounters, and this is extremely important in getting to cooperation. In the Repeated Prisoner's Dilemma, both players may be able to cooperate and reap the Reward. When they are likely to interact repeatedly, defection on either side is less likely compared to one-shot interactions. In game theory terms, the repetition of interactions changes the value of Reward (benefit, b, minus cost, c); it becomes $(b - c)/(1 - w)$, when w is the probability of meeting again. So when w is small (little likelihood of meeting again), R is still not very enticing; but when w is large, the Reward grows.

It is important in the Repeated Prisoner's Dilemma that players cannot predict the length of the game. If the game is known to be ten rounds long, each player will be tempted to defect in the tenth—but once that is apparent, it is really only a nine-round game, and the temptation arises to defect in the eighth . . . and so on.

The uncertain length of the game means that if one doesn't know when it's the final round, and if one is likely to meet again, then cooperation is more plausible.[14] The shadow of the future encourages cooperation, and we come to value reputation: the idea that we can be trusted to cooperate. Discrimination is a key element, for continually cooperating with a defector is a losing strategy;

no wonder we test each other for honesty and cooperativeness. Discrimination can become, even in nonhumans, a "social contract."[15]

In such a repeated Prisoner's Dilemma of uncertain length, what's best depends on the opponent's strategy. Based on extensive computer tournaments followed by analytical proof, Robert Axelrod has argued that a class of very powerful strategies has two components: (1) cooperate (be "nice") whenever you are the first mover in the game, and then (2) with nearly perfect regularity, mirror your opponent's moves. The simplest such strategy is tit-for-tat, in which you cooperate if you are the first mover, then do unto your opponent just what your opponent did to you in the last round: cooperate if he or she cooperated, defect if he or she defected.[16]

In symmetric two-strategy games, the order of the payoffs (P, R, S, T) determines the likelihood that players will cooperate or defect. In Prisoner's Dilemma, the Sucker's Payoff (cooperate with a defector) was the worst that could happen, and Temptation was the best under all circumstances.

Prisoner's Dilemma makes it easy to see why reciprocity is discussed entirely in the context of long-lived, repeatedly interacting, social animals, in which there are many unknown-length series of repeated interactions.[17] All the examples I can think of also involve organisms that learn, so that a continued "defection" by a particular individual seems likely to cost the defector: typically it will be detected and responded to. If A helps B repeatedly, but B fails to return that help, A can redirect his or her efforts, and other individuals in the group can see that helping B is a bad deal. In some species A can tell others—and the stage is set for tracking reputation, for discrimination, and for ostracism.[18]

One dynamic simulation of a group of agents playing Prisoner's Dilemma on a computer grid suggests natural dynamic cycles of defection and cooperation. Each agent on the grid has a rule or strategy that determines its next move based on its and its opponent's previous moves. An agent's strategy changes in two ways: through random mutations of rule components and through copying rules of particularly successful "neighbors." Always-defect agents dominate the early parts of the simulation. Eventually, through mutations, a cluster of tit-for-tat agents forms, gains strength by playing each other, and takes over the grid. Always-cooperate agents naturally mutate from the tit-for-tat agents and do as well in their interactions (with the tit-for-tats). But when the number of always-cooperate agents becomes large enough, some always-defect agents can invade and eventually take over the grid through their victories over always-cooperate agents. The cycle continues. The authors were particularly interested in what game parameters lead to longer periods of cooperation.[19]

Hawk-Dove and Other Games

Other games have different relative payoffs, and there is a rich array of strategies.[20] We can, in pure game theory, set the payoffs to whatever values we wish.

In the real world, additional considerations affect trust and cooperation: whether interactions are likely to be repeated, whether risks are high or low, and whether one is among kin or strangers. Cooperation can be a highly effective competitive strategy if help goes to kin and/or is reciprocated. For example, in the game Hawk-Dove (or Chicken), two individuals competing for a resource of value V (food, mates, territories) can act aggressively (Hawk) or avoid fights (Dove). Aggressive interactions impose a cost, C. When a Hawk meets a Dove, the hawk gets T (Temptation) = V (the entire value of the contested resource); the Dove gets S (Sucker) = 0. When two Doves meet, they split the resource ($Reward = V/2$). When two Hawks meet, they fight, and get P (Penalty) = $(V - C)/2$. The relationship between the cost of fighting and the value of the resource is crucial: if $C > V$, the payoff order is $T > R > S > P$ and the optimal strategy is a mix of the two behaviors.[21] If the cost of fighting is less than the value of the resource ($C < V$), this reduces to Prisoner's Dilemma.

Reciprocity and Its Relatives

Reciprocity may be direct or indirect (A helps B at some cost, B helps C, C helps A, and so on). Indirect systems, however, are not always stable, for at any moment, the balance of costs and benefits is likely to be uneven (A has helped B at some cost; B has not yet helped anyone). If individuals interact only rarely or occasionally, such indirect reciprocity is extremely vulnerable to cheating: perhaps you help me, but we may never again interact and I never reciprocate. If we are close relatives, my defection may still help my genes and some of yours. However, any individual who consistently helps nonreciprocating nonrelatives will have little effort left for kin.

Elegant examples of reciprocity can develop when risks are high and interactions are repeated. Vampire bats (long-lived, repeated interactors), for example, face uncertainty as they seek blood from livestock. If a bat fails to eat, it can be in danger of starving after as few as 60 hours. Successful bats tend to drink more than they need and share the surplus with unlucky neighbors. This system is facilitated by the fact that the same physical amount of blood makes a smaller hours-to-starvation difference to a well-fed bat compared to a hungry bat: the cost of helping is low and the benefit to being helped is great.[22] And remember the winter ravens from chapter 1: a juvenile responding to a "yell" from another juvenile who has found a carcass in an adult pair's territory not only gets something to eat but as a side effect helps make it impossible for the adults to drive off the marauding juveniles. Both the "yeller" and the responder profit.

Simple reciprocity, without additional considerations, is limited in its ability to explain much of what we see in large societies. When risks are high and individuals move about so that the next interaction is uncertain, expensive helping behaviors may occur only or primarily among kin, or individuals may mirror the behavior of others, for example, in a tit-for-tat manner: I'll start by cooperating, but if you default, I will, too. As the biologists Anne Pusey and Craig Packer pointed out, simple, clear-cut, direct behavioral reciprocity is probably far rarer in the real world than we had realized.[23] In many systems, cheating is discouraged by the third-party rule: punish not only cheaters but also anyone who fails to punish cheaters.[24] This is one of several cultural additions enhancing the spread of cooperative behavior.

Adding a Little Complexity

In all the games and models so far, an individual's payoffs (even if constrained, as in coercion) are what predict cooperation or defection. Kin, short-term mutualists, and long-term repeated interactors are likely to cooperate with each other within the group. Because natural selection favors genetically profitable behaviors, even our powerful human intelligence may have evolved in the context of cooperative competition for resources and mates and cooperative breeding.[25]

These examples are usually very simple, two-person games, and we face the same difficulty as we do in genetic models: to understand individuals in societies, we must build more elaborate models (though still as simple as we can get away with, yet still match what we observe). In the real world, some relatively simple additions to the two-person games may make a world of difference in promoting cooperation in large groups. First, do nothing unusual; remember what others did, and copy both your winning behavior and the winning behavior of others. Thus, copy common behaviors (they are likely to be profitable if they have become common), and copy behavior that has clear profitability (this could be judged indirectly as well: copy high-status winners even if the behavior you are copying is not obviously the source of the status). Second, watch others; do not offer cooperation to defectors or to those who give to defectors. Third, punish cheaters. This means that cheaters, in stable-membership societies, lose more than the game at hand (this strategy comes to involve reputation). Finally, advertise your future behavior ("commitment"), both negative (if you cheat me, I will make you very, very sorry) and positive (I am willing to make great personal sacrifices to belong to your group)—and exhort others to make expensive commitments as proof of future cooperation.

These relatively simple additions can lead to important influences: (1) strong and widespread norms, (2) strong socially imposed costs and benefits, (3) the punishment of cheaters (including those who do not enforce the norms), and (4) a low cost of imposing the norms. In fact, any behavior can become stable if these conditions are met; thus we can see great variation in social norms.[26]

It is clear that organisms in long-lived social species, including humans, are likely to do things that benefit others, either because the actor also benefits or because there is some likelihood that there will be future interactions. "Watchers" may assess us by watching how we treat others, so that our future costs and benefits with additional individuals may be affected. Cultural norms can become strong forces influencing how people will act. We still know little about how particular rules come to exist in particular situations, though we can say, I think, that in relatively small and very stable groups, the costs and benefits will be more predictable than in large fluid groups. It is probably no accident that happiness indices today are high in relatively small, homogeneous democracies like Sweden.[27]

The "Group Selection" Naming Muddle

You will remember core issues of individual selection from earlier chapters and the discussion above the first three columns of table 9.1: Darwin's original individual selection, sexual selection, and kin selection. There are also three forms of social selection: reciprocity (often with the complexities discussed above), coalitional selection, and cultural transmission; at least the last two can lead to cultural evolution. These and interdemic selection have no conflict with ordinary natural selection. Cultural evolution and coalitional selection can, under the right circumstances, lead to behavior that sets limits on the selfishness of individuals.

But reciprocity, coalitions, cultural transmission, and interdemic selection have at times confusingly been referred to as forms of "group selection"[28] even though they are quite distinct from Wynne Edwards' original benefit-of-the-group naive group selection, which occasionally surfaces in fields such as wildlife management, and even parts of demography. The use is probably justified in some cases, but it creates confusion.

The Global Group: Wynne-Edwards and Population Regulation

In the far right column of table 9.1 is the original (so far as I know) group selection argument. The biologist V. C. Wynne-Edwards made a truly distinctive argument about how selection acts on individuals in groups. He noted that groups have traits, such as dominance hierarchies, that are not simple statistical sums of individual traits; he argued that in any conflict between what was good for the group and what was good for the individual, the good of the group would prevail.[29] Thus, he expected individuals routinely to pay reproductive costs (incorrectly inferred) for the good of the group. In his view, group selection would swamp ordinary natural selection. The biologist Graham Bell put it succinctly: "group selection is a fallacious concept when it is held to cause the evolution of characteristics that benefit populations, or species, or communities, as a whole, without distinction of ancestry."[30]

Table 9.1. Summary of Selection Theories.

	Natural Selection			Social Selection			Group Selection	
	Individual Selection	Sexual Selection	Kin Selection	Reciprocity	Coalitional Alexander Irons	Cultural Transmission Boyd & Richardson Cavalli-Sforza & Feldman Lumsden & Wilson Durham	Interdemic Sewall Wright Sloan Wilson Hamilton Wilson	Wynne Edwards
Observations	Organisms seem well suited to their environments.	Individuals sometimes do extremely risky dangerous behaviors.	Individuals sometimes behave in ways which *seem* to cost themselves but benefit other individuals.				Gene frequencies differ across groups.	??
Hypotheses	Heritable variation occurs. Environments are not infinite; thus there is competition among organisms. Some variants will be more successful than others, and will leave more offspring. Over time, population proportions of the different variants will change.	In sexual species, the evolution of anisogamy (unlike gametes) favors (*ceteris paribus*) specialization by individuals into either mating or parental effort. The reproductive return curves for mating effort involve a high fixed cost (much expenditure before any return), resulting in high variance—which favors risky competitive behaviors.	Relatives carry genes identical by descent. Thus, acts that appear to cost the performer may, in fact, profit the performer genetically if: $$rb - c > 0$$ where b = benefits to recipient, c = cost to donor and r = degree of relatedness.	Individuals may profit genetically by assisting nonrelatives when the help is reciprocated. Conditions which must be met are (a) individual recognition, and (b) repeated interaction. Such reciprocity is most likely to evolve in species with long-lasting social units.	Within-group coalitions can impose their interest, constraining the actions of others.	Mechanisms of cultural and genetic intergenerational transmission differ. Cultural transmission and selection may favor traits that natural selection alone, acting on genes, might not favor.	The organization of individuals into groups can affect gene frequency. Conditions which must be met are (a) small viscous (restricted gene flow) populations; (b) close together (same selective pressure; can replace each other and (c) population turnover must be rapid, approaching the rate of turnover of individuals.	Groups have characteristics that are not simply the sum of individual characteristics. Group selection operates on the group and is thus very different from individual natural selections. If there is any conflict, group selection should prevail. Altruistic behavior should be common.

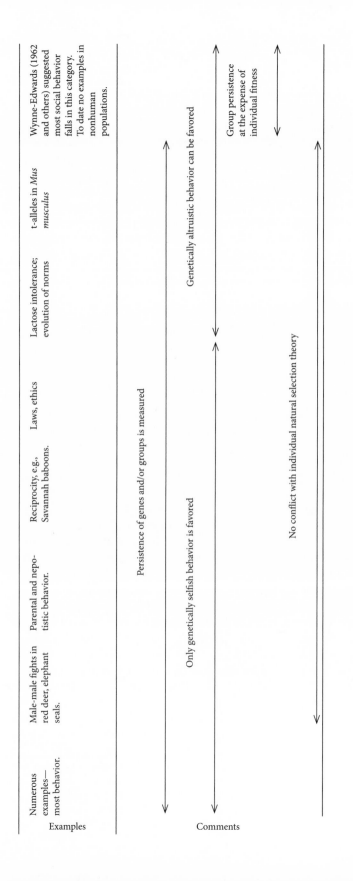

Examples						
Numerous examples—most behavior.	Male-male fights in red deer, elephant seals.	Parental and nepotistic behavior.	Reciprocity, e.g., Savannah baboons.	Laws, ethics	Lactose intolerance; evolution of norms	t-alleles in *Mus musculus*	Wynne-Edwards (1962 and others) suggested most social behavior falls in this category. To date no examples in nonhuman populations.

Comments

Persistence of genes and/or groups is measured

Only genetically selfish behavior is favored

Genetically altruistic behavior can be favored

No conflict with individual natural selection theory

Group persistence at the expense of individual fitness

To date, concerted work has uncovered neither logical nor empirical support for Wynne-Edwards's argument: the empirical evidence suggests that any system depending on individuals to perform routine genetic sacrifices for the benefit of unspecified "others" (e.g., nonrelatives, strangers) collapses quickly as the altruists are outcompeted and outreproduced by reproductively selfish individuals.

So why even consider such an arcane and incorrect argument? There are two reasons. First, even though Wynne-Edwards's formulation is biologically naive and easily dismissed after examination, it is widely accepted in parts of the social and management sciences today. It is also important to see that the *kind* of group can make a difference: groups comprised of related individuals are really kin selected, for example.[31]

Classic Interdemic Selection and Its Descendants

This theory, simply put forth by Sewell Wright, predicts that if certain conditions are met, *the structure and composition of groups can influence the frequency of various genes* (table 9.1).[32] For example, house mice (*Mus musculus*) are a well-known example of simultaneous conflicts within and among individuals and groups. Male house mice may have a gene, the *t*-allele, which causes sperm with competing alleles to clot up: the sperm are at war. Males who are heterozygous for *t* (and we would think would make 50% normal sperm, called +, and 50% *t* sperm) make 85% to 99% *t*-bearing sperm—that is, *t* wins.[33] But because males who are *tt* either die or are sterile, *t*'s victory within the individual in heterozygous males results in disaster for homozygous males. The group matters also: these mice hold harems, so if a female is in a (sterile) *tt* male's harem, she cannot leave and mate elsewhere. So selection is obviously working at multiple levels; the result is that the group structure lowers the frequency of the *t* allele.[34]

The *t*-allele is a rare exception, for obvious reasons. Within an individual, there are usually limits to the profit an utterly self-interested gene can garner. Think of a mutation resulting in cancer: if it kills its bearer, it too dies (although even here, cytoplasmic warfare and genes like the *t*-alleles remain lethal exceptions). The phrase "the parliament of the genes"[35] and the prevalence, for example, of suppressor genes reflect the fact that there are common interests shared by the total group of genes (keeping their bearer alive, healthy, and reproductive) that cause coalitions of genes to prevail over "outlaw" genes.[36]

You will note that the result of multilevel selection in *t*-alleles is a decrease in the deleterious *t*, In this model, the conditions under which deleterious alleles (such as one for genetic altruism) could theoretically spread are very restricted: small, viscous (i.e., largely inbred) groups, close to each other (so you can assume that the ordinary selective pressures are similar); occasional "mixing" followed by restricted interchange. Within each group, the more advantageous (selfish) alleles always increase at the expense of altruists; thus, the timing and type of mixing between groups are critical. Because differential success occurs at birth and death (and the group equivalents of these), and because the turnover of individuals within groups is typically more rapid than the turnovers of groups, a bias exists for selection to be more powerful at the level of individuals.

This model is logical but unlikely to be a strong force generating indiscriminate altruism; other explanations (below) are more likely.[37] In game theory terms, it is quite possible for "always cooperate" to lose within a group but for groups of "always cooperate" players to grow and spread more rapidly than other sorts of groups. So, you see that the catch to playing "always cooperate" is that this is true *only if no defectors sneak in*; this is a fragile system. As the anthropologists Richard Boyd and Peter Richerson summarized, versions of this model of group selection intended to show how altruistic genes can spread "clearly [are] not verified by the data concerning ethnic cooperation among humans."[38]

Coalitional Selection, Cultural Transmission, and Cultural Group Selection

WITHIN-GROUP COALITIONS

Cultural transmission exists widely (even in nonhumans), although cultural evolution has not been found for certain in nonhumans. These are clearly real phenomena. Cultural transmission exists at least at two levels: in within-group coalitions and across the entire group.

Individuals within a group may form coalitions to gain advantages. We find it useful to gather in groups working together to advance our interests. Think of laws, for example, which are inflictions of constraint on individual behavior by coalitions of others in the larger group.[39] Less formally, ethical and moral systems do the same thing (see above); for example, religious groups, like the military, can make demonstration of conformity, and sometimes costly commitment, essential to gaining entry. In many situations, advertising commitment can be important, from painful cicatrice scars (reflecting initiation into manhood in some societies), to military recruits' haircuts, to the "cheap talk" of bumper stickers (this last is an inexpensive, easily faked advertisement, and we give it little currency).

The fairly simply rule "copy those around you who are successful" (see above) can further advertise commitment and belonging, and promote group cohesiveness. Even in simple situations, copying is unlikely to be random.[40] Consider the Mukogodo (chapter 7); by any objective measure, they care for their daughters better than their sons. Yet they claim to prefer sons, because the Maasai, who hold higher status among the local peoples, prefer sons, and the Mukogodo claim to be Maasai (the Maasai often reject this claim). So the Mukogodo are "copying" (at least in their statements) high-status people. But their treatment of children reflects the selective reality that among the Mukogodo, daughters can rise in the world better than sons.

When groups consist of close relatives who share genetic interests, or when group members are threatened by external forces such as competing groups, individual and group interests converge. In a small group of hunter-gatherers, for example, sharing one's catch looks costly—but failing to share one's catch can literally be deadly. Coalitions of individuals within any group can change the costs and benefits of actions like sharing, especially when memory and punishment exist.[41]

OR ELSE!!

New work shows us much more about how cooperation can not only happen in humans but can persist and grow and become elaborated to third parties and the development of cultural norms. The pioneers of this approach to understanding complex human behavior in its culturally evolved context are Peter Richerson and Rob Boyd (along with several of their students), and Ernst Fehr.[42] Cultural transmission and social learning are crucial to human behavior.[43]

Cultural rules and those of natural selection interact; we have a *dual inheritance* of both genetic and cultural information, and we have biases for the sorts of things we learn most quickly. Because the rules for cultural and genetic transmission may differ, even under the same selective pressures, we find that different genetic equilibria can result from culturally transmitted versus genetically transmitted traits.

Cultural norms can be transmitted in many directions, and they can change direction far more rapidly than genetic transmission.[44] Perhaps even more important, unlike chimpanzees, for example, human mental representations can, over successive generations, become ever more complex; this means we can build on existing ideas to form new ideas; this allows us advances that are probably lost to most other species.[45] Cultural and genetic selection can interact, and culturally transmitted preferences can affect gene frequencies—behaviors can be favored in social systems that we wouldn't expect in the absence of social pressures.[46] So, indeed, genetically altruistic behavior by some individuals can, at least for a time, be favored (or forced, as in the coalitional selection above).

A very clever set of cross-cultural studies of three games gives insight into how this works. The Ultimatum Game has two players. Some reward is at stake, and player 1 proposes how the reward should be shared between the players. Player 2 can accept—in which case that is how the reward is distributed. On the other hand, if player 2 refuses, neither player gets anything. A "rational actor" economist would predict that player 1 would offer very little to player 2, but that player 2 would accept, because getting something is better than getting nothing. But, as we will see, that is very rare (if it exists), and there is considerable cross-cultural variation.[47] Here I will focus on this game's results cross-culturally.

Two more games were analyzed cross-culturally by Joseph Henrich and colleagues, and the results are fascinating; space does not allow me to dwell on them here. In the Dictator game, the first part of the game is the same as in Ultimatum, but player 2 has no right of refusal; thus positive (high) offers to share cannot arise from a fear of rejection. Finally, in the Public Goods Game, there is an intrinsic conflict between individual versus group good. A common approach is to play the game in two different forms and compare the outcomes. In each, a reward is at stake. In the Voluntary Contributions version, each individual has a stake and may (or may not) contribute to the common pool; at the end, the common pool reward is shared among all the players. In the Common Pool Resources version, the reward begins in a common pool, from which each player may withdraw up to a specified amount.

Henrich and his colleagues took these games beyond the "usual" experiments in WEIRD (wealthy, educated, industrialized, rich, and democratic) cultures to a cross-cultural set of experiments of about sixteen cultures around the world. The results are illuminating. First, there is no society yet studied that follows the pure "rational actor" rule. Second, there is considerable cross-cultural variation, and, within the ecological and cultural contexts, this variation makes sense.[48]

In the Ultimatum game as played in most modern Western cultures, player 1 tends to offer somewhere around a 50–50 split, and player 2 tends to accept. Does that reflect any sort of universally shared perception of what is fair? In a word, no. In summary, group-level differences in economic organization and degree of market penetration explain a substantial proportion of the variation across the societies studied; and the higher the payoffs to cooperation (the more important the social capital), the higher the "prosociality." As Michael Gurven noted, in many traditional societies, in which daily cooperation is likely to be viewed as a critical component of subsistence strategies, the game results are far closer to actual self-interest predictions than those found in the Western world.[49]

So perhaps it is not surprising that patterns of offers and rejections vary greatly, because what is important in each culture can vary. In some groups rejections are very rare, even when the offers are quite low (e.g., Kazakh living in Western Mongolia, the Quichua of Ecuador, the Ache of Paraguay, and the Tsimane of Bolivia). The reasons can differ, depending on whether they live intermingled with other groups, have coalitions, and more. At the other end of the spectrum, the Au and the Gnau of Papua New Guinea both make high offers ("hyperfair") but also usually reject those offers. One cannot understand these differences without knowing more about the culture. For example, the Au and the Gnau have a competitive gift-giving system (perhaps similar in some ways to Northwest Coast Native American potlatches), in which accepting a gift makes the recipient indebted to the giver. So it makes sense that very high offers are both made and rejected. Cultural forces indeed shape much behavior.[50]

In some cases, the size of the group matters; among the Hadza of Tanzania, people in smaller groups, who are potentially more dependent on other group individuals, tended to accept lower offers than people from larger groups. Among the Tsimane, in one of the most detailed studies, sex, age, and education made no difference in game behavior.[51]

In all these cases except Gurven's study of the Tsimane, only men participated; Gurven found no age or sex differences among the Tsimane. In Western cultures, many experiments have used undergraduate subject pools and have found a number of differences, for example, between majors (e.g., psychology versus commerce majors). With regard to sex differences, two findings are of interest. Remember that player 1 proposes a division of the reward, and player 2 can accept (in which case that is the division) or reject (in which case neither gets anything). When men were player 1, they offered higher (more generous) divisions when player 2 was a woman, and in another experiment, when player 2 was a woman, she chose higher minimum acceptable offers than predicted (this study was run in part to explore whether if women could be satisfied with less, they would suffer in salary negotiations, for example).[52] In a comparison of sex differences in the Public Goods, Ultimatum, and Dictator games, there was no evidence of systematic differences in men's and women's play when risk was involved; when there was no risk, women's choices were less individually oriented and more socially oriented.[53]

ALTRUISTS OR GOOD NEIGHBORS?

The bottom line? The more genetically costly a behavior is, the rarer it is likely to be. Yet the term "group selection" has been muddled to become almost meaningless; almost all the arguments above have been called group selection although they have little in common. It is too bad that the Wynne-Edwards definition has temporal priority; otherwise we wouldn't have this problem.[54] Culturally transmitted traits (like celibacy among Shakers) that are clearly costly tend to diminish over time; recently, I read that in 2013 there were three living Shakers still alive.[55]

But culture interacts with natural selection in myriad ways, and while the Shakers will go extinct, many other cultural practices continue to influence both our behaviors and our gene frequencies. We are probably using heuristic rules of thumb like "copy my successful neighbor," "hope to have a child if I'm doing better than my parents or peers," or "be especially demonstrative of my commitment." These will vary locally and be variously successful. Even after we have checked to be certain that we are correctly assessing costs and benefits, behavior we think is selectively stupid may take a long time to diminish and disappear.

We live in social groups; we see others every day, and there is cultural accumulation among humans that we don't see in other organisms ("Why, in my day, we . . ."). Cultural selection and coalitional selection clearly are important in human evolution. What I hope is that we can harness such cultural transmission to tackle the environmental issues we have created by our earlier, simpler, genetically profitable predispositions (chapter 15).

Cooperation and Free Riders

We expect, then, that genetically costly behavior will be uncommon, except when culturally mandated. Even if we can create groups of altruists who may last longer in dire straits than selfish, or simply reciprocal, groups, this advantage will persist only so long as altruists alone comprise the group and outsiders cannot also profit. If a cheater—a "free rider"—invades and takes benefits but does not do the costly help, the altruists face extinction.[56] And this is why, in human groups, third-party interactions can be so important: there may be not only rules to "punish free riders" but also metanorms to punish those who don't punish free riders.

Real-world examples of this difficulty abound, from nonhumans (lion males are free riders when lionesses hunt) to traditional societies to modern global climate change and acid rain. In many traditional societies, for example, the "fruits of the forest"—berries, mushrooms, wood, and so on—are common-pool resources: any villager can gather them.[57] If the village is small and close-knit, if villagers have lived there a long time, if exploitative cheaters are easily detected and punished, and if there are no external markets, the villagers' use is liable to be shared for subsistence and to be sustainable over long periods. But if the village has many transients, or if it is hard to catch and punish cheaters, or if anyone can convert forest products into lots of hard cash and move away, "take the money and run" defection is likely (see chapter 15).[58]

Today, we see new and larger-scale but similar problems. Ozone depletion, acid rain, global climate change, open-ocean whaling and commercial fisheries deterioration—all are common-pool, public goods, or open-access resource problems in which the costs of restraint have never been worth it. Can we make those costs worth it now? The difficulties of solving large-scale free-rider problems among nation-states are truly daunting. We can, for example, promote CO_2 emissions standards to reduce the degree of future global warming around the world; the difficulty is that such action requires some very local costs now for the sake of some quite dispersed future benefits—so it's hard to get cooperation (see chapter 15).

In today's large heterogeneous societies, individuals are not all subject to the same ecological or economic constraints. Cultural rules and the actions of others are important constraints in our environment. How we act daily in different situations to accomplish genetic selfishness varies greatly and can be strongly influenced by others, directly and indirectly, in the cultures in which we live, as we will see in the next chapter.

10

~~~~~~

Conflicts, Culture, and Natural Selection

Tragedy is when I cut my finger. Comedy is when you walk into
an open sewer and die.
—Mel Brooks

One death is a tragedy; one million is a statistic.
—Joseph Stalin

KATHERINE HARRISON was a witch. After the death of her rich and powerful husband, she was accused of witchcraft based on events occurring years
before. When accused, she was a rich widow who chose not to remarry, and she
had no sons to protect her property. She continued to be accused for years and
was even forced to leave her community, until, in 1670, her older daughter became engaged to a man from a powerful family. The future father-in-law, previously an attacker, became her protector, and the accusations ceased. Strong allies are important.

The demographic and economic particulars of witchcraft trials show a pattern that is a logical, if curious, example of conflict over resources and reproduction in a particular culture. The communities in which accusations of
witchcraft flourished were communities long torn by internal strife. Witchcraft
accusations often originated in property disputes. Despite that fact, women,
who owned almost no property, were three times as likely to be accused of
witchcraft, seven times as likely to be tried, and five times as likely to be convicted of witchcraft as men.[1]

Property disputes between men tended to go through the ordinary court
system, so why did property disputes involving women segue into witchcraft
accusations? Land was becoming a scarce resource in colonial New England,
being concentrated in few hands, and the circumstances under which women
could hold property were limited. Suppose a woman's husband died and she
had only young sons or no sons at all; men's routes to gaining the property she
held were limited: marry her or remove her. Two things appear to have mattered: (1) Did the woman have powerful male protectors? (2) Was she of marriageable age—or did she have low reproductive value?

Unless they were single or widowed, accused women from wealthy families could be fairly confident that the accusations would be ignored; women alone, with no powerful male protector, were likely to be accused, tried, and executed (and their land up for grabs).[2] Particularly if she had no sons and her daughters were the inheritors, a woman without a powerful male protector became very vulnerable. If she was young enough (of sufficient reproductive value), she would be courted; if she was older, she was likely to be accused of being a witch. Even a rich woman without a husband could be vulnerable: because of the way inheritance laws worked, widowhood did not bring women independent economic power.[3]

Most "witches," except in the heat of the Salem outbreak, were women who owned some property but had no strong male allies. The tales of witchcraft are rich and varied, but a significant pattern holds true: women who held resources alone and were not likely marriage candidates because their reproductive value was low were significantly more often accused, tried, and executed as witches.[4] Men's and women's interests were in conflict over resource allocation. The history of witchcraft is a fascinating set of interactions among genetic self-interest, coalitions over resources in that context, and cultural and historical particulars such as social norms. Moral systems are obviously far from immune to perversion by the strong to meet their own ends. I suspect, however, that with some thought we can predict much about the direction of such twists, even in quite particular situations.

COOPERATION, COMPETITION, AND GROUPS

Conflicts of interest are rife at a variety of levels: between men (often in the context of sexual selection); between women; between men and women (sometimes in the context of parental effort); between children. Much is informal, but we have formalized competition in many ways: sports (from kindergarten soccer to the Olympics), beauty contests, and more. How do the various forms of competition vary, and in what ways are they the same? Some of the same difficulties face groups of genes sharing a body and groups of humans sharing resources. In this chapter I want to explore how cultural rule-making and transmission might play out, and how group living has shaped the social nature of our intelligence.

WORKING OUT OUR CONFLICTS: MORAL SYSTEMS AND GROUP LIFE

The parliament of genes works because there really is a unity of interests once genes are stuck in the same body: what destroys other genes is also likely to be self-destructive. Among groups of individuals, as we saw in the last chapter, things are more complicated. Several forces appear to limit the degree of self-interest that could harm others. Kin selection, reciprocity, mutualism, and coalitional restraints such as laws and social mores can limit rampant individual self-interest.[5]

For an individual in a group, other individuals constitute a strong selective force; in many situations *other individuals* are the strongest predictor in the environment of the costs and benefits to any action. The evolution of prosocial behavior is an important, but still not fully solved, phenomenon often leading to cooperation within groups. Individual psychology matters, but so do extrinsic conditions. There are good arguments that culture, because it can be transmitted in so many ways and directions (below), differs from genetic transmission, which can only go down from parents to offspring. Cultural transmission yields greater scope than ordinary genetic selection for the evolution of large-scale prosociality. One problem is that, early on, at least, different fields interested in the evolution of cooperation ignored (or were ignorant of) insights from other fields; some modeling approaches have addressed that. Rob Boyd and Peter Richerson have created a nice synthesis of the issues involved in culture and the evolution of human cooperation. There are still, of course, considerable cross-cultural and even individual proclivities for prosociality.[6]

Coalitions of powerful individuals (revisit table 9.1) can make it very costly to display unbridled short-term self-interest. Laws, mores, taboos, public shaming—all are mechanisms to help subdue self-interest in the interests of tolerable group living (a cynic would say, in the interests of keeping your self-interest from impinging too much on my self-interest).[7] The social contract can be powerful.

The more we learn, the more likely it seems that the biological underpinnings of human group living strongly influence moral systems.[8] Years ago, the biologists Sewell Wright, William Hamilton, and Richard Alexander first identified the pressures of group living—and the resulting between- and within-group competition and cooperation—that underlie many of our cultural models. If our behavior toward our fellow humans is at all strategic, as behavior in general seems to be, there should be some common patterns (e.g., rules restricting cheating within the group) and considerable variation in other patterns (rules about who can choose marriage partners).

I should note here that (1) the naturalistic fallacy (imagining that what *is* reflects what *ought to be*; chapter 2) is a particularly dangerous trap in this inquiry, especially when we are thinking about humans; (2) some patterns are already predictable as being universal except in pathologies (parental and kin caring, not murdering innocents within the group); (3) men and women are likely to hold somewhat different beliefs in societies in which their roles, costs, and benefits are different; and (4) within-society differences seem likely when people's costs and benefits diverge; thus small kin groups will teach different and more uniform mores compared to large highly mobile societies. Perhaps because of the importance of small kin groups in our evolutionary history, there seems likely to be a bias toward retaining "small-group" sorts of rules.

Psychologist Lawrence Kohlberg structured a progressive list of moral "stages" he suggested people went through (from stage 1, "preconventional," in which "right" was whatever didn't get you punished, to stage 6, in which one upholds universal ethical principles as a matter of commitment and principle). He thought that moral sentiment developed toward stage 6 as a matter of human maturation, though some of us never made it. Not surprisingly, not only children scored relatively "early" on Kohlberg's scale, but so did adults who lived in small, traditional, kin-based groups with competition from outsiders.[9] Kohlberg's certainty of "moral progression" seems to me as shaky as the old idea of "evolutionary progression." The ecology of social behavior is far different when you live among your relatives and the same friends all your life than when you spend your life in a large and fluid society of strangers, and the rules will also be different.

Not only does social complexity matter, but, once again, what people say may not reflect how they behave otherwise.[10] It is easy to say that we believe in A, or B, or in environmental protection, or civil rights in China, but the real test is *what we will sacrifice to achieve those ends*. Our behavior, like that of other species, is sensitive to costs and benefits. We are likely to discriminate; while all of us are in favor (perhaps) of healthy children and a clean environment, what we are willing to pay personally to get some defined version of these goals is less clear. So conflicts of interest are rife, and males and females will differ in their goals and mode of conflict.

Male-Male and Female-Female Conflicts

Almost all within-sex competition can be viewed as a path to reproductive success (reviewed in chapters 3 and 4). For males, often being highly successful reproductively involves taking serious, potentially lethal risks. Most mammals tend to have polygynous mating systems in which males compete; in terrestrial species, males who must fight mature reproductively later than females (so they can grow larger), and because of intense competition and resultant injuries, males die earlier than females. If the male-male competition is aquatic or aerial, male mobility and maneuverability may matter, and males may be smaller than females (as in some seals and in raptors). Selection acts separately in the two sexes, so that the end result (who is bigger, for example) can vary.

Female-female conflict also has reproductive ends, but it is often more difficult to see—more subtle, perhaps.[11] Certainly choice of males in male-parental species is important (including humans). Learning "how to behave" so as to attract wealthy and powerful mates can also be a big part of cultural transmission.[12]

For example, in the 93 odd-numbered societies of the Standard Cross-Cultural Sample, from about age seven on, boys, but not girls, are taught to be aggressive and to show fortitude, whereas girls are taught to be industrious, responsible, and obedient (chapter 4). There was no significant difference in the degree to which either sex was taught to be competitive. So, while today we see many sex differences in the type and apparent intensity of competitiveness, in traditional societies, girls were taught as well as boys to be competitive. A detailed study of the Tsimane of Bolivia showed that Tsimane girls and women competed for social, mating, and food resources; importantly, the investigators found that women's focus of competition changed across the life span, reflecting the life history issues of tradeoffs of current versus future reproduction, and quantity versus quality of children.[13]

In the cross-cultural samples, too, one of the biggest causes of divorce was conflicts among co-wives—so female-female conflict within the family can be socially dangerous. In a patriarchal society, girls' competitiveness declined as puberty began. Cross-culturally today, Finnish, Israeli, Italian, and Polish girls, ages 8, 11, and 15, all use indirect aggression most commonly, followed by verbal aggression, with physical aggression least common. Curiously, from this life history stage onward, *physical* aggression (typically in the context of attracting the approval of boys in the group) can increase in some circumstances, sometimes to the extent of considerable violence and even murder.[14]

Male-Female Conflicts

Male-female competition tends to be most common in mated pairs, siblings, and in the families in which they are embedded.[15] As predicted long ago by the biologist Richard Alexander and psychologists Martin Daly and Margo Wilson, remarriages in which the woman brings children to the new relationship tend to be volatile, sometimes violent. Here is a case in which the reproductive interests of the man and woman differ both in traditional and in modern societies. Partner violence is more common when there are great asymmetries in power as well as divergent genetic interests. We also know that children in such remarriages, for example, are at far greater risk than either genetic children (of both adults) or adopted children, both cases in which the adults' interests converge.[16]

One potential area of male-female conflicts of interests is relatively modern, arguably a twentieth-century concern: gender equity (also called gender equality). I suspect questions on this topic would puzzle members of a traditional society. Today, we tend to think about gender equity in three areas: in job opportunities and equal pay for men and women; in political participation, from voting to office-holding; and in autonomy of reproductive decisions. Today, gender equity is spreading with socioeconomic development; it rises with industrialization and women's work and continues to spread in postindustrial countries. However, attitudes vary. Both the gender ratio (a United Nations Development Programme index of overall sexual equity) and the gender empowerment measure (which looks specifically at women's job and political opportunities) vary with the UNDP human development index, a measure comparing nations' human development index, a reflection of life quality. First-world nations, which are well off in health, standard of living, and education, lead in equity in both the gender ratio and the gender empowerment measure. The gender ratio also varies with a country's main religious persuasions: Hindu and Muslim countries have, to date, the least gender equity.[17]

I'M COMMITTED, ARE YOU? THE DECEPTION AND MANIPULATION PROBLEMS

What you and I are willing to pay for our own versus others' benefits (and how our willingness compares to other people's) reflects an important kind of social selection: the "commitment problem." Because we live in groups, have good memories for slights (usually better than our memory for praise), and because we, individually, do need help from our neighbors to thrive, we profit from discerning who is reliable, that is, who will give us what we need (when it's cheap, when it's expensive; if we reciprocate, if we do not, etc.). Our common strategies include exhorting others to demonstrate expensive commitment, requiring costly displays of future commitment by outsiders, and forcefully proclaiming our own commitment to the group.

In many subgroups in most societies, from military training to fraternity initiations, from financial dues to physical sacrifice, we require proof, typically expensive proof, that new members will do what's required when the time comes and will help others in the group. As in sexual selection (in which females are likely to require expensive, unbluffable displays of genetic quality or willingness to commit), in social selection we frequently require evidence of a willingness to be a good group cooperator, even—or especially—when that is likely to be expensive.[18]

Much manipulation and deception centers on commitment. The concept of "Machiavellian intelligence"—that we evolved to manipulate our cooperators and competitors—is highly relevant, and it represents a fascinating area for new work. When we leave a 20% tip in a restaurant, do we do so because we are altruists? Do we do it because we can use such "cheap talk," in game theoretic terms, to advertise our goodness?[19] Or is it because we have internalized such cheap strategies, because they are more convincing than having to think about when and how to advertise our goodness? Testing among such alternatives will be difficult but exciting.

INTERTWINING CULTURAL AND NATURAL SELECTION

Cultural transmission and the accumulation of cultural innovation are great complications, far more than simply two of several possible mechanisms to shape coalitions and conflicts and to foster socially acceptable behavior. Consider:

genetic inheritance can only be vertical and in one direction—down from parent to offspring. But cultural transmission has several possible routes: not only do we teach our own children, but we can learn from our children, our children can learn from nonparental adults (we formalize this in the school system; and perhaps teachers can learn from their students), and both our children and we can learn from others in our own generation.[20]

Cultural transmission is also initially difficult for Darwinian thinkers; a "meme," for example, is a conceptual replicator, an idea; we can count alleles, but often things are not quite so clear regarding memes (e.g., incomplete transmission, what amounts to "mutations" in their transmission). These related issues are still the topics of debate.[21]

Like most other human complexities, cultural transmission has antecedents in other species. It is true that many of these have very simple social lives, and I doubt that any of us would imagine that thinking in a biological framework alone could tell us what we need to know about human cultural diversity; our framework would be woefully incomplete. Yet other species certainly have intergenerational transmission that interacts with environmental conditions to change gene frequencies, and although the cases are surely simpler, perhaps they can show us something about how genes and environment can interact. What other species appear to lack is cumulative cultural evolution.

For example, optimal foraging theory postulates that foraging efficiency increases relative reproductive fitness. In ground squirrels, optimal foragers survive better and have more offspring than nonoptimal foragers. Foraging optimality is heritable: babies are more like their parents than like others in the population. However, learning is important. Heritability is about 60% genetic; the other 40% of the parent-offspring correlation in foraging optimality comes from babies foraging near their mothers and learning what to eat.[22] The functionally important facts are that heritable variation exists; that one can predict, in a specific environment, which strategies (learned as well as genetically transmitted) ought to result in an increased reproductive fitness for their possessors and an increased proportion of the possessors in the population; and that one can test and falsify these predictions. Humans, of course, are extreme in their reliance on culture as a means of responding adaptively to environmental pressures.[23]

Although many cultural practices appear superficially to work in ways that enhance the success of the practitioners, this is an area of inquiry in which much remains to be learned; we could use a deeply critical assessment of cultural practices, in addition to the excellent, more specific studies that exist.

Several groups of scholars have quantitatively studied the interactions of inheritance through cultural and natural selection. They have different foci, but some generalizations emerge.[24] When the two modes of transmission interact, the result can be an equilibrium in gene frequencies that differs from those expected from ordinary natural selection alone. Humans, like other learning animals, may have been influenced by natural selection to accept socially transmitted ideas that increase pleasure, reduce pain, reduce anger, reduce fear or increase cognitive consistency, and these pressures, arising from our group members through social as well as physical mechanisms, have shaped us.[25]

The course of cultural evolution is altered by the decisions people make as they learn which behaviors to adopt and decide whom to imitate. The anthropologists Robert Boyd and Peter Richerson and their students, in particular, have concentrated on the formal analysis of dual inheritance through culture and genetics.[26] They suggest two main decision-making forces: "guided variation" (because individuals acquire information socially, there can be rapid cumulative changes that may be adaptive) and "biased transmission" (people do not imitate others at random). Because changes can be rapid and cumulative, cultural transmission of memes has a particular advantage in environments that change moderately rapidly.

Cultural transmission is like genetic transmission in the sense that information about how to behave is transmitted from individual to individual, but because cultural transmission can be vertical, oblique, or horizontal, a belief or practice can change almost instantly. Nonetheless, vertical (parent-offspring) transmission remains important in cultural models; it is enhanced by assortative (nonrandom) mating in models of the transmission of, for example, sign language. Among African Pygmy hunter-gatherers, 80% of specifically identified skills were learned from parents. This is likely to be a typical pattern in many cultures, though empirical testing needs to be done. It has been found, for example, that among Stanford undergraduates there is strong parent-offspring transmission of political and religious affiliations.[27]

But problems remain. If culture were purely ideational (the piece of culture of most interest to many workers), we might have fewer problems. But people live in groups that experience conflicts, and culture contains material aspects. One general class of model incorporating this aspect suggests that, in general, individuals strive for rewards that have typically correlated with reproductive success. Culture, it seems, is not only the context within which we act, but also a tool for social manipulation. Recent work is progressing by recognizing individual conflicts of interest and thinking about cultural manipulation.[28]

Social and Lactic Acid Cultures

Cultural and natural selection interact in a potent way and can influence things we seldom think about. Consider the physiology of milk digestion. Around the world, there is systematic variation in how well adults can digest milk products. Many adults lack the ability to digest fresh milk; they have no lactase (which allows digestion of lactose), and so are intolerant of lactose. The anthropologist

William Durham, comparing alternative hypotheses, found a strong interaction between cultural practices and genetic inheritance.

The genes responsible for adult lactose absorption have evolved to high frequencies in populations that have a tradition of dairying and fresh-milk consumption and that live in environments of low ultraviolet radiation (high latitudes, where vitamin D and metabolic calcium are chronically deficient). Populations without dairying traditions and with low UV exposure have low lactose absorption (e.g., the Eskimo and Saami groups, who live at high latitudes with low UV but have diets rich in vitamin D). Similarly, populations with dairying traditions who live at low latitudes (high UV), as in the Mediterranean, have low lactose absorption. Milk-dependent pastoralists, in contrast, no matter at what latitude they lived, had high lactose absorption as adults. Thus, ecological conditions lead to a cultural practice that mediates selection on the ability to absorb and digest lactose.[29]

Religion, Government, and Sex Ratios

Culture can modify the most basic of biological phenomena. Worldwide, the average sex ratio at birth is about 105 male births to every 100 female births. But sex ratios at birth among orthodox Jews and during the recent one-child policy of China (where widespread cultural preference for sons exists) provide examples of cultural-natural selection interaction.

Among orthodox Jews, marital intercourse is prohibited during menstruation and for seven days thereafter, and the husband is not to masturbate or seek other sexual outlets. At the end of the seven days, the wife takes a ritual bath, and the couple is directed to have intercourse at that time and twice a week during the rest of the month, with the exception of men in unusual occupations that require the men's absence. There is additional advice if the couple wishes to conceive a son: intercourse should take place twice in succession.

It is difficult to obtain birth sex ratios for orthodox Jews independent of nonorthodox Jews, and conception biases are certainly difficult to measure, for the obvious reason that important parameters are difficult to control. Nonetheless several things are true. Y-bearing sperm, which combine with the egg to make an XY (male) fetus, are slightly pointier-headed (hydrodynamically faster) than X-bearing sperm; they are also smaller, with fewer resources to stay alive if the egg is not immediately ready. As a result, in humans as in most mammals, conceptions close to the time of ovulation tend to be male-biased. The orthodox cultural practice of abstinence for about 12 days per month, combined with frequent intercourse near the time of ovulation, appears to interact with biological biases in conception probabilities: sex ratios for Jews in a number of traditionally orthodox locations historically average 137 males / 100 females, whereas for nonorthodox Jewish populations and nearby secular populations, they average 105, the worldwide average.[30]

Historically, throughout much of Asia, a widespread preference for sons has existed, as reflected by the old Chinese proverb, "It is better to raise geese than girls." But as population numbers have increased worldwide, so has concern about population growth. In the early 1980s, the government of China instituted its one-child policy in an attempt to slow China's population growth rate. Couples were restricted to one child per family, with some exemptions.[31]

The cultural history of son preference has interacted with the limits on family size, and possibly with marriage preferences.[32] The proportion of families with only one child did increase, and the birth sex ratio became more male biased. An obvious first thought would be that perhaps couples about to have their first (and possibly only) child might engage in sex-specific abortion and infanticide, but this was not generally the case. First, the real concern over an infant's sex was in rural areas, where parents preferred boys to work in agriculture and therefore might have an interest in sex determination, but in these areas, the technology for sex determination was largely absent. In the cities where the technology existed, cultural son preference was not so pronounced. Further, the government decrees were seen by most people as perhaps "good" for the whole nation in terms of population limitation but also as running counter to one's own family's interests. Thus many people continued to have more than one child, even when that carried financial, and sometimes social, costs.

Occasionally, an effect could be seen in first births: in two provinces in 1985, the sex ratio of children in single-child families soared to over 129 males to 100 females. But these were exceptional: it was primarily in *later* births that the sex biases became most pronounced (fig. 10.1). In a nationwide study in 1989–1990, the sex ratio of first births was 105.6, right at the worldwide average, but the sex ratio of later born children depended on how many older brothers and sisters already existed. For example, the sex ratio of third-borns when there were two older sisters was 224.9/100, and the sex ratio for third-borns with two older brothers was 74.1/100. Some daughter-preference did exist when several older brothers were already born, and the "empty" spaces in figure 10.1 suggest that families do avoid some patterns of birth numbers and sex-ratios.[33]

Ironically, the effects of the one-child policy on fertility are absolutely trivial compared to some political (cultural) rules that were not aimed at fertility per se. Chairman Mao's efforts to "equalize" and to make intellectuals do farming and peasant work, for example, not only caused a variety of disruptions of families but also changed mortality and fertility rates dramatically. Analysis of these rates in Shanghai, a wealthy, densely populated urban province, Shaanxi, a rural province, and Hebei, a rural province with a higher GDP per capita since 1970 than Shaanxi, shows that many of his policies produced far greater impacts on fertility and mortality than any influences of the one-child policy. Government policies affected the three populations differentially; this was evident at both the aggregate and individual levels.[34]

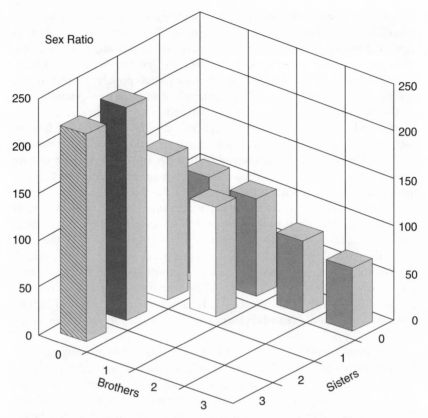

Figure 10.1. In the face of China's one-child policy during the 1980s, the sex ratio of births changed—though the impacts were seen not in first births, but in the sex ratio of later births, in response to the sex and number of existing children. Later births in families with more than one existing daughter were extremely male biased; later births in families with several sons were somewhat female biased. Curiously (see text), no impacts on fertility (or mortality) from the one-child policy were nearly as strong as those from the policies of Mao, which were not directed at fertility.

Religion, Fertility, and Celibacy

The interactions of natural selection and culture can be complex, and they operate at a variety of levels.[35] Ideas, attitudes—all can be transmitted between individuals as "semiautonomous units" and may, but need not, be adaptive.[36] Individuals in any society can choose to do things that are detrimental to their survivorship and reproduction—but whether such patterns will persist and spread, becoming and remaining the mode for longer than a generation or two, depends on how detrimental the behavior is. For this reason, in the coevolution of genes and culture, we do expect that, by and large, natural and cultural selection will interact to favor behaviors that benefit those who do them.

It doesn't surprise us, then, that most stable cultural institutions with a long history do not promote genetically costly behaviors among their members. Consider the positions of religious institutions on fertility: many religions, unlike the Shakers, have rules that urge the reproductive success of their members. Mormonism strongly promotes both financial striving and reproduction (God created all human souls long ago, and they must wait to be born, so having children is a kind act for these souls); polygyny was a long-time Mormon social norm, and it is still practiced in some locations of the United States today, despite federal laws against it. The Hutterites, who are used as the standard among demographers for maximum "natural" fertility, prescribe that all adults in a group must assist in the raising of children. This expanded and shared adult care (cooperative breeding), in small and close-knit groups, is quite possibly what allows the very high Hutterite fertility of 12 to 14 children per woman.[37]

There are other good reasons for religious cohesion. Time-consuming and costly rituals of many religions pose a dilemma for rational-actor economists. But perhaps different currencies are at stake. The behavioral ecologist Rich Sosis and the economist Bradley Ruffle have examined expensive religious rituals, testing the hypothesis that such rituals may lead to higher levels of cooperation. Using field experiments among Israeli kibbutzim, they found that, indeed, religious males (who are the main practitioners of collective ritual) were more cooperative than secular males and than either religious or secular females. They used observational data, which is important; there is evidence that people claim to be more religious and more cooperative than they really are. This, again, is a complicated issue.[38]

But what of the celibacy required of Catholic priests? While fertility is promoted for parishioners (even if indirectly, through abortion and contraception bans), men and women bound for religious life are required to be celibate, something that is an issue of conflict within the Church today. Here is surely a case of imposition of rules by more powerful group members on the less powerful.[39]

The rules of celibacy in the Catholic Church appear to have had their origins in property disputes. By 1139, under Pope Innocent II, priestly marriage was forbidden, but not until the dictates of the Council of Trent in 1545–1563 went into effect were married men no longer able to become priests (widowers, however, still could). If families had been small, it might have been more difficult to require celibacy of priests, but giving later-born sons to the Church often helped both the Church and the family in times of very large families, when too many sons complicated inheritance. Later-born sons were often destined for the Church (as appears to be true today: fig. 10.2).[40]

When celibacy rules were not absolute, they were typically easier on the younger and lower-status clergy, who had the least property open to dispute

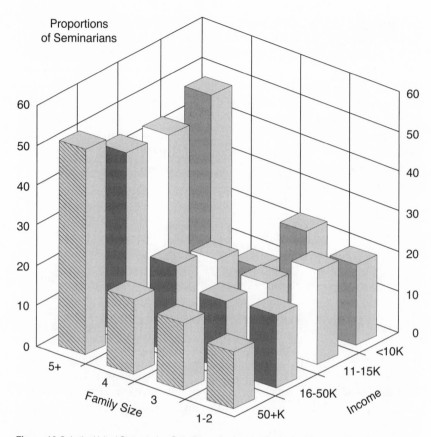

Figure 10.2. In the United States today, Catholic seminarians tend to be later born and from large families; some bias toward lower-income families is also apparent.

between the Church and possible heirs. Yet high-status men in the clergy could contravene the rules. Rank, as ever, tended to have its privileges: for example, Rodrigo Borgia, who became the Renaissance pope Alexander VI, not only was married but had several children by one of his mistresses, Vanozza de' Cataneis, whom he brought openly to Church celebrations. His son Cesare was his oldest son by Vanozza and, in the traditional manner, was destined from youth for a career in the Church.[41]

Today, with very small family sizes, the Church is having trouble recruiting priests in the United States. A major reason appears to be the stricture against marriage. For example, seminary enrollment in the United States peaked with Vatican II—but as disposable wages increased over the next few years, seminary enrollment plummeted.[42] A complex set of culturally derived behaviors—

giving younger sons of large families to the priesthood—once worked for large Catholic families. The trend today is still for (especially later-born) sons of larger, poorer families to become priests (fig. 10.2), but the interactions of devotion and Church requirements cause some proportion of only sons of devout Catholic families to become celibate.[43]

The complex and repeated interaction of culture and biology, combined with the rapid changes we engender, means that today the behaviors we see may or may not be adaptive. When we find apparently maladaptive behaviors, we have interesting further work to do: what will happen to the frequencies of these strategies? We cannot assume, for humans today, that any particular behavior is, or is not, adaptive.

LOGICALLY INEPT, SOCIALLY ADEPT:
THE SOCIAL CONTEXTS OF INTELLIGENCE

The inherent conflicts of group living—among kin, cooperators, and competitors—may well have shaped how we think as well as what we do (for example, consider the quite varied results Henich and colleagues got for the Ultimatum game cross-culturally in chapter 9). Most of the time, we think of our intelligence as something that liberates us from the exigencies—the selective constraints—of the environment. We can turn that consideration on its head and ask about the forces that may have shaped our intelligence. Our social environments, as well as our physical environments, have been central in shaping our intelligence, our cultures, and the ways we go about seeking resources.

The evidence suggests that we're not qualitatively different from other organisms in getting and using resources to survive and reproduce; we are quantitatively different: just more subtle and complex about it. We must beware of falling into thinking that somehow if it's conscious, it's different. As Robert Smuts noted:

> That choices made unconsciously in the midbrain normally have a biological function is taken for granted, but choices made consciously in the neocortex are often thought to lie in the realm of social values and culture, and to be free of biological constraints. The neocortex, however, is as much a product of natural selection as the hypothalamus, which means that it evolved because it served the most basic of biological functions: it helped human ancestors to perpetuate their genes.[44]

A strong argument can be made that our extraordinary intelligence, and much of our culture, derive from selection for ability to compete against *each other* (rather than against the physical environment), often as coordinated, cooperative groups.[45] Among the primates, big brains (which are costly and unlikely to be favored unless they confer an advantage) seem to be associated with both environmental and social complexity (for the hominins, see fig. 6.1). Primates with large home ranges tend to have larger brains; primates that live in larger social groups also tend to have large brains. For example, arboreal travel may require some cognitive map making, and toolmaking for foraging requires some intentionality.[46] Machiavellian intelligence, the smarts to manipulate one's social cooperators and competitors, clearly requires considerable complex understanding: to manipulate, to deceive, even to keep track of others' social lives.[47]

How to sort out the relative impact of these different forces? Neocortical enlargement seems to reflect the extent to which primate brains are specialized for intelligence, and the relative size of the neocortex can be compared. When the anthropologist Robin Dunbar did this comparison among primates, he found that measures of environmental complexity (range area, day-journey length, amount of arboreal fruit in the diet) showed no relationship to neocortical enlargement. In contrast, the size of the social group was correlated with all measures of neocortical enlargement. What's more, the ability to deceive conspecifics ("tactical deception," surely a measure of Machiavellian intelligence) was strongly related to neocortical enlargement. So, whatever the role of environmental complexity in shaping primate intelligence, it pales in comparison to the role of social complexities. In fact, scholars are leaning more today toward the view that communication may be best understood if we view it as evolving not only to transmit information but also to manipulate and deceive.[48]

Over our evolutionary history, then, other humans have been one of the strongest selective pressures molding our behavior. Conflicts of genetic interests, played out in social situations, may well have shaped our problem-solving abilities in special ways. We pride ourselves on our intelligence, but as any decision theorist or operations specialist will tell you, we are in fact lousy logicians. Daily we face political and procedural problems which have—and we know it—logical, intelligent solutions. Nonetheless, even when we know a great deal about the technology of solutions, we frequently find ourselves unable to agree on how to proceed, caught in "social traps"—that is, situations in which our decision rules might make social (but not logical) sense.[49] Perhaps we could profitably explore them as "evolutionary traps"; we can ask, in the context of our evolutionary and social history, whether patterns appear systematic in ways predictable from our ecology and our genetic conflicts.

The sorts of illogical things we do are legion, and they are consistent with an evolutionary history of (1) life in social groups and (2) environmental uncertainty that we remember and seek to understand (and impose a pattern on). We consistently interpret single or isolated events as predictive of a whole category, even when the event is portrayed as atypical or explicitly not representative.[50] We do this most emphatically if it represents something dangerous, noticeably different, or (especially) a social defection. We attribute someone else's behavior to his or her "character," but we explain our own behavior by circumstances (that way, we ourselves never behave badly, it was the circumstances, but those other rats . . .).[51] We are rotten at reasoning from the general to the particular, and we overdo reasoning from the particular to the general. In fact, as Robin Dawes, an operations research specialist, noted "a great deal of our thinking is associational, and it is difficult indeed to ignore experience that is association-ally relevant but logically irrelevant"—the vampire lore discussed in early chapters is an example. We also are likely to adopt wrong (sometimes blatantly wrong and illogical) explanations if they are proposed by someone "of status" in a group situation.[52]

We estimate the probable frequency of events on the basis of many things other than actual frequency data; yet logically, frequency is what matters. A famous example is the "compound probability fallacy." Consider some rare medical condition. Now, is it more likely if it is associated with something common but not correlated, like freckles? Well, no. But if asked about the likelihood of a rare condition, medical internists estimate it as more likely if another, more common symptom is listed with it. That is illogical: the combination of two events cannot be more common than either one alone. Nonetheless, illogical as it is, when we see the less common problem tied to a common problem, we tend to estimate the combination as more likely. We are apparently especially prone to this sort of thinking in the context of (perceived) dangerous or costly events and identifiable ethnic or socioeconomic groups.[53]

This is illogical associative thinking, and it looks as if *perceptions of individual interest in the context of a group have been an important force in shaping it* (as above). We are logically inept, but socially adept. One experience at being cheated, and we are likely to generalize to future interactions with individuals of the cheater's category. One dangerous event witnessed, and we fear it ever afterward. We remember and overestimate the occurrence of rare (especially dangerous or socially harmful) events and conditions. When we lived in small groups and interacted with the same people repeatedly, this may have been a reasonable predictor. Today, it means that if we are asked to estimate the relative frequency of murder versus suicide, most of us imagine murder as more common; it is not, but it's dangerous and dramatic and gets more media attention. We have more (usually vicarious) experience with it.

Daniel Kahneman, a Nobel Prize winner and frequent collaborator with Amos Tversky, has made many significant contributions to understanding how we think, although the issues of social currency have not been a major feature of his work. An important issue he raises is that we really have two systems of judgment and decision making. One (System 1) works on heuristics, giving us rapid opinions; this is the one that gets us into the sort of trouble I discussed above: one experience and we overgeneralize. He also noted, something important to our evolutionary background, that we "keep score," for instance refusing to cut our losses when that would mean admitting failure. And, for instance, if asked is the letter *k* likelier to be in the first position or the third position in a word? because it's easier to think of words beginning with any letter than in the third position, we have a predictable bias.

The second way we think, "thinking slow" (System 2), takes more mental effort and is more analytic. Because it takes more effort, many of us are apt (to try, at least) to slide into the intuitions of System 1. However, there are tasks and problems that can be done or solved only by System 2, and Kahneman gives excellent examples, making the point that we have a relatively sophisticated allocation of our attention to the two systems that has been honed by a long evolutionary history.[54]

Although we can certainly learn logic, we nonetheless typically solve problems, at least initially, in the context of our social history, and perhaps this is related to our predisposition to use System 1 if we can get away with it. And herein lies an interesting clue: some of our illogical thought makes sense if our history has typically involved competitors. One common solution seems to be not "what violates the logic?" but rather to "look for cheating against me." This could be simply because we are more stupid than we want to admit, but I doubt it. I suspect rather that, in our social history, we have been concerned with protecting our *own* rights under the social rules rather than protecting "rights" in general.[55] People solve concrete examples of logical dilemmas, perhaps especially if they are in the context of detecting cheaters, much better than the formally identical abstract versions of the problem.

For example, suppose you are a grocery checker. Every check for $50 or more must be endorsed by the manager. At the end of the day, which checks (stuffed in various ways in a drawer) will you examine, to be sure every check over $50 was indeed endorsed? If you think about it, you don't have to turn over any check that is for less than $50, or any check that has a manager's signature facing up. What you do need to do is turn over any check that is for $50 or more (face up) to see if it was endorsed. People do much better on this task than on a precisely parallel abstract task, the Wason selection task.

Taking this perspective helps us approach reasoning in new ways. Modern psychologists are, on the whole, pessimistic about teaching reasoning. And if one imagines that people solve problems using abstract algorithms and tries to teach in the abstract, the results will reinforce that perception. But people do seem to use rough cost-benefit rules in daily life, and if one teaches reasoning by tying abstract examples to problems people see as applicable to daily life, they learn much better.[56] Are there further patterns? Here is an area of inquiry ripe for cross-disciplinary sharing of concepts.

Undoubtedly, our evolutionary history has shaped our male and female brains in different ways.[57] This is apparent in many non–sexually oriented behaviors: spatially, women rely on "landmarks" rather than compass directions more than men (chapter 3); girls solve disputes more often by avoiding the context of the dispute while boys are more likely to argue about the rules (chapter 11). The very structure of our brains differs by sex, though it is important not to forget that our behavior toward our daughters and sons also contributes to gender differences in behavior. Few behavioral sex differences are consistently strong and clear, but in many traits there are strong trends (with overlap) for the sexes to differ. In the United States, girls tend to score higher on verbal tests, boys on mathematical tests; these differences increase as the children grow and learn (despite our best parental efforts) that boys are going to be the engineers and physicists. Once again, the interaction of natural selection and culture yields a rich diversity.

Our evolutionary history has shaped our intelligence in ways we would never suspect if we thought of ourselves as socially neutral logicians, as living computers. In fact, we quickly solve some problems that are extremely difficult and costly (in algorithms) for computers; but in very special, largely social ways, our logic is liable to be skewed. Though this is a fascinating proposition, our thinking in this area is in its infancy. So far, the main value of changing our perspective is that we have uncovered new ideas to explore.

11

Sex and Complex Coalitions

A Prince is likewise esteemed who is a stanch friend and a thorough foe, that is to say, who without reserves openly declares for one against the other, this being always a more advantageous course than to stand neutral.
—Niccolò Machiavelli

COALITIONS, like so many other phenomena, can be a reproductive strategy; and if this is true, male and female coalitions will tend to differ. Coalitions are not a widespread phenomenon; they occur in species that are long-lived, live in stable associations, have memory, and form friendships.

Dolphin society is (like chimpanzee groups and many traditional societies) a "fission-fusion" society, in which individuals travel and feed together, splitting up and reforming. But the sexes differ. Among bottle-nosed dolphins, females often swim with each other. A female might swim with her mother today, her mother and sister tomorrow, and her friends next week. Males, however, are a very different story. They travel in small and stable groups: you will always see the same two or three males together. However, sometimes you see mega-groups of two "usual" alliances swimming together; these mega-groups are predictable in their membership. What's going on?

When a female dolphin comes into season, she is a reproductive resource—but a male has a problem (as is true in deer, caribou—and many primates, in fact). Perhaps he can lure her away by dramatic aquatic display, or even kidnap her,[1] but how can he control her and prevent other males from fertilizing her? He might run into another group at any time. A single male simply cannot keep a group of males from taking away the female. What dolphin males do in the face of this predicament is form coalitions: the same two or three males will swim together, day in and day out, displaying together and cooperating in getting females.

But a two-male coalition can be displaced by a three-male coalition, for example. Dolphin males have a flexible system of stable coalitions and temporary supercoalitions (fig. 11.1). When group H (three males) has a female, group B (three males) may not be able to capture her from H; so they might leave, recruit group A (two males) and together steal the female. There are stable alliances at both the coalition and supercoalition level: males within each group cooperate; coalitions A and B cooperate; H cooperates with G and D, and so forth.

In primates, the cooperate-to-compete strategy is often starkly evident. Baboon males form coalitions specifically to steal females from other groups, much as the Yanomamö (chapter 7) form raiding groups to steal women from neighboring villages. In Barbary macaques (*Macaca sylvanus*), single males associating with a female frequently become the targets of male coalitions, though male coalitions also function to support kin in disputes, to offer reciprocal support, and to support nonrelatives when it is in a male's self-interest.[2]

Male coalitions were a clear reproductive strategy for the chimpanzees of Arnhem Zoo, which were studied by the primatologist Frans de Waal. The three big males (Nikkie, Luit, and Yeroen) played power politics for reproductive ends.[3] Sometimes the struggles were straightforward. Yeroen dominated, not only allying himself with males, but seeking the help of Mama, the dominant female. Luit (with some help from Nikkie) eventually dominated Yeroen, displaced him, and took over. Before he took over from Yeroen, Luit cultivated other males who had just won fights; but as soon as he displaced Yeroen, he switched to supporting less-powerful males who had typically lost, dissipating

Figure 11.1. Dolphin males travel in stable coalitions and occasionally, as here, two coalitions will form a supercoalition to gain access to females. (Photo by Andrew Richards.)

any centralization of power.[4] Yeroen cultivated Nikkie (who was the least powerful of the three males on his own), and they displaced Luit.

The rewards were clear: copulations with females. In the beginning, while Yeroen was the single (and most powerful) dominant male, he got about three-fourths of the matings, and almost all of the matings were with mature, prime females. So far, this is simply another example comparable to those in chapter 4, suggesting that dominance tends to be related to reproductive success for male primates.

However, Luit and Nikkie, neither able to displace Yeroen alone, formed a coalition and revolted against him. During the power takeover by Luit with Nikkie's help, Luit's share of matings jumped from 25% to well over 50% (while Nikkie went from perhaps 5% of the matings to over 30%). During Luit's dominance he (~37%) and Nikkie (~30%) together gained two-thirds of the matings. When Yeroen and Nikkie displaced Luit, both Yeroen and Nikkie gained while Luit's share declined to under 25% of the matings. In the next power takeover, Nikkie displaced Luit, and in his second year's tenure, gained over half the matings himself. Not only is dominance related to reproductive success in most species, but in social species, cooperative coalitions can be useful in gaining dominance and reproductive resources.

Coalitions, Resources, and Reproduction

Group living is a powerful force in human lives; our neighbors influence what we can do successfully. Much of our human behavior is coalitional. The anthropologists Lionel Tiger and Robin Fox commented that "the political system is a breeding system. When we apply the word 'lust' to both power and sex, we are nearer the truth than we imagine."[5] Not surprisingly, human sex differences are similar to those of other primates: if coalitions had reproductive ends in our evolutionary history, and if men and women succeed reproductively through different strategies, then men's and women's coalitions should differ.

While it is clear that men strive for power, it may not be obvious that men often use power and resources specifically as mating effort. But as we have seen in earlier chapters, across cultures men form coalitions to gain resources and use resources to gain reproductively, through formal reproductive rewards associated with status or simply through polygyny. It is probably true that *coalitions evolved as reproductive strategies and are likely to remain so*. Because reproductive return curves differ for mating and parental efforts (chapter 3), the potential reproductive gains, as well as the variation in success, are likely to be greater for men than for women. In our evolutionary history, men had more to gain from risky coalitions and high-stakes politics than women.

It should be no surprise that the payoffs differ for mating-effort versus parental-effort coalitions, and that defection (e.g., leaving a coalition) and open aggression will be more frequent in fluid, higher-risk, higher-gain mating-effort coalitions than in parental-effort coalitions. Parental-effort coalitions are much more likely to be longer term and relatively more stable than mating-effort coalitions. Coalitions among males supporting one another in dominance fights often involve nonrelatives and are likely to be much more fluid. When resources are sought to raise offspring, the struggle is likely to be individual or to involve close relatives or co-wives;[6] when higher-risk, higher-gain mating effort is involved, coalitions among nonrelatives are common. Men and women, normally expending as they do greater proportionate mating and parental effort, respectively, are predicted to form different sorts of coalitions at different risk levels and to be differentially involved in community and in wider levels of politics.[7]

Male-Female Coalitions

In all gametic sexual species, there is a male-female period of cooperation to produce offspring; these may be simple and temporary or life-long and complex. In environments requiring the parental effort of both parents in feeding offspring, males and females form lasting pair-bonds and cooperate in getting and distributing resources. Some male behavior will likely be directed toward mate guarding or other behaviors that increase the certainty of paternity. In polygynous species with coercive males or mate guarding, females may seek out powerful males as protection against other harassing males.[8]

Male-female coalitions can also involve mating effort, on the male's part at least. In some baboons, males "befriend" young females; when the female matures, the protector male is her likeliest mate. Even in fiddler crabs (in which females are relatively defenseless), male neighbors aggressively keep out male intruders from a neighboring female's territory but do not discourage females.[9]

The game Battle of the Sexes (BoS) reflects the tension in male-female coalitions nicely, because the two players have some convergence of desired outcome and some differences. The classic example (and the reason it is called BoS) begins with John and Mary wanting to spend the evening together. However, John wishes Mary would come with him to the baseball game, while Mary wishes John would accompany her to the Stravinsky String Quartet recital (Today, for political correctness, the game is sometimes called Baseball-Stravinsky rather than Battle of the Sexes; but the rules are the same). The players "want" to coordinate their behavior, for they want to spend time together, but they also prefer different activities.[10]

There is another common dilemma in other species, such as snail kites (*Rostrhamus sociabilis*) in the Everglades, which have converging interests in the success of their current offspring but also, because of the long mating season, have other mating opportunities. In game-theory terms, this battle—over mate desertion—is a form of Hawk-Dove or Chicken, not Battle of Sexes.[11] With snail kites, each partner, as the chicks grow, is likely to test the other, staying away a little longer on each hunting trip; if the partner is faithful to the nest, it is likely to be deserted.[12] Several things affect the likelihood of desertion: one's own status, availability of other mates, for example. In a bad year, with high water levels (that make less available the snail on which the kites depend), desertion is not likely (the cost of desertion is higher than the value of another mating).

In species in which a single parent is markedly less successful than two parents in raising offspring, the male-female coalition may persist, sometimes for the lifetimes of the individuals. When it does so, genetically monogamous breeding systems may result. In contrast, in any species in which one sex can raise offspring alone, the payoff for defection is high, and the cost of cooperating when one's mate has defected (deserted the nest for a new mating) is not devastating. When either parent can raise offspring alone, desertion is likely; the two parents will monitor the appropriateness of desertion (Can the other raise the offspring now? Will the other stay?).[13]

How Do Same-Sex Coalitions Differ?

In many mammalian species, same-sex coalitions exist relatively briefly. In dolphins and primates, however, long-term cooperation, in order to compete, is common. Individuals form coalitions of varying stability for reproductive reasons, and some coalitions, as in the Arnhem chimpanzees above, are sufficiently elaborate that they become indistinguishable from a broad definition of politics.

The basic differences between the sexes in the kinds of coalitions they form are further influenced by ecological factors. In many primates, males disperse from their natal group, joining a new coalition in a group of strangers.[14] In "male-bonded," or "male philopatric," species, in which males remain in their natal group while females disperse, male politics can become highly developed: groups of related males form coalitions; they have a power structure similar to patrilocal societies among hunter-gatherers. In "female-bonded" species, in which females remain in the natal group, matrilineal coalitions can likewise be powerful. Coalitions of females may assist males, resist male sexual coercion, forage together, and even harass subordinate females and female coalitions so that low-ranking females experience more reproductive failure than high-ranking females.[15]

MALE COALITIONS

Struggles among male primates for status, territory, or harem control often involve coalitions. The context is typically mating effort (above). Males may or may not be related, and coalitions may be temporary and fluid (e.g., with desertion from the group), particularly if they involve nonrelatives. Many species of primates have one-male groups; however, even in those species, male coalitions in the mating context occur occasionally. In red-tail (*Cercopithecus ascanius*) and patas (*Erythrocebus patas*) monkeys, several males may cooperatively invade a single male's group and mate with the females. In gorillas (genus *Gorilla*) and geladas (*Theropithecus gelada*), a subordinate male may share breeding rights with the dominant one for long periods.[16]

In most primates with multimale groups, males tend simply to be competitors—outside of coalitions in sexual selection, tolerant or cooperative relationships among males are rare (see chapter 4).[17] This is particularly true for species in which males leave their natal groups as they reach adolescence, so that adult males in a troop are seldom related. In male-bonded groups, cooperation is more common; in red colobus (*Procolobus*) monkeys, for example, only natal males are accepted into adult coalitions.[18]

In chimpanzees, it is clear that reproductive resources are at stake in male coalitions and political interactions, as in the Arnhem Zoo study.[19] Frans de Waal argued that "the unreliable, Machiavellian nature of the male power games implies that every friend is a potential foe, and vice versa. Males have good reason to restore disturbed relations; no male ever knows when he may need his strongest rival." Indeed, in the Arnhem study, unresolved male-male power tensions eventually erupted with lethal consequences, and one of the males was killed by a coalition of other males. The anthropologist Richard Wrangham noted that chimpanzees and humans are quite alike—and unusual—in having not only male-bonded groups but also intense male intergroup aggression. In humans we call the end result warfare (chapters 13, 14).[20]

Most data about coalitions and reproductive success are simply correlational. Recently, however, researchers tested the hypothesis that in free-ranging chimpanzees, cooperating in aggressive coalitions increased a male's reproductive success. Using fourteen years of genetic and behavioral data from Gombe (one of the best-known chimp study sites), they found that coalitionary aggression increased a male's chances of siring offspring and also of rising in the dominance rankings. Social network analysis lent depth to the analysis: the males who profited most from working in coalitions were those who had coalition partners who themselves did not form coalitions with each other.[21]

FEMALE COALITIONS

Female-female coalitions tend to center around parental effort and alliances against intruding males. Female chimpanzees in the wild are sometimes subject to prolonged and violent aggression and probably to forced copulations. For many years, there were no reports of female alliances to confront these attacks, as there are in bonobos (see below). Among wild chimpanzees in Uganda, females do sometimes form alliances in retaliation, possibly because, for whatever reason, these females are more gregarious than female chimpanzees in other locales. The Budongo Forest, where they live, is closely associated with human habitation and sugar manufacturing.[22]

Bonobos, or pygmy chimpanzees (*Pan paniscus*), are not only one of our closest relatives but also show an apparently unique and strong form of female coalition. Even though bonobos are a species in which females disperse and males stay in their natal groups (and thus adult females in a group are not related), female bonobos are skilled at establishing and maintaining strong bonds with one another; they perform a behavior called G-G (genito-genital) rubbing, which appears to help reassure and cement friendships in a variety of situations. Females control and share highly desirable food with one another (but not with males) and sometimes cooperatively attack and injure males. Food control appears to allow female bonobos to mature relatively early and, in the end, translates into enhanced lifetime reproductive success. Their coalitional power is remarkable in forming and maintaining relationships and in de-escalating potential aggression; it is also reproductively profitable. Females are receptive for longer periods than are common chimpanzees; copulations are common across groups of bonobos, and mating interference is rare. Most scholars working on bonobos attribute the relatively low level of male aggressiveness in bonobos to the strength of female alliances.[23]

In contrast, savannah baboons (*Papio cynocephalus* ssp.) in Amboseli, Kenya, which are often thought to be typical of species that form female coalitions, often have disputes. In about 5% of these disputes, a third party will intervene. Adult females strongly support close maternal kin, and Joan Silk and colleagues saw no evidence of trading grooming for alliance or of reciprocated support with nonkin. High-ranking females formed aggressive coalitions; when subordinates got into a dispute, support was more often given to the higher-ranking of the two. In sum, nepotism and individual benefits appeared to drive the form of female-female coalitions.[24]

Female coalitions in some species can have serious and bloody results. In a number of primates, groups of related females may cooperate to get resources useful to their offspring (while preventing others' access) or to harass subordinate females in reproductive condition.[25] In female-harassment species, less-dominant females have difficulty raising their offspring successfully. In fact, early descriptions of "allomothering" (a non-mother associating with an infant) often included incidents of female harassment ending in the injury or death of the infant.

So there are sometimes quite direct reproductive impacts of female-female cooperation and competition. Many examples of female reproductive coalitions involve preventing the success of others. In yellow baboons (*Papio cynocephalus*), mid- to high-ranking females about to ovulate or give birth form coalitions and attack low- to mid-ranking females, suppressing their ovulation or causing spontaneous abortion.[26] After a lower-ranking mother gives birth, her infant is likely to be roughly handled by more dominant females during its most vulnerable first three months—again, by those in reproductive states most likely to gain from the death of a competitor's infant. Dominant bonnet macaque

(*Macaca radiata*) females, often working in coalitions, also handle the infants of subordinate females roughly. Further, they differentially direct such handling toward female offspring of subordinate females, forming a sort of "corporate matrilineal hierarchy." Female mate choice tends to increase the coefficient of relatedness within the matriline, consolidating the coalition. Harassment of infants of other lineages may be a method of decreasing the size and strength of competing coalitions. In at least some baboons, we are discovering that female dominance struggles may have reproductive costs for the aggressors as well (chapter 4).[27] Remember, however, that female mammals are limited in the maximum number of offspring they can raise. The variation in reproductive success of females is less than that of males, so the impact of dominance on reproductive success for females is typically less marked than for males.[28]

In common chimpanzees and macaques, female coalitions can influence male competition, because female coalitions come to the aid of preferred males in power struggles (yet another reason for males to seek friendships with females). In chimpanzees, female coalitions are quite stable and overlap with the pattern of social bonds.[29] Female chimpanzees interact much more with other females than with males, and the frequency of aggression among females is about one-twentieth of that among males. Females also show far fewer submissive and other reconciliation behaviors. Although females do interact with males and sometimes enter into aggressive encounters in support of a particular male, mostly male and female life are relatively separate spheres. Overt strife is more characteristic of males. This generalization appears to hold true in both wild and captive populations of chimpanzees.[30] In the wild, furthermore, while the male patterns appear to be similar to the Arnhem Zoo population, the females appear to be even less sociable and less involved in bonding than the Arnhem females.[31] Once again, the ecological conditions favoring group formation have a ripple effect, influencing social patterns as well as simple dispersion.

Male chimpanzees are both more aggressive and more social than females because they are more political, particularly in the wild. But males have at stake access to matings with estrous females and the possibility of fathering the majority of offspring in any mating period. For females, reproductive success is not a matter of matings but of access to resources such as food for raising offspring and being able to protect one's offspring. For female chimpanzees, a strategy of avoiding competition by dispersing, becoming solitary females with their offspring, can work—but it does not widely lead to coalitions, negotiations, and politics.[32]

In gorillas, female kin are friendlier to and more tolerant of one another than they are with nonkin; coalitions involve kin against nonkin. In a number of species, there is evidence that females form coalitions to intervene in aggressive encounters (against either males or other females) on behalf of maternal kin far more often than on behalf of other individuals, and to incur greater risk for maternal kin.[33] Reciprocity does occur among unrelated females, although it does not appear to be intense. Coalitions among unrelated females are rarer overall than coalitions involving unrelated males, and the context of female cooperation is typically parental effort. Nonkin coalitions are often directed against lower-ranking females within the group and involve little risk for coalition members.[34]

In nonhuman primates, then, there are important sex differences in the formation and function of same-sex coalitions. Male coalitions largely serve mating-effort competition; male rank and male coalitions tend to be fluid, and unrelated males may form coalitions. Female coalitions involve parental effort such as nursing, foraging, and the support of particular individuals—kin and friends (reciprocators)—as well as the harassment of competitors in some species. Female coalitions, though they may involve both kin and nonkin, appear to be typically more kin-biased than male coalitions.

Both male and female coalitions in nonhuman primates appear to form in the context of reproductive gain, either through heightened mating or parental success or through hurting competitors' success. The major sex difference is that the potential reproductive impact of coalitions is greater for males: in any season, a male's success can range from zero to many, while a female's success can range from zero to one or two. For males, cooperation with nonrelatives can be highly profitable, and male coalitions, more often than female coalitions, involve nonkin and high risk.

Informal Human Coalitions

Coalitions are central to human life and society, are potent forces both within societies and between groups—and clearly can be more complex and flexible than other primate coalitions. I will not look at formalized coalitions, such as organizations, but only at the costs and benefits for individual alliances (including within organizations).

Among other primates, one's likelihood of winning appears to influence whether one seeks a coalition partner. In humans, this may also be true. In a gaming setup of groups of three, male and female participants were told that they had a certain "strength" level; they further knew the strengths of their two other players. When an individual was strong (had a high probability of winning all the points without help), the participants tended to compete alone. People also tended to show a threshold for switching to a coalition—some judgment of the strength they needed to win alone; when they chose a coalition partner, they tended to choose those defined as strong.[35]

Remember that in Prisoner's Dilemma (chapter 9) the payoff for defecting was better than the alternative no matter what the other player did. In related games like Hawk-Dove (= Chicken), a tension exists. For snail kites, the payoffs for deserting if your partner stays are the highest, but cooperate-cooperate is next best. Being a sucker to a defector pays better than the payoffs if both desert. So the temptation to desert is real, but the risks are also high, and the loss for defect-defect mutual desertion is high.

If one could count on cooperation—if one could trust—all would be rosy. As the economist Ken Binmore has pointed out, this has led to a common fallacy: because the payoffs from defect-defect are arguably the worst possible, isn't it not only immoral, but even foolish, to be "rational" and defect in games like this?[36] This fallacy of ignoring that we can't trust that others with different interests will cooperate is sometimes called the Fallacy of the Twins. When you hear it invoked, you have a clue that the author is falling prey to the fallacy "What if everyone were to behave like that?"—without recognizing that strangers (and even mates with outside options) have no basis for cooperation. This, as Binmore notes, misdirects attention to the outcomes only if everyone does the same thing. Since hawk-hawk (defect-defect) has a worse payoff than dove-dove (cooperate-cooperate), one could be led to the mistaken conclusion that surely everyone should—*will*—act identically and cooperate.[37]

But neither player can count on the other, so the players act independently. In Prisoner's Dilemma, the payoffs for defecting if your opponent cooperates are the highest of all. If the players were identical twins, we would predict cooperation, and indeed (chapter 2) twins and other siblings typically act more cooperatively than nonkin. Similarly, individuals who interact repeatedly in contexts in which the number of interactions is uncertain cooperate more than strangers.[38] This means that trust is another dimension of coalitions and cooperation. Cooperation is more difficult without trust, obviously, though it can exist.[39]

Coalitions and Communication

Important clues to the dilemmas of cooperation versus defection in group decisions come from a recent series of experiments by Nobel laureate Elinor Ostrom and her colleagues, who manipulated the payoffs for defecting, how much people were allowed to talk about decisions, and how much they could discuss penalties for cheaters.[40] The games had the same structure as "commons"—resources to be allocated among group members.[41]

"High-endowment" games, in which it paid handsomely to defect (take more than your share), had more defection and lowered everyone's "yield" (profit).[42] But even in high endowment games, communication mattered: the more people could communicate and the more they could punish cheaters (though what worked best were graduated sanctions, not "grim triggers"), the more they cooperated and the better their payoff.[43] Communication—even once—among participants in these experiments could significantly increase cooperative behaviors.

In low-stakes commons, the average net yield was 35% if no communication was allowed; when repeated communication was allowed, the yield rose to 99%. When the payoffs for cheating were higher, people behaved differently:

in high-endowment commons, the average net yield increased from 21% (no communication) to 55% (one-time communication) to 73% when communication was allowed repeatedly.

Just being able to talk should, in contemporary noncooperative game theory, should make no difference—but it does. Further, when the participants had the right to choose a sanctioning mechanism, even with only a single opportunity to communicate, groups achieved a net yield of 93% (compared to 21%); when the costs of fees and fines were subtracted, the yield was 90%. Thus, while there are no easy answers to solving large-scale ecological "commons" problems, there are some clues and some useful strategies to try, if one can monitor for cheating, identify cheaters, and punish them—all these are examples of coalitional selection (see also chapter 10).[44]

In an intriguing extension of these experiments, Ostrom and her colleagues eliminated face-to-face communication; people knew each other only through a computer network and only as numbers.[45] If a strategy had to be unanimously accepted, eventually someone would propose, and others accept, an even-handed allocation to all seven people. However, if it only took a majority (four of the seven players) to implement a strategy, a common outcome was that someone would quickly pick three others in the group using the numbers of their computer terminals. Proposals tended to be for computers 1, 2, 3, and 4; or 4, 5, 6, and 7; or 1, 3, 5, and 7. So they used an abstract logic to make these proposals and propose a one-fourth-each allocation. This formed a coalition that effectively shut out three of the seven people in the group.

COALITIONS AND THE ECOLOGY OF TRUST

As you would guess from the game outcomes discussed above, cooperation is likely when cooperators can trust that others will continue to cooperate. In non-humans, evidence is accumulating that only rewarded cooperation is likely to continue.[46] Conditions that make it easy to exclude individuals within the group, as in the computer-commons game above, promote very specific coalitional cooperation, which may constitute "defection" against the rest of the group.[47]

The issues of cooperation, discrimination, and payoff show up in places we would not expect from traditional paradigms. Just as in other species, cooperation in humans begins among kin, for defection is less costly genetically. How far it is extended beyond kin depends on the costs and benefits. For example, among the Ache (who share much food and are often called egalitarian), when a man dies, his small children are likely to die: since he can no longer share, other men do not share with his widow and children (chapter 7).[48]

Does teaching trust vary cross-culturally? If, for example, cooperative strategies can function to reduce the risk of failure (e.g., food sharing), then we expect such strategies in environments in which food return is uncertain and frequently low. Though we still do not expect indiscriminate cooperation, external extremeness and uncertainties seem likely to promote some kinds of cooperation.[49]

Cross-culturally, as population density increases, the degree to which children are trained to trust others decreases somewhat. In societies with communities of more than 5,000 people, children tend not to be taught to trust others.[50] This is hardly surprising; as the proportion of strangers—those who are neither friends nor relatives—increases, the less fruitful trust is likely to be as a first approach.[51] As the risk of infection from serious, life-threatening pathogens increases, children are less strongly taught to trust others (chapter 4). Here, the issue is not the social implications of "stranger," but the risk from any individual.[52] Patterns of marital residence—and the relative strength of male-male versus male-female coalitions—are strongly related to the intensity with which children are taught to trust. In patrilocal societies, in which related groups of men co-reside, children are less strongly taught to trust others, compared to other living arrangements.

When internal warfare (among villages within the same society) occurs more frequently than every ten years, children are taught to trust others much less than in societies in which such warfare is rare;[53] no significant pattern emerges with the frequency of external warfare. These patterns carry over into other social behaviors: when trust inculcation is low (as in polygynous societies or societies with frequent internal warfare), boys are trained to be more aggressive and competitive (chapter 6).

To move beyond simple kin coalitions and small-society reciprocity and conflict requires either trust or a rather Machiavellian approach to fail-safe devices.[54] International trade specialists have hoped that trade might create coalitions and develop trust.[55] Societies trade when they have complementary needs and perhaps the need for repeated interactions. In the societies of the Standard Cross-Cultural Sample, trade is more common when temperature and rainfall conditions vary greatly. Hunter-gatherers trade less than others; fishing and agriculture are slightly positively associated with trade; and trade increases strongly with the importance of animal husbandry. Despite these patterns, the inculcation of trust is not at all related to the importance of trade nor to any of the subsistence types.

SEX AND HUMAN COALITIONS

Once again, the fact that males and females evolved to get and use resources somewhat differently in reproduction influences the patterns we see in coalitions. Two very different conditions seem likely to lead to strong male-female bonds.[56] In harsh environments in which a man and a woman can form effective resource coalitions, they are likely to do so, and the coalitions will follow confluence of reproductive interests (e.g., husband-wife).[57] Such male-female coalitions center around garnering resources for offspring and are likely to be strongest in situations in which male and female interests are identical, that is, in genetic monogamy. The evolution of monogamy is something of a puzzle; in the cross-cultural data, more than 85% of societies are polygynous. Many years ago, the biologist Richard Alexander suggested that as societies grew larger, even the most powerful men might find it difficult to maintain polygyny. Recently, scholars have explored the cultural evolutionary hypothesis that monogamous marriage has been favoured by cultural evolution because of its group-beneficial effects—promoting success in intergroup competition.[58] In suppressing intrasexual competition and reducing the size of the pool of unmarried men, normative monogamy reduces crime rates, including rape, murder, assault, robbery, and fraud. By assuaging the competition for younger brides, normative monogamy decreases (1) the spousal age gap, (2) fertility, and (3) (perhaps) gender inequality. By shifting male efforts from seeking wives to paternal investment, normative monogamy can increase savings, child investment, and economic productivity. By increasing the relatedness within households, normative monogamy reduces intrahousehold conflict, leading to lower rates of child neglect, abuse, accidental death, and homicide.[59]

Brother-sister coalitions may exist in polygynous societies in which husbands live apart from wives and the certainty of paternity is low; a mother's brother (the maternal uncle) teaches his sister's sons. Male-female coalitions are likely to be relatively weak in many polygynous situations in which the male's reproductive interests are identical to those of all his wives and children and overlap with the interests of any particular wife inversely with the number of wives.[60]

Male-male coalitions are associated with status competition and with resources that can be more effectively obtained and protected by groups of males.[61] Thus, heritable land of lasting value and some large game or livestock (especially if used for brideprice) are often the focus of male-male coalitions. At their extreme, these can become warfare (chapter 13). Such coalitions are likely to be among brothers and are thus associated with patrilocal residence,[62] but more or less fluid coalitions of this sort can arise among men of various relatedness and among nonrelatives in many societies. These male-male coalitions may exert considerable power and control significant resources. Male-male coalitions exist in spite of the conflict of reproductive interests of the men involved—but only if each member can gain sufficient resources or influence to better his position compared to operating alone. The fluidity of such coalitions may be related to this factor.

Female-female coalitions, like male-female coalitions (and unlike male-male coalitions), tend to operate in the familial sphere and are seldom powerful outside the household. Co-wives, even if sisters, have less convergent interests than genetically monogamous husband-wife coalitions, and in fact a major cause of divorce cross-culturally is conflicts among co-wives (the level of conflict is lower when wives are sisters). Female-female coalitions may arise among female relatives or co-wives; most appear to function for the exchange of information (e.g., location of good foraging spots), child care, and subsistence-related work. The resources women gather are used primarily for their offspring and family, and significant female control of resources is rare. These coalitions are seldom significant beyond the household boundaries; even female solidarity groups tend to be among relatives.[63]

For a woman, the potential conflicts of interest over resource distribution in the household seem likely to be distribution to her relatives, distribution to her female reciprocators (related or unrelated), distribution to her husband's relatives, distribution to her husband's reciprocators, and distribution to his other wives in a polygynous situation. The use of resources for the man, the woman, and their joint offspring are unlikely to be a source of conflict.

All these sex differences—in the potential reproductive risks and returns, in the usual degree of relatedness of cooperators, and in the sphere of activity—seem likely to be reflected by differences in men's and women's behavior. Consider the options open to the following individuals if they are dissatisfied: a monogamously married woman, a polygynously married woman, a man in a coalition of other-related and unrelated men. A man's options depend on his resources and his connections in other coalitions; he can fight for power openly within the coalition, try to manipulate the rules so that the coalition operates to his own advantage, or leave, changing coalitions.

Even though a woman's natal family may have some importance as support, neither of the women is likely to gain by changing coalitions. Open conflict within families is seldom successful for a variety of reasons, and avoidance of conflict is common.[64] A monogamously married woman loses the only ally whose reproductive interests are virtually identical to hers, and if she remarries, she must begin reconstructing the same confluence of interests, sometimes a difficult task.[65] A polygynously married woman also has few options and must operate in a coalition with a lesser confluence of reproductive interests with her husband. Such a woman may control few resources, cannot easily find a new coalition, and is unlikely to gain by open confrontation within the family; she cannot easily gain status and resources in a second, more desirable marriage.[66]

When do women compete directly for resources? When do women's coalitions function more like men's coalitions? There seem to be two such conditions, both more prevalent in Africa than elsewhere. In complex matrilineal or dual-inheritance societies, a powerful woman, while not increasing her own number of children, may increase her grandchildren by passing wealth and power to her son, as in the Ashanti (chapter 12). Even when such systems are changed by contact with industrialized societies, traditions of women's independence and power, though they no longer yield reproductive gains, may persist and thrive, as in West Africa today.

In sum, the sexual dimorphism in human coalitions follows from the arguments of chapter 2 (mating and parental return curves) and parallels that in other species to a surprising degree, particularly in light of cultural diversity.

Coalitions in the Play of Boys and Girls

The two sexes begin to diverge in their coalitional behavior early in life (chapter 4). Girls, but not boys, are more likely to rely on adults to settle disputes, both in the United States and cross-culturally.[67] Similarly, boys, when challenged, tend to respond directly and aggressively, whereas girls give signals of "disengagement" to possible conflict.[68] Even in nursery school, girls compete by cooperating rather than by confronting.[69]

Today, in Western nations, sex roles are far more similar than in most societies, but important sex difference remain in competitiveness and striving. The psychologist Piaget found girls more "tolerant" in their attitude toward rules and more willing to make exceptions.[70] In general, boys play more often than girls in large age-heterogeneous groups; they play more competitive games; and their games last longer than girls' games. Boys' games last longer, apparently not because they are more complex or less boring, but because continual disputes over rules arise, and boys engage in conflict resolution. In a wide variety of settings, boys tend to cover more space in their play, to play more roughly, and to be "more chaotic, more disorganized, and less neat."[71]

In one study, boys were "seen quarreling all the time, but not once was a game terminated because of a quarrel, and no game was interrupted for more than seven minutes."[72] It almost appeared that the negotiation of rules (politics?) was as important as the game itself; certainly it was more constant— whatever the game, boys argued about the rules. For girls, the occurrence of a dispute tended to end the game, in a sense sacrificing continuation of the game for continuation of the relationship, as if no options for negotiation within the coalition or for changing coalitions existed. Put bluntly, starting early in life, boys get more practice than girls in negotiating their best interests.

These differences are understandable if, in evolutionary history, women have enhanced their reproductive success by cooperating in the familial sphere with female relatives and co-wives—that is, in situations in which they could not gain through open conflict or by attempting to change coalitions. Men, on the other hand, have enhanced their reproductive success by cooperating to get greater resources and power with both related and unrelated men—situations in which open assertion of dominance (with greater risk) may frequently gain. The patterns we see suggest that different coalition strategies may indeed have led to reproductive success for the two sexes.

12

Politics and Reproductive Competition

Politics—strife of interests masquerading as a contest of principles.
—Ambrose Bierce

HUMANS, like other primates, move almost imperceptibly from "coalitions" to "politics." Arguably, coalitions represent part of the give-and-take of politics. We have come far from our starting consideration of the ecology of sex differences in reproduction. But a variety of current issues have precisely the characteristics that make this logical stretch worthwhile: intelligence and simple rules, interacting with environmental and historical particulars, create outcomes of increasing complexity and scale. If we can understand how men and women profited reproductively from differing resource strategies in the past, perhaps we can follow the emergence of complexity and begin to make sense of likely differences—and how to separate those from issues of equity—today.

The primatologist Frans de Waal's definition of politics as "social manipulation to secure and maintain influential positions" is a biologically useful one. The art of politics, as defined this inclusively, is older than humanity itself. De Waal, describing his work on chimpanzees at the Arnhem Zoo, commented that "whole passages of Machiavelli seem to be directly applicable to chimpanzee behavior" (and surely the review in earlier chapters of this book bear this out).[1] All chimpanzees appear to strive for power and influence, but there are major differences in the ways male and female chimps strive and in the ways they use their influence and power. There are also important interpopulation differences. In chimpanzee society, the final reproductive goal of political strife is clearer, perhaps, than in human societies, but it may be useful to examine human politics from this broad behavioral ecological perspective.

Here, I want to follow the kinds of coalitions male and female mammals are likely to form, up to that most rarefied of coalitions—formal politics. In most countries today, women are rare in national and international politics (Norway, India, the United States, Germany, and England are recent exceptions), even when there are societal and legal mechanisms to mandate that women as well as men officially have access to power and control of resources. Women may spend most of the money and outnumber men in professional jobs, but they hold relatively few top-level positions.[2]

In some societies, politics and reproduction are overtly interwoven. Among the Tiwi of northern Australia, for example, there is an extreme form of gerontocratic polygyny.[3] Older men monopolize women; the difference in age between spouses in primary marriages is 23.6 years, which is quite different from the 2- to 7-year age difference we think of as common in our society.[4] Men get wives through demonstrating various skills and negotiating with other men, trading favors, and female relatives—a kind of politics. Highly polygynous men are old and have demonstrated skill and power.

Even if coalitions have been central in the evolution of social behavior, the connection to complex modern phenomena like international politics may be far from obvious. Yet I am suggesting that both politics and war (next chapters), if defined in a biologically functional way, are outcomes of highly developed coalitions and have their roots in prehuman history.

Not only primatologists like de Waal have viewed politics as the seeking of rewards. Disraeli claimed that politics is nothing more nor less than "the possession and distribution of power": power over others is clearly an issue in behavioral ecology. That this is not a new idea is reflected by the wealth of informal definitions of politics, including:

> Nothing but corruptions; the madness of many for the gain of a few. (Jonathan Swift)
> The conduct of public affairs for private advantage. (Ambrose Bierce)
> The possession and distribution of power. (Benjamin Disraeli)
> Economics in action. (Robert M. La Follette)
> The science of how who gets what, when, and why. (Sidney Hillman)
> Nothing more than the means of rising in the world. (Samuel Johnson)

To be sure, these are informal and ironic definitions; any connection to power is hardly obvious from most formal definitions of the word (and few nonbiologists, except perhaps Henry Kissinger and François Mitterand, openly relate power and sex).[5] Because most definitions concern the formal practice of politics in nation-states with powerful official positions gained by election or in societies with political inheritance, they obscure the functional significance of political behavior. They lack a consideration of *why* we have evolved to strive to gain influence.

Under the formal structure, however, two phenomena in all of politics are particularly interesting to a behavioral ecologist: coalitions and contests about resource control and status.[6] Sometimes these are relatively straightforward matters of local "horse-trading" of influence and support; sometimes the situations are more complex.

MEN, WOMEN, AND POLITICS CROSS-CULTURALLY

With notable exceptions such as Angela Merkel, Hillary Clinton, Indira Gandhi, Gro Brundtland, and a few others, national/international politics in the developed world is largely a man's game. In contrast, comments abound that in tribal and band societies women had significant influence and power. Whether this is true depends largely on one's measures of power and influence. Cross-culturally in traditional societies, though women's rights vary in many domains, men hold power almost uniformly in politics. About 70% of societies cross-culturally have only male political leaders; just over 7% have both sexes, though men are more numerous and more powerful.[7]

In very few societies are women coded as equally powerful as men (and men are always more numerous).[8] Women can hold kin leadership posts and have more say and influence than men in 6 of 93 societies. Both sexes have roughly equal influence in 4 other societies.[9] Women can attend and participate equally in community gatherings, though they may be segregated from men, in over half the studied societies. While women may widely have an informal "voice" in community affairs, formal political influence is rare for women, both within the kin group and in the community at large.

We think of men's and women's interests as converging most strongly in monogamous societies, but women are not more likely to hold political office in monogamous societies,[10] in societies in which men are absent, in those with biased sex ratios, in particular subsistence bases, or in many other conditions I can find—I think women as leaders are too rare to see any pattern. Recent (April 2014) data from the Inter-Parliamentary Union show approximately 18% of United States members of the House or the Senate are women, thus ranking 84th out of 147 places. Of the countries for which there were data, Rwanda, Andorra, and Cuba rank high (~49% to ~64%); Canada has about 25%; the Scandinavian countries have 40% to 45% women in their parliaments.[11]

The rarity of women as major political figures is consistent with the hypothesis that men have evolved to make reproductive gains from striving in coalitions that are relatively large scale, but it is important to predict the exceptions as well as the general case. The very rarity of women's political power makes thorough statistical analysis difficult. The societies in which women are coded as holding political power in the odd-numbered societies of the Standard Cross-Cultural Sample occur on several continents, are settled or migratory, are patrilocal or matrilocal, can be highly polygynous, and include subsistence forms that are generally not associated with women's ability to control resources (e.g., animal husbandry and agriculture). Four of the nine societies are African, suggesting a possible geographic bias, but it is important to remember that African societies comprise over a fourth of the half-sample. In fact, the geographic association appears weak.[12]

There is one potential pattern with women's large-scale political activity: descent rules. Five of the 9 societies in which women are said to have important political power are societies with matrilineal or double descent, precisely those societies in which women's power is predicted to enhance their son's (not their own) reproductive success. Over half the cases of women having political power occur in one-fifth of the sample, the matrilineal and duolineal societies. Of the 18 matrilineal and double-descent societies, 5 have women in political office (we would expect 2); of the 56 patrilineal-, ambilineal-, and bilateral-descent societies, 4 have women in political office (7 would be expected). Thus, despite the difficulties of recording descent systems in great detail, there does appear to be an association between descent systems and women's ability to hold public office.

There may be virtue in examining in more detail those societies in which women are coded as having significant power, to see if there are patterns in the *type* of offices and power women have and what the obvious independent variables are. As you will see, there are problems with the coding of women's power in some of the nine societies; in an effort to recognize women's participation fully, three societies were coded as "women hold power" when the reference was a mention in conversation or a rumor.[13] When these societies are excluded, no statistical result is changed. No factor other than descent has even a marginally significant relationship to women's ability to hold political office, and descent is significant.

Nama

The sociologist Martin King Whyte, in a cross-cultural study of women's status, coded the Nama of the central and northern Kalahari as having women in political positions, but women were neither as numerous nor as powerful as men.[14] If anything, this is an overstatement. The Nama were patrilocal and patrilineal; thus we would not predict they would have women in political office. There were few goods to inherit. Men's weapons were usually buried with them, and women sometimes gave their ornaments to their daughters.[15] In their small family-based groups, there was no organized system of government; old men simply had the greatest influence over daily affairs.[16] In the northwest, where bands were larger, each band had a recognized chief, usually hereditary in the male line. The heir succeeded only after he passed through the puberty ceremonies and killed his first big game.

If the office fell to a man while he was still a minor, there was no regency except among the more northern Naman group of the Nama. One supposed example of female political power: the successor to the office, under normal circumstances, is the eldest son of the last chief, and as such he is usually accepted without question by the tribe. If he is a minor, his father's brother or some other near relative in the male line acts as regent, although one instance is recorded, in the case of the Gei//Kuan, of a woman reigning on behalf of her young brother until he came of age.[17] Thus, while the code is technically accurate, most Nama groups had no political offices. Only one example of a woman holding power is recorded, and the real force of women in politics was nothing like that found in other societies coded as having female politicians.

Mbundu

The Mbundu, in south-central Africa, were polygynous and patrilocal but had a double system of inheritance. Land descended patrilineally, and movable property was inherited matrilineally. In one sample village, there were twenty-seven (socially) monogamous marriages, nine men had two wives, and the headman had four wives and a concubine. Other households were occupied by formerly monogamous widowers now living alone.[18]

The village headman functioned as both priest and legal authority. Succession as headman was patrilineal, from father to son or from elder to younger brother. Villages were grouped into twenty-two chiefdoms, but almost half of these chiefs were tributaries of one of the more powerful chiefs. Chiefs had religious, legal, and external-affairs duties. The report that women could hold political office apparently arises from the following:

> The Sambu kingdom has long specialized in the training of medical practitioners and witch-doctors. The tradition is that it was founded by a woman who seems to have come either from Bailundu or from the stock from whence came the Bailundu royal family at an even earlier date. From the family of Wambu-Kalunga, in addition to Ciyaka, came those who set up the kingdoms of Elende, which had a woman ruler early in its history.[19]

Thus the Mbundu, like the Nama, have a historical tradition of only one woman having held political power at some time, and there is no evidence that women routinely held public office. In fact, the ethnographies are explicit that public office is patrilineal. Women's public activities and power in these two societies are quite clearly different from, for example, the Ashanti.

Ashanti

The Ashanti were polygynous, matrilineal, and avunculocal or virilocal, with wives occupying separate quarters.[20] Both women and men could own land, but there were complexities in property inheritance: a woman could inherit only from a woman and a man from a man. Political power was associated with the "stool." In each political unit there were two stools: the chief's and that of the queen mother. The classic study on the Ashanti states that "the recognized seniority of the woman's stool is no empty courtesy title. In fact, but for two causes (physical inferiority of women, menstruation and ritual avoidance) the stool occupied by the male would possibly not be in existence at all."[21] Menstruation involved ritual avoidance: this meant that premenopausal women would be barred from war. Several postmenopausal queens did historically accompany armies to war.

The queen mother had considerable say in the choice of the chief, and her veto could not be overruled. After the chief took office, her place was at his left hand wherever he went, except to war (and there, too, if she were postmenopausal). She alone could rebuke the chief, his spokesman, or his councilors in open court. She could address the court and question litigants. She did not, however, attend court when menstruating. Each queen mother had the right to choose the chief's senior wife and to replace her if she died. The senior wife, or her daughter if the senior wife was old, became regent when the chief went to war.

The queen mother thus had direct political power, influence over coalition formation, and indirect power through her choice of senior wife. Further, because the senior wife was likely to have little or no political training or experience, and no knowledge of customary law or court procedure, the queen mother often assumed these duties, deciding cases with the full powers of chief. The queen mother was also entitled to a share of the court fees.[22]

Matrilineality influenced women's importance. As Rattray commented: "A king's son can never be the king, but the poorest woman of royal blood is the potential mother of a king."[23] A queen mother's resources, which were accumulated through her reign, could be used to enhance her sons' abilities to get wives in this polygynous society.

Rattray's comments about women's power and the clan system of the Ashanti are poignant in the light of historical development. Europeans had never recognized the women's stool or the queen mother's power. Noting the comments of other authors on the "troublesome" activities of the queen mother (while not recognizing her position), Rattray asked the Ashanti about her and found that they recognized that women held no such positions among the European visitors, so they assumed that Europeans held women to be of little importance. Rattray argued that the queen mother could be an ally, but Europeans' treatment of her was guaranteed not only to alienate her but to be destructive as well. Later reports seem to have proved him correct.[24]

Bemba

The Bemba, who occupied the high plateau of northeastern Rhodesia (Zimbabwe), considered themselves warriors. They practiced shifting cultivation and traded relatively little.[25] Elephant hunts were a source of wealth to chiefs. Descent was matrilineal, and marital residence was at least initially matrilocal. The Bemba were polygynous, but the degree of polygyny declined; by 1951 it was rare to find more than one man per village with two wives, in part because of women's resistance to polygyny.

A striking Bemba political feature was a centralized form of government under a hereditary "paramount chief."[26] The paramount chief ruled over his own country and also acted as overlord to a number of territorial chiefs, who were drawn from the immediate family of the paramount chief. Some sisters and nieces were chieftainesses with authority over villages. The chief's mother had a territory of her own and was important in tribal councils. Succession to office was matrilineal. When a male chief died, the office passed first to his brothers, then to his matrilineal nephews, then to the children of his sisters' daughters. A female chief was succeeded by her sisters, maternal nieces, and granddaughters.[27]

Rome

The Roman Empire represents a very different level of societal organization and technology than most other societies in the sample. Much more information is available on certain aspects of women's and men's political lives. Since early in the empire, wives, who legally were considered wards of their husbands, were in fact emancipated. Also quite early, marriage shifted from a purely parental arrangement to a requirement that a woman's consent be obtained. Married women could, by Hadrian's time (he was emperor from AD 117 to 138), write their own wills. Upper-class women were well educated.[28]

Women were active in society, and their titles, privileges, and distinctions were apparently as closely graded as those of men.[29] A woman generally shared her husband's station, but emperors occasionally gave women themselves consular rank, if they were relatives and not married to consuls, and rarely might let them retain their status if they then married lower-status men. Such women had great privileges, even though they may simply have been individual privileges.

Women in formal political bodies were unknown, except for the *convenius matronarum*, an ancient guild of religious origin that became the empress's maids of honor. This body determined allowable fashion and etiquette: what costume was appropriate for women of various rank, who had precedence and the "right of the kiss," who ought to have sedan chairs, and whether they should be adorned with silver or ivory, who could have gold or jeweled shoes, and so on—primarily issues of fashion rather than real power.[30]

Women had considerable personal latitude under the Roman Empire; they were active in public social life and in a sort of women's senate. But they did not hold senatorial appointments of the same sort as men, and the women's senate made decisions solely on matters of etiquette. This description arises from the restricted sources allowed in the Standard Sample analysis; if one looks further, the impression that women's political power was largely illusory is strengthened. Though wealthy Roman women may have had considerable personal autonomy, their formal activities were largely honorific, and their influence on public policy came primarily through informal influence and association with powerful men.[31]

Marquesans

The Marquesans, of Polynesian origin, lived on a series of islands in the central Pacific. The tribe was strictly a localized group without a hierarchical political structure. Many anthropologists treat data on the Marquesans with caution, arguing that they are a society too changed by contact to trust reports and that by the time anthropologists arrived, the old customs had been swept away. Marquesans are well known for being polyandrous, but the circumstances are important. The adult sex ratio was male-bised, although authors disagree as to the intensity of bias.[32]

A household's status depended primarily on the number of men available to work for it, building houses and manufacturing items (men produced 85% of manufactured work); the discrepancies reported by ethnographers regarding the sex ratio probably related to whether chieftains or ordinary men were interviewed. The eldest child was considered the heir, and all sons who were not first-born children were expected to find alliances as secondary husbands. Eldest-born sons tried to marry attractive women; secondary husbands were acquired by offering an attractive wife to share. "Average" households comprised a household head, a group of other men, and a single wife. While officially all husbands had equal access, one ethnographer commented that "the first husband ran things and distributed favors, although it was to his advantage to see that his underlings were sexually satisfied so that they would work for his house and not wander off with other women." Polyandry could arise in a number of ways: two men co-marrying one woman; a young man moving in with the family of a young woman—if things worked well, the couple moved to their own house and a wealthy man might join them; or an older man might marry a young girl who had several lovers. Polyandry thus could cut across social classes.[33]

Polyandry appears to have been an outgrowth of male-male alliance in this way: status was achieved by the work of men. Men were common, women were scarce. Most women had at least two husbands some time during their life. The chief's household had not only several husbands but also several wives. When a family became powerful, it could challenge the current chief; to avoid this, the chief might arrange a marriage between his eldest son and the first-born daughter of the rising family, or he might adopt their eldest-born son. Such adoptions and marriage alliances required wealth, and wealth required manpower. It is unclear how skeptical we should be about this argument, and there is no written analysis of the conflicts of interest arising from such arrangements.

Women appear to have had some choice in sexual arrangements, but their accession to public power was rare—and anthropologists' reports leave us uncertain about just what "rare" meant or whether women chiefs had the same duties and perquisites as men chiefs. The relative impact of chieftainship on men's versus women's reproductive success is also a matter of speculation; however, women could not gain through direct reproduction, and if male chiefs had more wives than others and could control other's access to their wives, men at least had the potential for direct reproductive gain through the chieftainship.

Saramacca

The Saramacca of Guyana were matrilineal. Polygyny was general and residence was not patrilocal. Kin clans had ties and claim to particular areas, with fishing, hunting, and forestry rights. Lacking any centralized political system, the villages and clans were relatively autonomous. Lawmaking rested with the chiefs and the councils (*krutus*) over which they presided; law enforcement was the task of other officials, the *bassias*.[34]

Most villages had at least one woman bassia; she was responsible to the village head for the behavior of the women. In many tasks, she organized the women and was responsible for seeing that work got done. If two women were involved in a dispute, she informed the village head and transmitted his decision if the affair was not of "sufficient gravity" to warrant calling the old men together. When anthropologists asked if women talked in the big council, they were told "No, not in krutu. But they talk plenty at home," suggesting informal influence.[35] Nonetheless, since bassias had duties in council, and women were bassias, presumably the women bassias were active in the councils.

In theory, women bassias had "equal power" with male bassias; but the ethnographies make it clear that women's and men's spheres of influence were separate. Since the Saramacca are coded as women having "equal power" as men, it is important to note that their power was restricted to women's matters.

Montagnais

The Montagnais (Naskapi), in and around the Labrador Peninsula of North America, were a sub-Arctic hunting people, living in low density and hunting either in family groups (in the southern part of their range) or communally (in the less populated northern areas).[36] Marital residence was sometimes matrilocal, sometimes patrilocal.[37] They gathered in the summer but were dispersed for the rest of the year. One southern informant suggested that a chief was irrelevant—it would be absurd to depend on the authority of a chief hundreds of miles away.[38] The authority of the chief was more developed but still weak in the northern areas, where the Montagnais hunted communally, than in isolated forest groups.

Political organization was loose. There was usually a chief, and if the band was large, a council of older men assisted him. Political units were loosely associated with territories and not strictly kin based; there was no political organization beyond the band. In the early days, chieftainships, when there were any, were patrilineally inherited. In 1927 a government Indian Act required the election of chiefs, though many bands did not bother. One ethnographer reports that chiefs were chosen on the basis of personal characteristics: a man must be an excellent hunter with a well-developed sense of responsibility toward the others in the band and (because of the interactions with the Hudson Bay Company) a shrewd negotiator. No chiefs were bachelors; in part, informants felt this was because a large kin group made for a strong support base in attempting to attain the chieftainship.[39]

One informant, Joseph Kok'wa, remembered that Tommy Moar's mother had been a chief in the Nichikun band for several years, but Tommy Moar responded that, while she had been held in high regard by her tribespeople, she had not actually been a chief.[40] The issue of women and public office is thus cloudy in the Montagnais. An informal system of advice seeking was common. Tommy Moar's wife, Maggy, for example, was highly regarded, and both women and men submitted their problems to her judgment and tended to follow the suggestions she offered. Thus women could have considerable informal influence in the "community" sphere, though it is not clear that they were ever "chiefs."

Creek

The Creek, once ranging throughout the southeastern and southwestern United States, were warriors, traders, and agriculturalists, growing corn, beans, squashes, and other crops. Descent was matrilineal.[41] By about 1850, inheritance rules usually depended on parents' wills, but intestate cases were relatively equal with regard to both sex and birth order.

The Creeks had a nobility, but the chieftainship did not attain the level of power among them than it did, for example, in the Natchez. In the eastern part of the Creeks' range, particularly near the eastern Siouan area, women were frequently chiefs. Though this was an uncommon occurrence toward the west, some women in the western tribes did become war leaders. After the end of the Civil War, chiefs were elected.[42] But consider this description of the annual assemblies of the Creek nation: "Only the chiefs of the warriors are admitted there; the subordinate chiefs who are present are intended to serve the others, but they have no voice in the deliberations. The women are charged with the duty of preparing the necessary food and drink for the assembly; the subordinate chiefs go to fetch the provisions and place them in their turn in the grand cabin for the members of the assembly."[43]

Women sometimes were awarded the rank and title of warrior, but in at least one case, a woman's bravery was rewarded by conferring a war title on her son. It is unclear how frequently women chiefs were war chiefs, though it is evident that they could hold this office. We know nothing else about the duties and power of women chiefs.

WOMEN IN POLITICS: WHEN DID IT PAY?

Cross-culturally, women have seldom held public political office or been politically powerful in any formal sense. In many societies, of course, women are active in the "informal influence" sphere and within the kin group; these effects are harder to measure. In the societies in which women do or did hold public positions, it is frequent that, as among the Romans and the Saramacca, they held a position of power over women's affairs but had only informal influence on men's issues. For the Nama and the Montagnais, public offices were not always relevant, because the societies were so dispersed and at such low densities; nonetheless (e.g., among the Montagnais), strong women could have considerable informal influence. Among the Nama and the Mbundu, women did not hold office; there was only one reported instance in each tribe of a woman's importance in the political sphere. Among the Marquesans, women rarely ascended to office. Creek, Bemba, and Ashanti women, though they held power less often than men, appeared to be able to make the same decisions as men.

The societies in which women's decision-making capabilities centered on the same issues as men's were complex, matrilineal, or double-descent societies in which heritable resources existed.

Some societies not in the Standard Sample may prove instructive. The Tchambuli, or Chambri, in Papua New Guinea are well studied, but conflicting reports exist.[44] Margaret Mead originally suggested that women dominated men and that this society represented a role reversal compared to Western ideas about sex and power. From this she inferred that our ideas of masculine and feminine roles in power issues, for example, were culturally determined and culture specific. Since this was a polygynous and patrilineal society, reports of great female power were unexpected (but reasonable if power were a fluke of culture and had no ecological payoffs).

Later studies found that women and men essentially operated in separate spheres. Men were indeed thought by the Chambri to be more aggressive than women, and issues of female dominance existed only with relation to women's matters—they never crossed the sex boundary. Thus, among the Chambri as among the Saramacca, women might dominate other women about women's matters only.

Concern over Galton's Problem, the possibility of bias introduced because neighboring societies may share practices, raises the issue of whether there might be an "African" bias in these results. The association between African location and overt political power for women, though statistically insignificant, suggests further exploration could be rewarding. Other African societies may be instructive. Among the southern Bantu, there existed an institution called the "female husband." Economic and political power were the concomitants of female husbandship for women in Bantu society.[45] Among the southern Bantu, in which women could hold political office, female political leaders were expected to become female husbands. Women gained socially from becoming autonomous female husbands, but the impact on their reproduction is far from clear. If powerful "sons" from female husbandship returned gains to her genetic kin, there may have been indirect reproductive gains as well—but that is not discernible from the existing data. Among the Shilluk and the Nyoro, women who inherited or achieved high political status were forbidden to marry—a clear and direct reproductive cost.[46]

The most important relationship with women's overt political power is the descent system. I think this is related to the evolutionary ecology of resource gain and reproductive success. In the few societies in which women wield substantial public power, as opposed to informal influence, there is no evidence of clear, direct reproductive gain. In fact, in some examples, there is a conflict between political and direct reproductive gain for women. In some descent systems, women's power may accrue to their sons, who then reap a reproductive benefit (as above, in the Ashanti);[47] in such systems, *women gain by increasing the number of their grandchildren.* In other systems, women's direct gains are not clear, but when power gives access to substantial resources, women in power similarly can make gains through their sons, or nepotistically.[48]

Once again, the sexual dimorphism in usefulness of resources and power in reproduction is critical. Men appear to seek overt political power for direct reproductive gain (e.g., wives, for owed reciprocity), whereas women seek resources for themselves and their offspring. Sometimes this is accomplished through indirect or informal influence and nepotistic gain. Most commonly, the amount of resources controlled by women is sufficient to support their family, but rarely, as we have seen, particularly in matrilineal societies, women may exert public power to the gain of their families. While women hold relatively few high-level public positions, they are major consumers; the relevant cross-cultural codes are for controlling the "fruits of own, both, and men's labor." Even today, in both business and politics, women are underrepresented.[49]

There is real point to the old joke about the man who tells his friend, "I leave all that day-to-day minor stuff to my wife: what we'll have for supper, what kind of car to buy. I make all the important decisions: whether to intervene in Ukraine or impose a carbon tax in Europe."

Men and Women Today: When Does Politics Pay?

High political positions carry monetary and informational rewards, which may last beyond the position itself (think of the former politicians who are professors, university presidents . . .). High political status also typically has sexual-access rewards as well. Even a cursory look at the list of those caught or convicted in going too far (money in the freezer, inappropriate sexual advances to employees of both sexes . . .) suggests that the rewards today are great enough to engender great risk taking. The sex ratio is heavily male biased, but the data are difficult to analyze, because some transgressions are single sexual follies, and some are repeated abuse of power for monetary and sexual gain.

Women have evolved to use resources differently than men in reproductive matters, and this difference has had an impact on their political strategies and degree of influence. But this is surely not a justification for the rarity of women in public life today (see above). It may be true that women in Western societies, for example, have comparatively great overt power, but it is also likely that for both men and women, the exercise of formal power may well have less impact on reproduction than in smaller societies. Thus, it is interesting that during Western European history, as societies moved toward more monogamy (reducing the potential rewards of power for men), politically powerful women seemed to become more numerous (e.g., Mary Tudor, Elizabeth I, Mary of William and Mary, Queen Anne). Men may still reap a reproductive advantage from political power, compared to women, through remarriage and children born to mistresses.[50]

Further, it seems clear that laws in both preindustrial and modern societies are far from neutral with regard to reproductive competition, and conflicts of interest exist between men and women. The accession of more women to positions of real public political power has the potential to change the balance of power on such issues.

13

~~~~~~

Sex, Resources, and Early Warfare

[These bloody feuds] are for the sake of the women.
—Jivaro informant

THE PRUSSIAN MILITARY strategist Carl von Clausewitz said war was simply "the continuation of state policy [politics] by other means." Although warfare gives rise to some of the strongest, most tightly knit and potent international coalitions in modern times, that is surely not its evolutionary context, for states are a relatively modern phenomenon, whereas organized intergroup conflict among competing coalitions is older than humanity itself. Conflicts of interest, if not coalitions in open aggression, are universal among living things, and certainly lethal conflict exists in many species. Thus, it makes sense to begin our inquiry by focusing on simple conflict, asking: Over evolutionary time, what have been the ecological contexts of conflict and killing? Why and when are there sex differences? What were the costs and benefits to the individuals involved?

From our perspective, two phenomena matter: the reproductive impacts for individuals of fighting and killing (including formal war), and the potential conflicts of interest among different individuals involved in any conflict. It is obvious from earlier chapters that conflict, which is clearly risky, will center on items of real reproductive importance: mates, or status and resources when these lead to mates.

In general, risky conflict is unlikely over trivial resources. Warfare in traditional societies *was* risky. In traditional societies, it was far more lethal and pervasive than scholars had thought.[1] In contrast to most countries fighting in modern wars, a very high proportion of men fought and mortality rates could be extreme. You will remember that among red deer, about a third of the mortality in adult males came from injuries received in fights.[2] In a number of traditional societies, the proportion of adult male mortality could range as high as about 42% (Yanomamö) to 60% (Jivaro). Among the Turkana, a nomadic pastoral society in East Africa, fighting could be quite large in scale—and while individual warriors suffered the risks, the benefits were collective. Cooperation was maintained in this risky endeavor by community-imposed serious punishment of free riders: fines, corporal punishment.[3] Compared to these figures, mortality in the twentieth century for European and American servicemen was trivial; casualties could be high, but only a portion of the adult male population fought.[4]

Lethal conflict looks, at first glance, as though it should decrease reproduction—certainly it's hard for dead individuals to reproduce. In fact, lethal conflict is like infanticide and delayed reproduction: in specific environments, for some individuals, risky conflict is a high-stakes gamble in which there is some probability of winning greatly increased reproduction, versus losing (death or disability). Were this not true, these behaviors would remain rare. Evolutionary novelty (see chapter 2) is an important complication for humans. Certainly modern war is characterized by rapid changes in technology, and it is likely that the original driving force of reproductive reward no longer exists (though why do generals wear all their medals to important events?).

Because of this complexity, we need to ask the following questions: (1) Was there previously a reproductive advantage to engaging in lethal conflict? (2) Is there currently a reproductive advantage to engaging in lethal conflict? and (3) Are the proximate triggers leading to conflict "unhooked" from any previous selective advantage?

Resources and Conflict

Within groups, risky, damaging aggression is surprisingly widespread in many species. Fights occur both over direct reproductive resources (territory, mates, etc.) and indirect reproductive advantages (elimination of competitors or competitors' offspring). Typically, deaths arise from fights over mates or resources for getting mates, from infanticide, and from cannibalism.[5] The ultimate costs and benefits are reproductive, although the level of risk varies with the kind of aggression. The risk to an infanticidal killer, often a male (e.g., lion or lemur) taking over a harem and killing the offspring of his reproductive predecessor, is probably small. Not surprisingly, infanticide is more common than killing adults.

Potentially lethal fighting among individuals of similar age and status is typically a male endeavor and occurs primarily among adults of the same population rather than among groups from different populations. That is, homicide is more common than war.[6] In a game-theoretic approach, combatants assess each other's resource-holding power: "The stake played for is infliction of loss of resource-holding power, and is determined by the fitness budgets of the opponents. This defines a critical probability of winning . . . for each combatant, above which escalation (fighting) is the favorable strategy . . . and below which withdrawal is favorable." Escalation, then, should occur only when the *absolute probability of winning minus the critical probability of winning* is positive for each combatant. Someone always loses; if information were perfect, the loser-to-be would never attack and would always withdraw if attacked.[7]

Within-population groups can also fight, although this is more common in humans than most other species. The classic film *The Ax Fight* is reminiscent of Bernd Heinrich's *Ravens*: an apparently inexplicable event (a fight among Yanomamö men) becomes clear after learning a great deal about the participants. In *The Ax Fight*, as the anthropologists are unpacking upon their arrival, a chaotic fight breaks out across camp; they grab cameras and film what they can. Later, identifying individuals, the roots of the fight appear to arise from kinship disputes; this was one of the earliest empirical tests of kin selection. Or was it? There might have been a correlation between social structure and kinship. And in some other groups, such as Lamaleran whale hunters, kinship is not a strong organizing principle. Using matrix regression, the anthropologist Michael Alvard showed that genetic kinship was the primary organizing principle in the Yanomamö ax fight. So there can be considerable variation in how important kinship is in any particular case.[8]

Throughout military history, bluff has been a major way of keeping an opponent's information imperfect; think of Toussant L'Overture marching the same company around the block several times to fool his opponent Leclerc into thinking he had a large army. There is a long prehuman history to bluff; in many species, displays involve deception and bluff that make a potential combatant seem bigger and stronger, advertising that an attack might prove costly.[9] Conflict is most likely between potential combatants who are similar in status or power; in such a situation, the exact probabilities are hard to "calculate." In red deer, for example, subordinate stags are unlikely to escalate a confrontation—the risks of serious injury are too high. Depending on the costs, the stronger combatant may press an attack.

In some primate species, with the elaboration of coalitions among individuals, we see intergroup conflict: groups of varying size may separate from the main population on foraging trips, and if they encounter smaller groups or lone individuals from another population, they attack, exploiting the uneven balance of power. Chimpanzee raids are like this; male chimpanzees negotiate alliances within groups (chapter 12) and use those alliances to coordinate attacks against males of other groups. Intergroup, rather than interindividual, conflict is likely only in long-lived, social species; such group aggression is not qualitatively different (the context, benefits, and costs are similar) compared to individual fights, but it is more complex, involving cooperation among individuals. The rewards are reproductive. In many species, females change groups as spoils of war for the winners.[10]

Potentially lethal conflict is only likely when the possible reproductive rewards—mates, status, resources for mates—are high. Mating effort, not parental effort, is the context of most potentially lethal conflict: if the offspring is dependent on its parent, a live parent is crucial. Thus, within mammals, males, more often than females, will be in a position to gain from risky, possibly lethal fights (chapters 3, 4). Sexual selection in competition over mates, and kin selection in infanticide and intergroup conflicts, are the context.

WHY WOMEN WARRIORS ARE RARE

The United States may be unusual in not having a major "warrior queen" icon in either its history or its mythology.[11] From Britain's Boudica (or Boadicea), said to have risen up against invading Romans, to Joan of Arc to Queen Jinga of Angola fighting off Portuguese invaders, many cultures have at least a myth of important women warriors. In truth, of course, as the historian Antonia Fraser has recounted, women warriors are rare, and royal women, though often called upon to carry symbolic weapons, rarely or never fought like warrior kings. Having women in conflict roles is still controversial. Currently, the only countries allowing women in combat are the United States, Denmark, Sweden, Finland, Canada, New Zealand, Italy, Germany, Israel, Serbia, Switzerland, and Taiwan.[12] Why?

Because the two sexes have different payoffs for risk taking, aggressive coalitions tend to be sexually dimorphic in mammals. In most coalitional species, compared to females, males tend to form coalitions that are riskier, more fluid, more aggressive, and more often among nonrelatives (chapters 4, 11). Females' conflicts typically center on getting food or parental resources or on infanticide, while males' conflicts are likely to focus on getting mates. This means that the reproductive profit from fighting may be many times greater for male mammals than for females. Further, mating effort conflicts, usually among males in mammals, are more likely to escalate to lethal proportions than conflicts over food, for example. This difference in risks and returns, of course, is what prompted Darwin to treat sexual selection differently from "ordinary" natural selection, even though functionally it is identical (table 9.1).[13]

In humans, too, men's reproductive success varies more than women's in most cultures. So it is not surprising that aggressiveness is one of the most consistent sex differences across cultures; for example, most homicides are committed by men. Women's politics and conflicts over resources tend to be at the familial and neighborhood level, whereas men's conflicts tend to have broader scope; this too is hardly surprising. Cross-culturally, men can often make enormous direct reproductive gains when they acquire power, status, and resources (chapters 4, 7, 11, 12). In bridewealth societies, rich and powerful men use resources to buy more wives or to buy younger wives of higher reproductive value, who typically are more expensive than other women.

In the few societies in which women play high-stakes power games or go to war (rather than simply wield informal influence), they show no immediate, clear reproductive gain.[14] Throughout evolutionary history, men have been able to gain reproductively by risky warfare; that is how heroes gain status and access to women. Women seldom have been able to gain. Nonetheless, women have, although rarely, fought with or accompanied men to war. From at least the time of Alexander, who ruled in 336–321 BC, women traveled and sometimes fought with their men; in Alexander's time, when campaigns were long and women might have children before the campaign was over, the children were legitimized after the soldier completed his duties. As the historian Richard Holmes noted: "During the seventeenth, eighteenth, and nineteenth centuries, women occasionally passed themselves off as men and fought in the ranks of infantry and cavalry regiments."[15]

As noted earlier, it isn't that men are bigger and stronger than women. In primate species, and in human societies, social complexities so outweigh the impact of physical size that size alone is a poor predictor of success.[16]

As in politics, the critical factor behind our sex differences is that resources and power have different reproductive utility for men and women. Across cultures, three common conditions—patriliny (inheritance by sons) combined with patrilocality (sons stay home) and exogamy ("marrying out")—foster men's, but not women's, confluences of interests in war. Further, men can make reproductive use of large-scale resources more than women can. Related men, but not women, tended to live together and may fight together.[17]

WAR: RUNAWAY SEXUAL SELECTION?

Human war can become more complex and varied than intergroup aggression in other species, largely as a result of the development of technology (which itself is a product of intelligence). But the roots of lethal conflict clearly lie in sexual and kin selection. In view of these roots, the development of technology to today's superlethal levels raises an important question: is war an example of runaway sexual selection?

Sir Ronald Fisher, a mathematician who made significant contributions to biology, noted that "remarkable consequences" follow if females exert a strong preference for extreme traits in males. As Darwin noted, two influences are important in sexual selection: initial, sometimes considerable, advantages not due to female preference (e.g., the advantage of large antlers in combat for red deer); and any additional advantage conferred by female preference. The first advantage reaches some limit: for example, a deer with monstrously large antlers is unlikely to win a fight. This second advantage can appear "whimsical," but female preference centers on male advertisements that are expensive and hard to fake, which leaves considerable leeway.[18] Females are forcing males to show their condition.

The intensity of female preference will continue to increase through sexual selection so long as the sons of females exerting the preference have any advantage over other males, and as the daughters show the same (or greater) extreme preference.[19] Fisher noted: "The importance of this situation lies in the fact that the further development [of the favored trait] will still proceed, by reason of the advantage gained in sexual selection, *even after it has passed the point in development at which its advantage in Natural Selection has ceased*" (italics added).[20]

Thus, when immediate reproductive gains are so great that they outstrip the countering pressure of ordinary natural selection for survival, lethal traits leading to extinction can arise in sexual selection. When, as among the Yanomamö, men with great warring skill have more wives and more children than others, sexual selection can be very powerful.[21] Even in modern industrialized societies, in which participation in wars (and other risk-taking behaviors) may be "unhooked" from the advantages given by sexual preference, if sexual preference still exists for "war heroes" or if there are other proximate rewards, previously linked to selective advantage, risky behavior may still be common.

OTHER BIOLOGICAL APPROACHES TO UNDERSTANDING WAR

One early biological approach to understanding war was Konrad Lorenz's book *On Aggression*. Lorenz argued that aggression is an "instinctive" drive favored by selection. Although he specified that this did not make warfare unavoidable, others have inferred some sort of genetic basis (indeed, "instinctive" does imply a genetic basis to many), rather than a flexible response to ecological conditions, in which genetically identical individuals might act differently, depending on circumstances. Lorenz also argued that because humans lack lethal weapons in their simple physical makeup and rely on tools, they also failed to evolve reliable inhibitions against killing one another.[22] But this predicts neither the occurrence of aggressive behavior—what conditions (e.g., reproductively important conflicts of interest) are most likely to precipitate aggression—nor the constraints (e.g., individual costs and benefits).

Hopeful Assertions

Perhaps in response to just such arguments, well-known and well-respected scientists, in a statement (May 16, 1986) for the International Society for Research on Aggression, argued from the fact that there is no evidence for a specific allele for aggression, that warfare was "biologically possible, but . . . not inevitable, as evidenced by its variation in occurrence and nature over time and space." Here lies an assumption: if there were an allele for aggression (as has been found among males of one family recently), warfare would be inevitable—the genetic basis might dictate something.[23] This view moves from patently true statements about the nonevidence of any special alleles for "warring behavior" to generalizations that "biology does not condemn humanity to war, and that humanity can be freed from the bondage of biological pessimism and empowered with confidence. Just as 'wars begin in the minds of men,' peace also begins in our minds. The responsibility lies with each of us."[24]

Certainly war is not "biologically determined"—but it *is ecologically influenced*. Such an approach fails to come to grips with the ecology of war—the circumstances in which aggression profited the individual or lineage genetically, and today may also profit corporations in what Dwight Eisenhower called the military-industrial complex. And as we learn more about the complex interplay of genes, ontogenies, and environments, today it seems naive to imagine that anyone would postulate "an allele for aggression."[25] This approach is hopeful but remains insufficiently specific or predictive.

Echoing Lorenz, one argument posits that humans and chimpanzees have particular "Darwinian algorithms" that govern coalition formation and predispose both species to warfare.[26] These psychologically imposed structures have certain characteristics: cheaters must be identified and excluded or punished; and participants are rewarded or punished in proportion to the risks they take and in proportion to their contribution to success. Each coalition member has an impact on the coalition by regulating his own participation in the coalition and by the actions he takes to enforce the contract on other members. However, all of these specifications seem likely to be true for numerous other coalitions (hunting dogs, female lions, etc.), so the argument is not very convincing that humans and chimpanzees are unique in having the appropriate algorithms for warfare. In fact, we cannot eliminate the possibility that any evolved psychological mechanisms are the *result*, not the cause, of the patterns we see.[27] Finally, this argument is not a true alternative to the ecological argument. If, as I suggest, lethal conflict exists because individuals and families have profited from assuming the risks of lethal conflict under specific conditions over evolutionary time, then we expect proximate psychological mechanisms related to aggression. But, so far as I know, we have yet to demonstrate any specific ones.

The Ecology of War: Uncertainty and Payoffs

Warfare, like a number of other human social patterns, appears to be related to environmental unpredictability: cross-culturally, warfare is more common when people in a society perceive unpredictability in the environment, when particular sorts of resources unreliably become (or threaten to become) limiting.[28] A strong predictor is the threat of weather or pest disasters.

An important empirical study raises another consideration of ecological influences. The anthropologist Napoleon Chagnon, working with the Yanomamö (chapter 7), found that the frequency of warfare differs among villages in an important way. In the rich lowlands, where the potential payoffs are high, villages are relatively large (300–400 people), warfare is highly developed and chronic, and men are aggressive both personally and in raids. In the foothills and mountains, plant productivity is lower, villages are smaller, and warfare is far less frequent; there seems to be little point to risking one's life.[29]

The behavioral ecological hypothesis predicts strong sex differences and reproductive payoffs to male warriors; the environmental uncertainty hypothesis and Napoleon Chagnon's work suggest that the frequency of warfare should vary with ecological conditions. Societies with more warfare also encourage boys to be more aggressive, tougher, and show more fortitude, perpetuating the pattern. This is just the way we would expect sexual selection to affect warfare.

The Demography of War: Too Many Males?

There is also some evidence that a male-biased operational sex ratio can predispose groups to war. This, too, makes ecological and evolutionary sense. Remember that in many societies, capturing women from other groups is a main purpose of war. Also, young males are societies' most violent members. We have sex-ratio data for few societies, but a trend is clear: societies with lots of young adult males are likely to see strife.[30]

INTERGROUP CONFLICT IN OTHER SPECIES

Red deer stags, elephant seal bulls, and mountain sheep (*Ovis canidensis*) rams all fight individually over females and other reproductively important resources. But there are true coalitional conflicts in other species—both within and between groups—that are more similar to human conflicts. Three social carnivores show intergroup aggression. In gray wolves (*Canis lupus*), family-based packs (both male and female) occasionally invade neighboring packs' territories and attack residents; in one study, intraspecific conflict accounted for 43% of wolf deaths not caused by humans. Among spotted hyenas (*Crocuta crocuta*), which, like wolves, live in territory-holding family groups, intruders into a clan's territory are likely to be attacked and killed; smaller clan subgroups patrol the territory boundaries, confronting other "patrols." In lions, which also live in groups (prides) based on a group of related females and one or more associated males, interpride encounters occur, but lethal injury is rare. When invading males are attempting to take over a pride, there may be lethal injuries (not only adult but infant), though once a resident male gives up his reproductive rights, aggression typically stops. Males who have just taken over a pride are likely to commit infanticide, as in several other species. Vervet monkeys, both male and female, defend territorial boundaries, males fighting males, and females, females.[31]

Aggression occurs widely in primates, including in baboons, new world monkeys, lesser apes, and group-living prosimians. In many primates (e.g., vervets), a behavioral challenge (e.g., territorial incursion, conflicts over a specific resource, including females) generates a defensive response, and groups of individuals from different local populations fight.

Among apes, male-male coalitions may approach the complexity of human politics (chapters 4, 11). And in primate intergroup aggression, we see for the first time a level of complexity and coordination that approaches the warfare patterns of preindustrial societies. The complexity of group raids means that intelligence is important.[32] In chimpanzees, humans, and perhaps gorillas, there are regular cooperative raids by breeding adults against adults of neighboring groups.[33]

In chimpanzees, such raids are probably low cost because groups vary in size—one can hide when big parties are encountered and can attack smaller parties with little risk. Adult male chimpanzees make aggressive forays into the ranges of neighboring groups, sometimes fatally injuring defending males. Females are semisolitary, using a core area but often traveling outside that area. Adult and subadult males are more gregarious and travel more widely. Total community size ranges from 20 to 110 individuals, but temporary groups range from 1 to 20 animals; they have fission-fusion group formation. Because group size fluctuates, it is not predictable how many conspecifics a group might encounter. Further, in chimpanzees it is more common for females to transfer from their natal group, whereas males are likely to remain among their relatives—and the costs and benefits of risky fights are different if one fights among relatives rather than nonrelatives.[34]

So males in traveling groups may profit from attacking smaller groups when they encounter them and capturing females when they can. Although male-male cooperation and the benefits of risk taking are enhanced by groups of related males living together,[35] it is not a requirement. In gorillas, both sexes may leave the natal group. In lions, males leave the natal group whereas female relatives remain—yet lion males engage in intergroup lethal conflict.

Females join males in potentially lethal intergroup aggression in a few species: some baboons, some monkeys, wild dogs, wolves, and mongooses.[36] Each of these last three species has a socially monogamous, extended-family structure in which male and female costs and benefits are more similar than in polygynous species. These mixed-sex battles also tend to be at family territorial borders; females do not participate in mating-effort "raids" of the sort described above for male chimpanzees.

In nonhuman vertebrates, then, most aggression, both intra- and intergroup, has a reproductive cause. Among primates, groups of males may fight in ways that resemble ambush attacks reported in traditional human societies. Male-male coalitions are frequent among relatives but also occur among nonrelatives. Males frequently come into open conflict over access to females and over con-

Figure 13.1. Men from two Yanomamö groups form a temporary alliance to raid a third village. (Photo by Napoleon Chagnon.)

trol of resources useful in attracting females. While most of the examples involve males, there are female aggressive encounters in a number of species. Within their group, females may work in related coalitions to attack reproductive competitors or the offspring of reproductive competitors.[37] And in a few species, such as vervets, females may join males in territorial disputes.

Peacemaking in Other Species

In most species, visibly giving up causes an attacker to stop. In chimpanzees, an attacker may continue in the face of submissive gestures. Remember that in the Arnhem Zoo study (chapters 4, 11), at least one male coalition had lethal consequences. Of all nonhuman species, it is among chimpanzees that we see the most elaborate peacemaking. Frans de Waal suggests that the clear-cut dominance hierarchy provides a ritual format for reconciliation, which often follows a behavioral confirmation of formal status.[38]

We humans have devised elaborate mechanisms, even institutions, for ending conflict. Even in the simplest societies there are important roles for third parties. Intergroup conflicts appear to have been both frequent and selectively important in our evolutionary history—and to resolve them, we mastered new complexities that involve groups of individuals who may not only have disparate interests and incomplete information (or have misinformation) about one

Figure 13.2. Among the Yanomamö of northern South America, men who achieve the status of *unokai* (roughly, "revenge killer"), (A) have more wives and children; (B) do so earlier in life (* indicates $p < .05$).

another, but who may not know one another or be able to predict reliability or probability of default.

CONFLICT IN PREINDUSTRIAL SOCIETIES

What we call warfare in preindustrial societies is indistinguishable in context and function from much intergroup aggression seen in other species; it differs only in scope. Cross-culturally, 60% of the societies for which data exist engage in warfare at least yearly. Most attacks in traditional societies are ambush attacks, which are often well-coordinated to take advantage of the element of surprise and often have numerical superiority. A description of such attacks would differ little from the description of chimpanzee raids.[39]

Individuals in preindustrial societies travel and work in small parties of varying sizes, like chimpanzees, and attacks and escalations by larger groups can be fairly low cost.[40] Further, patrilocal and patrilineal societies are very common— and this means that groups of related males live and work together, influencing men's costs and benefits in ambushes and warfare. These patterns are repeated in otherwise divergent societies around the world.

Among the Yanomamö of South America (chapter 7 and above; figs. 13.1, 13.2), the reproductive context of men's warfare is clear. Men who participate in revenge raids and ambushes have more wives and more children than others, and men who avoid warfare suffer reproductively. Yanomamö men who have killed on a war party raid are accorded the title of *unokai*, and a man's performance in war parties affects him reproductively. War parties are small, from 2 to 20 men, and tend to comprise related men. It is true, nonetheless, that ecological conditions change men's costs and benefits; only in the rich lowlands is aggression rampant.

Although there are mystical aspects, most war parties arise from disputes about reproductive matters. Men may choose to avoid joining any particular warring party, and war parties may turn back, often as the result of a prophetic dream. Nevertheless, if a man avoids several possible opportunities or behaves in ways perceived as cowardly on the raids, he becomes the butt of jokes, and other men may begin to make sexual overtures to his wife. Once a man establishes himself as unokai, he is likely to have more wives and more children than non-unokai (fig 13.2). Thus there are clear reproductive advantages among the Yanomamö for men who participate in war parties, and particularly for men who kill.

Among the Jivaro of northern Peru and eastern Ecuador, as among the Yanomamö, there is no stratification. During times of peace, there is no chieftainship. When wars erupt, older experienced men who have killed many men and captured many heads are chosen as war chiefs. No Jivaro can be chosen if he has not killed. Bloody feuds, reported as functioning to obtain women, are frequent and follow familial lines; adult male mortality from fighting (above) is among the highest known.[41]

In North America, the Blackfoot Indians were known throughout the nineteenth century as formidable, aggressive warriors. Blackfoot warfare centered on capturing horses (for bridewealth) from neighboring tribes. Most parties comprised fewer than a dozen men for reasons of stealth, though raids of up to fifty men could occur. Many of the most active raiders were men from poor families ambitious to better their lot. Even sons of reasonably well-off families needed more horses than their fathers could give them if they were to marry and set up their own household. Horse-raiding parties were led by experienced men with a good war record and good judgment. Participants were volunteers, and though the leader might be a mature man in his thirties, most of his followers were in their late teens or early twenties. Occasionally, a childless woman would accompany her husband on a raid. Unless prior arrangements had been made (e.g., for equal distribution), each man could claim the horses he had led out of camp or the range stock he had captured. Bitter arguments could occur over ownership, and it was the leader's job to settle these. Some leaders gave the horses they themselves had captured to men who could claim no horses. A leader's generosity helped him maintain a popular reputation, facilitating the recruitment of future followers. Successful raiders either paid their horses for a bride or gave them to relatives, most commonly to their fathers-in-law or brothers-in-law.[42]

Among the Meru of Kenya, livestock were used for bridewealth; to marry, a man needed to accumulate sufficient livestock (preferably cattle) to purchase a wife. Men fought to gain livestock and status, and the military cycle followed the seasonal pastoral cycle. Negotiation of bridewealth was carried out by the families, principally male relatives of the bride and groom, and interfamilial alliance was an important function.[43] The father of the warrior transferred five specified items—a cow, a bull, a ram, a ewe, and a gourd of honey—to the father of the bride. To sweeten the deal, additional gifts of these and other items were negotiable but could not replace the basic five. Further symbolic gestures of alliance were involved (e.g., the warrior's father would send beer ahead of the actual brideprice, then the bride's father would initiate a general beer-drinking fest to which the warrior's father was invited); however, a main point of the transaction was clearly economic. A portion of the bridewealth was often kept back by the warrior's family, often for years. War was the principal method of gaining livestock, and warriors were expected to earn bridewealth as well as increase familial wealth and status.

Rules and traditions of warfare among the Meru facilitated cattle stealing. In individual conflict between warriors, for example, a warrior could save his life by declaring that his opponent could take his cattle; this was accepted as a declaration of surrender. All captives, female and warrior, were redeemable for livestock. Married female captives could be kept as concubines or wives; unmarried and uncircumcised females were taken as "daughters" and later traded for brideprice. Among the Meru, as among the Yanomamö, men clearly gained reproductively by establishing themselves as successful warriors.

The broad cross-cultural data suggest that these societies are fairly representative, in that reproductive matters lie at the root of war in most traditional societies.[44] Women (abductions, failure to deliver a bride) were causes of warfare in 45% of societies in one major study of 75 traditional societies. Material resources specified as useful in obtaining a bride were causal in another 39%, and in about a third of these, ethnographies specified that richer men obtained more wives than poorer men. In only 12 of 75 societies was there no immediately obvious connection between warfare and men's direct reproductive striving. Other studies have found that 75%–80% of wars involved land (clearly useful in establishing a family), and that adultery and wife stealing were major sources of conflict. Similarly, in the societies of the Standard Cross-Cultural Sample, women were captured in 66 of 158 societies; in the vast majority of these cases, women were married or kept as concubines by their captors. In societies like the Maori, in which there is no direct association between warfare and women or the resources directly used to acquire women, the warfare patterns still appear to reflect conflict over resources useful for the family line.[45]

It is easy to see a link between small-scale societies' warfare and reproductive and familial payoffs. Even in the Judeo-Christian heritage, women were a valued profit from warfare. In a cogent review, John Hartung examined the centrality of rape and forced marriage in biblical warfare.[46] Consider Moses: "Now, therefore, kill every male among the little ones, and kill every woman who has known man by lying with him. But all the young girls who have not known man by lying with them, keep alive for yourselves" (Numbers 31:17–18). Or the biblical injunction:

When you go forth to war against your enemies, and the LORD your GOD gives them into your hands, and you take them captive, and see among the captives a beautiful woman, and you have desire for her and would take her for your wife, then you shall bring her home to your house, and she shall shave her head and pare her nails. And she shall put off her captive's garb, and shall remain in your house and bewail her father and her mother a full month; after that you may go in to her, and be her husband, and she shall be your wife. Then, if you have no delight in her, you shall let her go where she will; but you shall not sell her for money, you shall not treat her as a slave, since you have humiliated her. (Deuteronomy 21:10–14)

Interesting here, of course, is not only the proposition that forced marriage with a female captive is sanctioned, but also that the captor must wait a month—sufficient time to establish that his captive is not pregnant by another—before he can "go in to her."

The warriors themselves in traditional societies often have a very clear and straightforward perception of these costs and benefits. Napoleon Chagnon, who worked among the Yanomamö for decades, reported this conversation with a Yanomamö friend:

"Who did you raid?" he asked.

"Germany-teri."

"Did you go on the raid?"

"No, but my father did."

"How many of the enemy did he kill?"

"None."

"Did any of your kinsmen get killed by the enemy?"

"No."

"You probably raided because of women theft, didn't you?"

"No." (Chagnon reports that this answer disturbed his friend.)

"Was it because of witchcraft?" he then asked.

"No," I replied again.

"Ah! Someone stole cultivated food from the other!" he exclaimed, citing confidently the only other incident that is deemed serious enough to provoke men to wage war.[47]

14

Societal Complexity and the Ecology of War

Four things greater than all things are,—
Women and Horses and Power and War.
—Rudyard Kipling

Is it possible to conceive of life without force?
—Charles de Gaulle

IT IS NOT SURPRISING that the functional relationships of warfare are clearer in small, simple societies than in large politically complex ones. The transition from traditional-society warfare to the complex multinational warfare discussed in treatises on military history may seem almost unfathomable, but we must explore it if we are to understand whether modern warfare is functionally different from tribal warfare.

The military historian John Keegan's description of Alexander the Great suggests that even in large hierarchical armies, as during Philip's and Alexander's rule in Macedonia, personal characteristics, kin-group size, and the ability to inspire loyalty and reciprocity still were crucial to a man's success in warfare.[1] The Macedonian kingship was elective; Alexander had claim to the succession as the eldest son of the king's acknowledged wife, but if he had not been bold and eager for battle, he would have found it hard to press his claim.

Macedonia was an imperial power, and the Macedonian army was large, diverse, and hierarchical. It included cavalry and light cavalry, light infantry, and specialized troops—archers, siege artillerymen, engineers, surveyors, and supply/transport specialists. Soldiers were neither a tribal war band, as in the Yanomamö, nor were they conscripts; they were recruited from a variety of social classes. Nonetheless, there was a central, inner core of warriors whose relationship to the leader was close, often a blood relationship.[2] It was important that Alexander consistently led his men, fighting by their sides, performing dramatic feats of courage and leadership. While courage has probably always mattered to men's fates in war, in Macedonian times, heroism became a prerequisite for leadership of large, complex organizations.

Similarly, Prithvinarayan Shah's ability to forge the modern state of Nepal owed much to his strategy of rewarding individual men's interests and thus commanding their loyalty. What is now Nepal comprised several small kingdoms; geographical constraints divided the area into small, self-contained units. Peasants' lives were hard. All land belonged to the state; those who worked the land typically paid half of the harvest to the state, as well as giving compulsory unpaid labor—even though few of the valleys were very fertile. Gaining freehold land was the only escape. With only perhaps 8000 to 10,000 men and with serious logistical problems, Prithvinarayan Shah drew men of the hill tribes, notoriously pragmatic and unlikely to follow others' dreams, into his ambitions, offering land grants for services as a way for followers to break out of the cycle of agricultural poverty.

Shah was a real leader. He spelled out in concrete terms the individual advantages for his soldiers, and showed them the advantages of a farther-reaching group goal. In addition, throughout the long campaign, in negotiating with his enemies, he offered substantial rewards to anyone converting to his views and was usually successful. No other individual had been able to overcome the fragmentary relationships and form a modern nation from the region's tribes.[3]

Leaders in war were likely to become leaders in peace (though not always equally effective in both roles). Indeed, war has been proposed as a mechanism involved in the very formation of states. Warfare has been argued to be a necessary, if not sufficient, condition for the formation of states, and there are clear relationships between political complexity and warfare patterns.[4]

War in traditional societies requires individual striving and centers on reproductively important resources. At all levels above the simplest ambushes, warfare involves organization and opportunities for gain through leadership; successful war leaders are likely to be good manipulators of others, and they accumulate an armed following. The path from war leadership to political leadership may be short but profitable: the risks of political leadership are usually less lethal than those of war leadership. Nonetheless, the bold aggressiveness that can make a war leader great may not be the most effective coalition-building strategy in times of peace.[5]

I think that recognizing the importance of reproductive interests in lethal conflict can resolve some apparent discrepancies among earlier models of warfare and the rise of states.[6] Reproductive competition is a major evolutionary selective force underlying lethal conflict; warfare is a principal mechanism and may be waged in the name of women, bridewealth, revenge, agricultural lands, new territory, or any devised reason.

Not surprisingly, as societies become more complex, so does the scope of the problem.[7] In warfare involving hierarchies of power (i.e., rank and specialization; probably all but tribal ambush warfare), risk is correlated with prior status and/or rank. Since at least the Middle Ages in Europe, disenfranchised or low-status males have gone to war in positions of greatest risk. Sons of Portuguese nobles, for example, would take three-week crusades to nearby, relatively safe locations, while sons of poor families went to Jerusalem, often dying there.

Among the Cheyenne, in North America, men who controlled more re-sources and had larger familial networks became peace chiefs; such men tended not to assume the risks of war. Instead, men who had no relatives (often orphans) became war chiefs; they could gain the proximate rewards of status but could not turn status into reproductive gain because they had to remain unmarried. The Cheyenne case appears to be an example of intragroup conflict of interest, between peace chiefs and war chiefs, with kinship networks and power on the side of the former. It would be extremely interesting to learn the history of the Cheyenne condition. Were the first peace chiefs (political lead-ers) originally successful war chiefs who then discovered how to avoid the risks of leading war parties without losing status? Taking extraordinary risks with no hope of reward should be rare, seen only when an individual's other options are severely constrained, or if he becomes convinced of some over-whelming benefit.[8]

If we put together basic sex differences in aggression and ecological influ-ences on "payoff" for aggression, some predictions follow. Truly lethal endeav-ors—such as the suicide missions of some Japanese in the Second World War, the extreme risks taken in jihads in the Middle East, and the lethal attacks of 9/11—should correlate either with very low status of the men or with promises of gain otherwise unmatched. And this pattern is important: in developing na-tions, the proportions of young males in many populations are increasing and will continue to increase; meanwhile, in the developed world, young males are expected to decrease by 4 million.[9] As I noted in chapter 13, as the proportion of young males increases in resource-limited environments, the potential for violent solutions in both local and larger conflicts seems likely to increase.[10] Nonetheless, recently, religious and other inculcation can be convincing even to well-off young men, complicating the picture further; several of the 9/11 and current attackers were wellborn and well educated.

There is a persuasive argument that even today kin selection still may influ-ence much conflict.[11] When societies become very large and are comprised of diverse groups, intragroup or ethnic conflict once again becomes important. If, as I have argued above, lethal conflict arose in the evolutionary context of re-productive striving, international wars, like World Wars I and II, are likely to be simply epiphenomena—although national leaders may use tactics of referring to proximate cues that are important in evolutionary time, such as familial terms to promote patriotism. Strife over local resources seems most likely to escalate into lethal conflict; thus it is not surprising that even today civil war and ethnic strife are far more common than international war—and more com-mon and potent than when this book was first written. Of the eleven conflicts in which the United States has played some role since 2001, nine are/were local insurgencies or conflicts. Only In Iraq and Afghanistan did the United States play a major initial role of invasion to overthrow a current government; and insurgency remains a dominant force in those countries today.

GREEK HOPLITES: EARLY "WESTERN" WARRIORS?

The Greeks may have provided the first real shift (in the West, at least) away from the small-band, guerrilla ambush attacks that were the usual pattern in preindustrial societies. Indeed, the Greek situation shares some characteristics with modern warfare, but it also shows some important and perhaps underappreciated differences. The rise of the hoplite—a heavily armed and armored infantryman—in Greek warfare during the second half of the seventh and the sixth century BC had important political as well as military consequences.[12]

The hoplites fought in phalanxes, and each man's life depended on the behavior of his neighbor in the battle line. For the most part, the hoplites were small farmers. As with the Meru, Greek warfare followed the agricultural seasons. The constraints of vineyards, olive groves, and grain fields meant that battles were fought during the summers by local farmers to gain or protect local property. In most city-states except Sparta, there was little combat specialization and very limited drill. Men were vulnerable to the draft in any summer from their eighteenth to their sixtieth year, and in any battle the majority of men were likely to be over thirty. Men of rank fought as ordinary hoplites among their less noble neighbors, and men fought with others of their tribal affiliation. When columns were decimated, they were not immediately reconstituted, but someone simply took the place of a man killed—usually a friend, relative, or neighbor. Men fought alongside their kin and their neighbors, with whom they held many interdependent relationships, and survivors of a battle were likely to know all who were killed. Men testified to the numbers of battles others fought, a sort of mutual recommendation system that fostered intensive mutual interest. Coalitions beyond the battlefield were strengthened.

Thus, while these conflicts involved hierarchy and trained warriors in organized combat rather than tribal ambushes, individual men were still fighting to protect or gain their own land resources, and they fought with their families, tribes, and neighbors. Commanders fought in the front lines with their men, suffering almost certain death if defeated. A relatively high proportion of victorious generals also died. In contrast, in most modern wars, commanders have moved farther from the front lines and share fewer risks with their troops, with some notable exceptions in the American Civil War.[13]

THE ECOLOGY OF RENAISSANCE WAR

European warfare in the Middle Ages, up until the mid-fifteenth century, involved many small-scale territorial wars and local, powerful men; it involved reproductive and familial gains and losses—a sort of violent housekeeping. Until perhaps 1450, knights fought largely as individuals and in small coalitions. Fights often started over revivals of old family claims to previously lost estates: "Political

Europe was like an estate map, and war was a socially acceptable form of property acquisition." Although particulars differed, familial economic interests were important. Noblemen, the landed gentry, had strong vested interests in waging war, not only for territory defense or acquisition, but for the spoils and riches to be gained. Later-born sons and sons of poor families, with less access to resources and titles than wealthy elder brothers, tended to end up in high-risk warfare. In Europe, from Roman times until Charles the Bold's military ordinance of 1473, differentiation by insignia in larger armies reflected social status.[14]

Just as among the Yanomamö, kinship mimicry was used to manipulate others: men called important people by kin terms to create or emphasize alliances. Military adventurers often became *fratres jurati*, or sworn brothers. When William the Conqueror invaded England, Robert de Oily and Roger de Ivery—unrelated men—were fratres jurati. After receiving honors for battle, Robert gave one to Roger as his sworn brother. Fifteenth-century *compagnies d'ordonnance* engaged the monarch's personal interest, were a route to high administrative office, confirmed disputed titles, and paved the way for handsome plunder of ransom profits. Successful warriors could gain higher status in peaceful times.

The introduction of guns permanently changed the nature of war, its conduct, and conditions. At some level, of course, it is fair to say this of any technological development, any new war tool: horses, crossbows, and so forth. But gunpowder may have had an unforeseen impact that fed back into the nature of selection on men at war. Gunpowder led to larger and costlier armies, with more support personnel (e.g., masons to make balls of stone if the iron ones ran out). Once, conflicts were limited by an individual's ability to exert sustained physical effort; now they lasted longer. Sieges became more common. Now, guns could be poked through holes in fortifications. Perhaps most important, guns changed the requirements for soldiers: it took far less skill and fewer resources to fire a musket than to train as a longbowman or cavalryman.

The use of guns further separated the risk bearers from the profit makers in warfare. A new group emerged: weapons makers, who had complex skills and political power and who could profit from lethal conflict without the risk of engaging in it.[15] The military-industrial complex was born.

Increased costs of the new technologies meant higher taxes. Although field armies may never have drawn on more than 5% of the population, recruiting was voracious and taxes became an important issue in all but the most local wars.[16] Mercenaries became prominent. Efforts to recruit the landed gentry were not too successful, for predictable reasons: because percentage quotas, rather than individuals, were drafted (except in the case of personal indentures), anyone who could buy or litigate his way out of the draft passed the burden on to someone poorer.

Early in the period, substitutes were likely to be younger sons of the nobility, but later they became the unskilled, the poor and hungry, and even those on the run from criminal proceedings. In 1572 the Venetian commander Giulio Savorgnan reflected on why his troops had enlisted: "To escape from being craftsmen, working in a shop; to avoid a criminal sentence; to see new things; to pursue honour—but these are very few. The rest join in the hope of having enough to live on and a bit over for shoes or some other trifle that will make life supportable." Similarly, in 1600, Thomas Wilson assessed the English forces as comprising chiefly cottagers and copyholders but also those who are "poore, and lyve cheefly upon labor, working by the day for meat and drinke and some small wages."[17]

Renaissance wars, then, came to be fought by mercenaries, not those defending or seeking lineage resources. The standing army came into being after the Renaissance, and power passed largely to others, not to those who fought. Further, mercenaries had impact on civilians during peacetime: unlike knights, who were likely to return home, or paid soldiers, responsible to their employers, mercenaries during peacetime were "the responsibility of no man and they consequently became bandits."[18]

Over time, then, the potential individual reproductive gain of warfare fostered technological innovation, and then the technology of war changed the nature of war's profits. The balance of benefits seems to have shifted, for most actors, from a familial, resource- or status-building strategy, as in the Yanomamö, the Meru, and Europe of the early Middle Ages, to a more stratified situation in which the poor and disenfranchised began to shoulder the risks and costs of warfare, as in Portugal and even in Vietnam. More expertise was required for the manufacture than the use of common weapons, further separating the risks and the costs for warriors. Even today it would be difficult to disprove an accusation that individual fortunes and family empires are sometimes built from war—but it is done typically by the manufacturing entrepreneurs, not by the warriors. Of course, it's hardly so simple; rewards and booty remained important resource advantages for some men.

The Behavioral Ecology of Modern War

It is probably true that past correlations between warriors' behavior and reproductive success are weak or nonexistent today, though we lack the necessary data to know. But aspects of men's behavior in modern wars, and of the organization of fighting forces, suggest that (1) proximate correlates of reproductive success due to risky and aggressive behavior still exist in modern wars, and (2) successful leaders organize field units in ways that play on the past kinship structure of warring groups.

Training

Many features of men's training for warfare mimic the proximate cues of both kin groups and close reciprocity. New recruits in many armies undergo forced transformation to uniformity (GI haircuts, uniforms). Training emphasizes communal values, often by using kinship terms; training is aimed not only at the obvious skills, but also at ensuring cohesion, inciting hostility, enforcing obedience, and suppressing mutiny.[19] Recruits are likely to be called "son," "boy," or "lad." This paternalistic language goes far beyond basic training and has done so for centuries. The "sworn brothers" fighting together in the Middle Ages engaged in reciprocity bolstered by kinship mimicry.[20]

David Hackworth, one of the most-decorated of military men, argued that much U.S. combat failure in Vietnam arose directly from a few causes, including poor training that broke the important patterns above.[21] Instruction was given by returned short-timers who had not wanted to go to Vietnam in the first place; instructors' dissatisfaction often led to war stories rather than to training when supervision was absent. Most significant, men were not trained and put into combat in units, with their mimic of familial structure and their strongly developed reciprocity. Instead, they were sent individually into combat in a strange land (the finishing preparation for this jungle war was often done on snow-covered fields). Without the required skills (e.g., none of Hackworth's trainees knew what to do when a gun jammed) and with their dependence on strangers, many died.

Pre-battle Exhortation

Successful commanders from time immemorial have played upon the major themes of gain: individual wealth, social reciprocity, kinship gains, and sexual success. Thus Du Guesclin, a French knight of the mid-fourteenth century, mixed penitential motives with profit, when he urged his recruits: "If we search our hearts, we have done enough to damn our souls. . . . For God's sake, let us march on the pagans! . . . I will make you all rich if you [follow me]!" Interestingly, Hugh of Caveley, an English knight fighting with him, responded by invoking kinship and reciprocity: "Bertrand, fair brother and comrade, mirror of chivalry, because of your loyalty and your valor, I am yours, I and all these here."[22]

Shakespeare captures the essence of this strategy, when he has Henry V exhort his rag-tag collection of men before Agincourt, calling them brothers even while highlighting their class diversity:

> We few, we happy few, we band of brothers;
> For he today that sheds his blood with me,
> Shall be my brother; be he ne'er so vile,
> This day shall gentle his condition:
> And gentlemen in England, now a-bed,
> Shall think themselves accurs'd, they were not here;
> And hold their manhoods cheap, whiles any speaks,
> That fought with us upon Saint Crispin's day.
> (*Henry V*, act 4, scene 3)

Henry invoked sexual selection, status-seeking, kinship mimicry—all to stir his men (many of whom were the poor and ragged) to do their utmost. The folks at home are sometimes similarly exhorted even today. In the news headlines of the Middle East conflict, headlines have reinforced kinship images: "Shipmates become like brothers in Gulf pressure-cooker."[23] And men who serve under extreme conditions can come to value their comrades literally as kin. In *Band of Brothers*, the story of E Company, a rifle company exposed to face-to-face combat for long periods of time, Stephen Ambrose recounts both the valor and the sacrifices men make under these conditions.

Interrogation and Resistance

Group structure and within-group reciprocity are crucial in fostering resistance to interrogation.[24] During the Korean War, for example, the most successful North Korean interrogations followed the breakdown of the prisoners' group structure. Reliable group structure—the presence of dependable comrades-in-arms—contributes not only to stalwart behavior in battle, but also increased ability to resist interrogation after capture. More recently, since incidents such as Abu Ghraib, a number of countries appear to have begun training personnel to resist such "interrogation" techniques as waterboarding, and others. The top ten methods used by the United States of softening up prisoners to confess include some classics and some new forms: isolation, sleep deprivation, sensory deprivation, sensory bombardment, forced nudity, sexual humiliation, cultural humiliation, exposure to extreme cold, exploitation of phobias when known, and waterboarding.[25]

Insurgency and Terrorism

Technological advances have fostered newly dangerous ethnic conflicts, as noted above. As the lethality of weapons increases, a small number of people can threaten the political stability and lives of ever larger numbers of people. Major powers today shy away from escalation of international conflicts likely to lead to nuclear engagement. As local, usually ethnic, conflicts become more important, the danger posed by small groups increases.[26] Actual death counts may not be as high as in the two world wars, but as, for example, school-girls are kidnapped or shot, everyone in an affected area feels unsafe.

Reviews of contemporary terrorism and low-intensity conflict suggest that despite exceptions like the Red Brigade, most conflicts in fact originate as local ethnic or religious conflicts of interest—e.g., Rwanda, Northern Ireland, the West Bank, Bosnia, the Philippines, the Liberian civil war, Nigeria, Kenya, Cambodia, Somalia, Afghanistan, Yemen and Mali, and more. Conflicts like these probably are intensified by the sorts of pressures prevalent throughout the evolutionary history of warfare: genetic lineages in conflict, which have expanded to become ethnic regional conflicts.

It is probably not irrelevant that in many successful (long-lasting) terrorist groups, the leader assumes a paternalistic role, and the group structure mimics that of families. Even a review of larger-scale conflicts suggests that ethnic and racial components are still important after World War II—far more important than ideology or nationalism. Many insurgencies speak clearly to this point.

The real danger is probably that small, originally local conflicts, because of technological advances, can wreak international havoc, and major powers can then be drawn into the fray (as appears to have been the case in many of the U.S. involvements since 2001). For example, U.S. funding for Special Operations Forces, which are typically involved in low-intensity interventions in local conflicts, quadrupled between 2001 and 2011.[27] There is real potential for major powers to be drawn into confrontation through (originally) local conflicts, and balancing such conflicts can impose high costs and real risks.

Deception

In other species, aggression is frequently accompanied by advertisements that exaggerate an individual's prowess (chapter 15); some primates also deceive by using "neutral" postures to get close to a target. In human conflicts, there are probably not qualitative differences, but it may be that there are important quantitative differences in the importance of deception compared to that in other species. We are capable of subtle deception; this may make conflict more, not less, likely—because conflict is more likely when at least one contestant's information is faulty.[28]

In human arms races, secrecy, deception, and misinformation are perhaps strategically useful, even if bluff doesn't always work. As a result, policy makers and government officials must often oppose citizen concerns.[29] This situation is complicated by the fact that leaders and their advisers play a double game—both laterally, with other national leaders, and vertically, with their public—and the influence of the public on leaders' policies is maximized only periodically (in the United States, every four years)—plus the public receives imperfect, manipulated information.

Women and Disadvantaged Men in War

From the engagement of relatively few, typically related, men in combat over resources that were directly related to their lineage's success, armies have grown and become hierarchical, composed of unrelated men and men with quite different backgrounds and wealth (although there tends to be stratification in rank with education and wealth). The result is an increasing divergence of the actors' interests. From related men who might squabble over this bride or those cows, we now have administrative groups sending others to fight. Maintaining discipline and loyalty in the face of unequal payoffs can be tricky. This problem lies at the very root of the transformation of societies from often highly polygynous states to (at least prescriptively) socially monogamous states.[30] In polygynous societies engaged in large-army warfare and conquest, formal reproductive and resource distribution schemes are common. As warfare technology changed and armies became larger, status differentials increased; no longer, as among the Yanomamö or even Alexander's armies, were the spoils of war a relatively simple reflection of individual courage and skill. More high-status men opted out of conflict, and more low-status men were recruited or drafted. The latter, from at least medieval times to the Vietnam War, suffered higher casualties than their richer competitors.[31]

Yet the fate of disadvantaged men is not so simple; if it were, the only question would be why they ever serve. Just as the costs and benefits for any individual of living in a group must be weighed against the costs and benefits of living alone or in another kind of group, the question must be asked: did these men, on average, fare better or worse than if they had remained civilians? I know of no data. For those who survived, it is possible that they did, in fact, fare better than if they had not served, through the status of wearing a uniform, just as in older high-risk, high-gain endeavors.[32] Further, in many wars, less well-trained men (e.g., high school education or less) have been taught skills (mechanics, etc.) that can help them get a better job back in civilian life.

For at least 400 years, women have accompanied men onto battlefields; typically they cooked, nursed, cleaned, and supported their men. In the First World War, the Russian forces had one all-female combat unit. The first U.S. woman to enlist was Loretta Walsh in 1917, but it was not until 1948 that a law was passed that made women officially able to be part of the military, and not until 2013 that women were offered the same combat opportunities as men. Just over 15% of the 1.1-million strong U.S. military is composed of women. Today, countries besides the United States that allow women as combatants include Australia, Canada, Israel, New Zealand, Norway, Sweden, and Sri Lanka.[33]

WAR AND REPRODUCTIVE SUCCESS TODAY

In societies like the Yanomamö and the Meru, a man's lifetime reproductive success was likely to be closely correlated with his performance in war. These two examples are not atypical; in many societies, reproductive rewards for valor were standard.[34] Both direct reproductive rewards and heightened status or privilege before the law were common rewards for leadership and success in war. As armies became larger and more stratified, and as direct formal reproductive rewards for performance disappeared, the situation became less clear. For many, there were obvious risks—but no longer obvious reproductive rewards.

For disenfranchised men today, with little or no chance of success in peacetime environments, there is possibly a correlation, but it is hard to test.

Has the reproductive nature of lethal conflict really disappeared? Rape is a well-known but little-studied and seldom-quantified concomitant of warfare and perhaps of the male-dominated military itself. In the Bosnian conflict, for example, there were reports of rapes of Muslim women by Bosnian soldiers: "We are seeing the same pattern repeatedly, of Chetniks telling women, 'It is better to give birth to Chetniks than to Muslim filth,'" said Mahir Zisko, executive director of the war-crimes commission. "Another statement that recurs is: 'When we let you go home you'll have to give birth to a Chetnik. We won't let you go while you can have an abortion.'"[35]

With the increased numbers of women serving in the military, opportunities for sexual harassment and rape are exacerbated. There are three main types of conflict-related rapes: rape of an enemy girl or woman, to terrorize them, their families, or other women; rape of male prisoners; and internal sexual harassment of women military by their superiors. A study in 2010 found that 22% of men and 30% of women in Eastern Congo reported conflict-related sexual violence. In the Sudan, a UNICEF advisor was quoted in 2004 that rapes continued unabated. The documentation of humiliation and sexual harassment of male prisoners at Abu Ghraib is well known. The point here is not that these are horrifying exceptions, but that sexual harassment, including rape, continues to be a weapon of war, and sometimes even of spreading one's genes (above).[36]

Proximate and Ultimate Causes of War: Evolutionary Novelty

In evolutionary terms, warfare is predicted to become and remain common only in circumstances in which the net fitness of warriors is enhanced. Throughout the history of conflict, in humans as well as in other species, reproductive profits have been associated with the risks of lethal conflict. With the elaboration of war and the increased pace of weapons development, selective outcomes have become less tied to individual actions and characteristics. Those with the most to gain from warfare have come to suffer lower risks than those with little to gain.

We may well have unhooked the reproductive rewards from the behavior, so that lethal conflict is now counterselective for actual warriors and driven only by proximate cues. Now, perhaps (though I know of no data), war may not profit anyone directly involved in the conflict—yet the driving cues remain. With this in mind, consider briefly some of the proximate causes of warfare assigned in modern conflicts. This subject is far too broad for a detailed analysis here; specific causes are multifarious, and others deal with it in more detail. There are, however, some major patterns that are worth reviewing briefly.

Oft-cited causes and correlates of war include the number of contiguous neighbors, economic conflicts (comprising approximately 29% of wars from 1820 to 1949), territorial disputes, ideological differences, and misperception or distortion of information (e.g., in wars in the twentieth century). Pacifying influences have been judged to include shared ethnicity, common government, recent alliance, and extended deterrence (when military strength is sufficient in the short term). Recent work suggests that more quantitative analysis is needed and that analysis must separate, for example, local from international terrorism, and terrorism from large international conflicts.[37]

Leaders' personalities can have a potent influence on events. Even in the huge number of specific causes of wars in modern times, the "ecological" categories are still rather limited: conflicts arise over resources (economic or territorial) and are less likely the longer and deeper are common bonds (kinship or reciprocity); open conflict is often precipitated by faulty information.[38]

There is, however, a behavioral ecology of war from our evolutionary past. The multiplicity of proximate correlates in modern warfare does not mute the importance of the kin, ethnic, and male competitive forces. There have been over 100 small-scale wars since 1945, resulting in 15 to 30 million deaths. The vast majority of these are between local groups; few are international. A review of recent atlases of war—or perusal of this morning's paper—reinforces the importance of essentially tribal conflicts of interest.[39]

Can Evolutionary Theory Help Avert Arms Races?

If potentially lethal conflict has very old evolutionary and ecological roots, can learning about these roots help us avert arms races? It is difficult to tell. Understanding the evolutionary background of lethal conflict and arms races may simply help us to understand that costs and benefits, deception and misperception, may involve currencies other than the immediate and obvious ones. Whether this could help us mitigate modern arms races, I don't know.[40]

In a major work on the kin selection roots of warfare, after eloquent analysis, two authors are reduced to calling for "some form of world government, some management force that might stabilize the most immediate threat to humanity— nuclear destruction."[41] The entire work, however, is an acknowledgment that the power of in-group amity and out-group enmity would likely force any such world government to be a conquest state, a chilling prospect.

There are hints, however, that even in the absence of power hierarchies or hegemony, certain costs and benefits may promote cooperation.[42] The reproductive profits may be largely gone, but the proximate drivers remain. Writing of just such problems, the biologist Richard Alexander found that "it seems to place me in a camp of those who see mutual deterrence as the basis for peace, even though I doubt that either self-extinction or massive destruction can be prevented indefinitely by deterrence alone."[43] He then suggested that our finely honed social and predictive intellect might provide several partial brakes on arms races—for example, if sufficiently numerous and powerful individuals or groups perceive that no matter who wins the confrontation, we all will lose, and their power in internal social/political coalitions may allow them to force some solutions.

The difficulty is that the brakes are weakened by the dilemma a biologist would call the "levels of selection" problem (chapter 9): because natural selection largely fosters genetically selfish behavior, long-term costs and benefits to groups are discounted compared to immediate, short-term costs and benefits to the individual, family, and reciprocators. The larger and less related, the more diverse, the group, or the farther in the future costs and benefits must be calculated, the greater will be the discount. Thus, given a short-term gain in status or tax base for a local constituency versus an unspecifiable risk of nuclear warfare sometime in the future, we do not predict restraint. Of course, modern war is not the only issue for which our large-scale activities are problematic (chapter 15).

Any successful attempt to foster peaceful behavior must change individual costs and benefits in the proximate sense; yet currently many social institutions and rewards mediate strong status pressure on individuals to enter into wars.[44] This sounds simple enough and has parallels in economic approaches, but there are serious problems: (1) there are no longer simply a few easily identified, powerful groups or coalitions with focused costs and benefits, so it is hard to figure out how to direct any such proposed manipulations; and (2) now, to an extent previously unknown, small groups can control highly destructive devices. When these individuals are unaffected by the sorts of costs and rewards humans have evolved to recognize, we call them terrorists. Our technology is sufficiently advanced that even such small groups can wreak havoc, killing great numbers of people.

15

~~~~~~

Wealth, Fertility, and the Environment in Future Tense

Desire is the very essence of man
—Spinoza

No species has ever been able to multiply without limit. There are two biological checks upon a rapid increase in number—a high mortality and low fertility. Unlike other biological organisms, man can choose which of these checks shall be applied, but one of them must be.
—Harold Dorn

IT IS EASY TO IMAGINE that our evolutionary past is remote, unconnected to our lives today, and of interest only when we think of traditional societies or ancient history. Yet, as we saw in the last few chapters, our evolved tendencies interact with today's novel environments. Cities, no less than rain forests and savannas, *are* our environments, and there is evidence that selection still acts on humans.[1] Just as soldiers in modern warfare display remnants of behaviors from past times, so do we all in our daily lives, as we live, work, and raise our families. Some physical and social aspects of our current environments are evolutionarily novel, largely the result of our past and current actions; and we may well have disconnected our behaviors and tendencies from their ancient impacts (love of sex, fertility, contraception). Even so, our evolved tendencies interact with our current environments—and our future depends in part on the results of those interactions. Our evolutionary past strongly influences our ecological future.

In chapter 14 we saw that our evolved traits affect our behavior in warfare even today. This is even more true about the paths that multiply our evolved resource-acquisition tendencies into modern regional and even global problems of population, environment, and sustainability. Further, warfare, and the resource issues I tackle in this chapter, are clearly related; each can lead to the other and, for example, to the extinction of other species.

Three strands of argument are central to this book; they contribute, I think, to the curious resource situation in which human societies find themselves today. First, in past environments, successful resource striving led to successful reproduction: more resources led to more children, particularly for men. Furthermore, in most species and in prehistorical and historical human societies, females who failed to acquire some threshold amount of resources, like unsuccessful males, failed to reproduce.[2] In our past, this pattern has been clear, at least as reflected by the demography of traditional societies (chapter 7).[3]

A shift began in the nineteenth century with the demographic transition, in which wealth led not to *more*, but to *better-invested*, children: fewer, but with each being more costly (chapter 8). Yet even when monogamy was the rule and people married late, as in the demographic transition in nineteenth-century Sweden, men with rich resources were more fertile than others (chapter 8). In the evolutionary history of all living things, including humans, getting "more" has always been reproductively more profitable, especially for men.

Second, our intelligence and sociality have allowed us to continue creating novel technologies and environments; this has resulted both in heightened efficiency of resource extraction and accumulation and in a phenomenal spreading and interchange—we affect others across time and space in truly new ways. (In songwriter Tom Lehrer's words, "The breakfast garbage that you throw into the Bay, they drink for lunch in San Jose." [4]) This brings the levels of selection problems of chapter 9 into stark relief. Our individual and corporate actions can affect populations far away, but it is not usually for their sakes that we do what we do.[5]

Finally, for humans in particular, an additional complexity exists: we have intensified the importance of heritable wealth. In essentially all societies studied, parents try to manipulate resources in ways that preferentially help their own children, whether the important subsistence base is hunting and gathering, horticulture, pastoralism, agriculture, industry, or postindustrial resources.[6]

Together, these strands (the success of intense resource striving, intelligence, and cooperation when useful, and a growing shift to ever-more-expensive offspring) have changed the selection pressure on family lineages. Today, in developed nations, sheer fertility—number of offspring—is not a good predictor of lineage success in many environments. Although we evolved to be fertile, to help our offspring survive, and to enhance their reproductive success,[7] there are obvious trade-offs involved, and the most effective lineage-persistence strategy will differ with environment. If a Yanomamö chief can marry earlier and more often than others but cannot control his children's survivorship, then high fertility is a successful strategy. But because competition from other humans has always been a key selective pressure on us, when intense this competition may lead to an advantage of not *more* babies, but *better-provisioned*—and thus more consumptive—babies.

There is evidence that family size may be, in both traditional and modern societies, an adaptive response to the cost of per capita investment required for children's success.[8] In the modern developed nations, for example, expensive education may be required to gain a successful job, and parents will, when they

A. Routes to familial success *ceteris paribus*

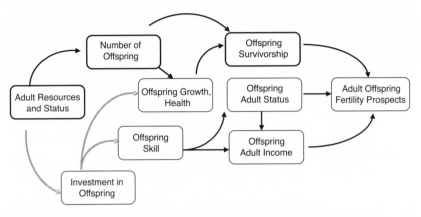

B. Routes to familial success in competitive environments

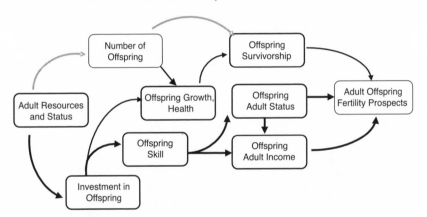

Figure 15.1. Successful parental-investment trade-offs (numbers vs. size vs. care) will differ with environmental conditions. Black lines indicate strong influences; gray lines show less important pathways. **(A)** All else equal, individuals who control more resources will produce more children and are likely to have more grandchildren than others. **(B)** But when offspring face highly competitive environments and it is difficult to get established and reproduce (e.g., high density in many nonhuman species; expensive training to get a well-paying job in many human societies), the most successful paths to lineage success over time are likely to involve very high investment in each child, with concomitant lowered fertility—essentially creating a demographic transition. (Modified from Low 2013a.)

can, invest more in each child (fig. 15.1). And this is related to another, sometimes more subtle, change: the nature and scope of women's work. In the United States in 2012, for example, 64% of the female population ages 15–64 were gainfully employed outside the home. This increase in women's work has had a huge impact, as we will see, on the value of women to men as mates, which once was simply reproductive value.[9]

If only superbly invested children succeed in reproducing, the trade-off between numbers versus success of offspring becomes acute. In other species, we typically see only the endpoint. The impact of density on degree of competition was first explored in the contexts of islands, for which the biologists MacArthur and Wilson suggested that when organisms first arrive on an island and there are few competitors, selection favors productivity (high fertility) even if foraging, for example, were not highly efficient—but then as the island becomes crowded with competitors, selection favors efficiency and investment (e.g., larger seeds), with consequent lower fertility.[10]

In humans, other factors besides density can affect the degree of competition one's offspring face, but it is clear that across all kinds of societies, parents act to endow their children as well as they can—to make them as competitive in the particular environment as possible. Sometimes the investment is in schooling, as in Thailand, sometimes in market training—the particulars vary, but the general pattern is becoming clear. And in the United States in contrast to most of our past, highly educated women have fewer (and perhaps better-invested) children than less well educated women (fig. 15.2).[11]

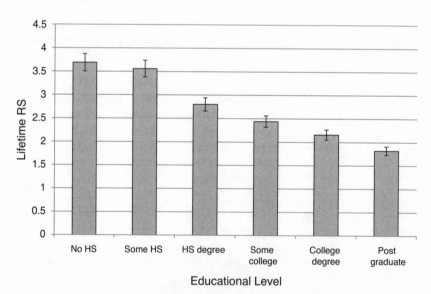

Figure 15.2. Data from the Panel Study of Income Dynamics suggest that highly educated women in the United States have lower fertility than less well educated women. There are also effects of when (in what decade) a woman was born (e.g., women with little education who were born in the 1920s had higher fertility than similar women in later cohorts; Anderson and Low 2003, Low et al. 2002), but this general pattern prevails. RS = reproductive success; HS = high school.

Genealogy matters. Data suggest that, under many conditions, stepparents do not invest so intensely in their stepchildren as in their own children, thus strengthening the hypothesis that genealogies matter. Further, when a man remarries, if he is paying child support, he is less likely to sire more children in his later marriage.[12]

These three strands suggest that we will face new problems as growing, and increasingly consumptive, human populations interact with environmental quality and stability (or fragility). The interactions of the strands will, I fear, affect our very ability to continue to live as we wish. We already have hints in the news headlines: global climate change, with its increase in the frequency of extreme weather events (extreme cold as well as extreme heat, tornados, hurricanes, droughts), increases in atmospheric CO_2 levels, worldwide precipitous decline of ocean fisheries on which some national economies depend . . . how ironic that our evolved tendencies to favor our own through increasingly clever ways of extracting resources have led us to such large-scale problems.[13]

While the proximate causes of all of these concerns differ greatly, at a deep level *we humans have created the difficulties simply by doing well, perhaps too well, precisely what we have evolved to do*: garner and consume resources every more cleverly, be fertile, give as much as we can to our children, and not look too far ahead.[14]

In 1992, the "Northern" (developed) and "Southern" (developing) nations squared off at the Rio Conference. Although the language was polite, each group accused the other of making the earth unable to sustain human life as we know it: "your fertility is too high" versus "we could afford our fertility if you were not so profligate with world resources." In 1996, the same issues were raised again, with little progress and no resolution. Yet, as figure 15.3 shows, both sides are right: wealthier nations do have much lower fertility than less developed nations (fig.15.3A; not shown, they also have an older age structure), but they also consume more global resources per capita (fig. 15.3B). And the most successful family-planning programs internationally are those that emphasize that by controlling the number of children, the family can be "better off" (can consume more goods). This is, in most developing countries, a positive thing, but note that it does increase per capita consumption.

At the heart of this ongoing debate lie conflicts, in the family and between the sexes, over resources, fertility, and power—the issues of this book—multiplied to population levels. Human population is at an all-time high, per capita consumption is higher than in past eras, and wealth and power are unevenly distributed. Uneven resource control and power are older than humanity, of course, but the combination of high population and high consumption is relatively new.

The stage is set for real-world ecological impacts of our evolutionary history. Suppose lowered fertility in some environments produces greater lineage success through fewer, better-invested children (fig. 15.1B). What is likely? Per capita consumption will rise and costs (e.g., of extraction and transportation)

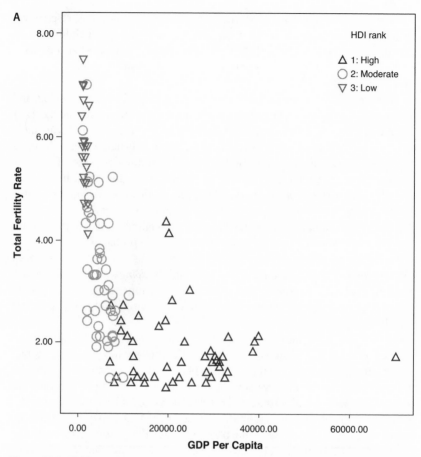

Figure 15.3. (A) Women in better-off nations (HDI = Human Development Index Rank, a multivariate reflection of quality of life, calculated by the United Nations Development Programme 2007 (http://hdr. undp.org/en/data) have lower lifetime fertility than women in less-developed nations, as measured by the total fertility rate (the average number of children a woman would have if she lived through her entire reproductive period and at each age had the average number of children).

are likely to be externalized; both have typically been effective competitive strategies in the past.[15]

In a scenario such as this (arguably the First World versus Third World situation), *lowered fertility alone produces no solution to the population-consumption dilemma*; sustainability requires more than that.[16]

Part of the reason the Rio Conference was so disappointing to many was this: a population's total impact on its environment arises not simply from the number of people, but also includes consumption rates *and* the impact of consumption technology: *Impact* = *Population* × *Affluence* × *Technology*, or *IPAT*.[17] The

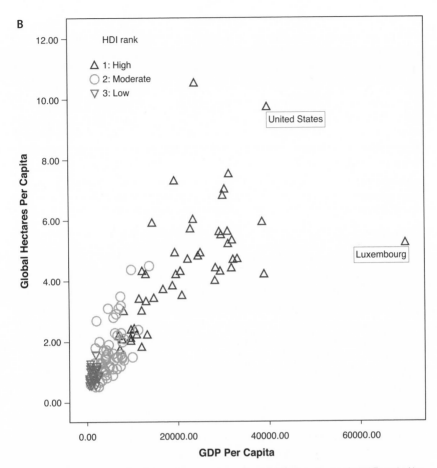

15.3. (B) However, per capita consumption, as measured by global hectares/capita (Global Footprint Network 2007, http://www.footprintnetwork.org/en/index.php/GFN/) is also much higher in wealthier nations.

relative importance of each part of this relationship differs for developing and developed nations. Although the particulars vary, ignoring *either* population numbers or per capita consumption will mislead us.

WEALTH, FERTILITY, AND CONSUMPTION TODAY: EMPIRICAL DATA

Analyzing fertility relationships today is complicated for a number of reasons. First, because so many conditions are evolutionarily novel, there is a possibility that whatever patterns we see will be emergent phenomena, influenced by proximate cues once—but no longer—related to selective pressures. Second, societies are large, containing quite diverse subcultures whose members may face varied pressures. Third, "wealth" and "status" today can mean a multitude of things (and those meanings can differ for men versus women, or for people in different subgroups of a large and heterogeneous population). It can be difficult to be certain that relationships we think we see are not simply covarying along with other factors.

Perceived status is (as we would expect from our small-society evolutionary history) based on those we know, not large statistical aggregates—so broad patterns of wealth and fertility might no longer exist, even though we are doing what we evolved to do. That is, a nation is not "a population" in biological terms: there are many populations likely to exhibit varying patterns in any measure. Certainly we can learn from differences among nations, but we still only learn a part of what we need to understand. For example, today we confront growing wealth inequality and fertility within nations, as well as across nations.[18]

So we need within-society comparisons. Even then we may have the wrong form of data; we usually have aggregate data, and these are often inadequate to answer these questions. Remember the Swedish patterns of chapter 8.[19] Aggregate data for the nineteenth century in Sweden show no significant relationship between status and fertility, but the individual and family lineage data show considerable, complex impacts of resources, especially land, and status on men's total fertility and women's age-specific fertility. Individually based lineage studies in several societies have analyzed the effects of cohort (when you were born, reflecting historical factors), wealth, and markets on human fertility (chapter 8). In the best empirical studies on modern societies so far, the results are mixed. Several show continued relationships between some measure of status and lineage continuance; in others, once-evident wealth-fertility differentials have disappeared in younger generations.[20]

Both theory and empirical data suggest a complex and locally varying picture (as in chapter 8). It appears that (1) children are an economic cost that is less defrayed today than in some earlier (e.g., agricultural) societies; and (2) heritable resources matter to the competitive success of children in most modern societies as well as in traditional ones; unless a family's wealth is boundless, fertility must decline as per capita investment in children goes up.[21]

Demographers, like behavioral ecologists and evolutionary anthropologists, are wrestling with these questions of wealth and fertility. The Nobel laureate Gary Becker treated families as a neoclassical economic problem; Richard Easterlin suggested that a man's perception of his relative wealth would predict fertility and introduced wealth-dependent preferences into the models. The demographer John Caldwell noted that in many societies, children can defray at least some of their economic costs. However, some assumptions that might seem obvious to behavioral ecologists are lacking among at least some scholars:

a recent economics review came to the conclusion that "in short, there is no explanation for why Americans still want children"![22]

Despite our human preoccupation with preferences and proximate cues and despite all the complexity, some generalizations emerge. Fifteen of 22 tests of the Easterlin hypothesis (perceived relative wealth promotes fertility), using micro- and macrodata in a variety of cultures, support the hypothesis. A person's marital status and number of existing children make a difference in people's deliberate fertility decisions. It also matters whether a government offers pronatalist tax breaks or maternity payments, or taxes fertility. Working to increase wealth represents harsher trade-offs for women than for men (more below). Finally, some additional complexity is introduced by simple preferences. For example, individuals whose religious beliefs prohibit birth control may be likelier than others to have children.[23]

No wonder patterns are complex! People who perceive themselves as well-off tend to want more children than their competitors; but these perceptions (at least so far) do not translate into strong positive wealth-fertility patterns across large heterogeneous groups.

Wealth, Women's Age-Specific Fertility, and Women's Life Paths Today

A strong predictor of high fertility is an early first birth, and one of the best predictors of low fertility and slowed population growth is delayed fertility: having children later in life than others. The strongest correlates of delayed fertility, in both developed and developing nations, are women's participation in the workforce and women's education. Women face harsher trade-offs than men, because most modern resource-garnering activities are less compatible with child care than, say, gathering (which itself reduces women's child-rearing capabilities), and declines in fertility may arise in part from women's choices in the face of these trade-offs.

What trade-offs do work and schooling represent for women? Time and opportunity costs? Wealth versus fertility?[24] There may be multiple and complex reasons for earlier versus later fertility, and to understand what relationship these patterns have (if any) to evolved patterns requires that we can rely on the accuracy of the data.

In life history theory, the strongest predictor of age at first birth is life expectancy at birth: if life expectancy is low, births are predicted to be early.[25] When women's lives are examined through this lens, which is so important to the issues of fertility and consumption, age at first birth does increase with life expectancy (fig. 15.4A), but curiously, the "best fit" of age at first birth to life expectancy is *not* among the short-life-expectancy nations, but among the wealthiest, most-developed nations (fig. 15.4A); clearly cultural factors in human populations have a huge impact. It is also clear that life expectancy at birth has not been constant (fig. 15.4B): there are nine major patterns of change in life expectancy

from 1955 to 2005, reflecting wars, HIV, and a variety of other causes. So neither assumption is met. The major hypothesis, however, is clearly supported: age at first birth increases, and population growth slows, with increased life expectancy at birth. Further, age-specific fertility is lower at all ages when life expectancy at birth is higher (fig. 15.5).

Those are transnational data; the picture can differ within nations, as noted above.[26] Fertility varies among populations within nations. Sometimes we are eager to "explain" patterns through our biases: for example, many widespread preconceived notions ("welfare payments increase women's fertility"; "it is a particular ethnic group driving this pattern") may simply not be true. There may be, e.g., particularly after child-contingent welfare reforms as in Great Britain in 1999, increases in fertility, particularly among low-education women in couples.[27] In the United States, "wealth" is not tidily related to women's fertility: it looks as though it affects total fertility differently than the timing of fertility.[28]

Both men and women today work to gain resources; in the mate market, men's value to women is still largely resource value, but women's value to men now includes resource value as well as reproductive value. How does men's income versus women's income affect fertility? Work generates income, work costs time, and lost time affects men's and women's fertility differently. One study of fertility of modern Swedish women and of white and nonwhite U.S. women sorts out covariates like time versus income (though not across generations). Some major evolved patterns are apparent (see endnote), although considerable complexity exists in transitory (and possibly nonselective) economic and fertility preferences.[29]

Women's fertility-productivity trade-offs are perhaps even starker in industrial nations today than in the traditional societies of chapter 7. Yet when women's work experience is controlled for, women's wages have several positive effects on fertility, though they differ among the groups of women (increased first births for Swedish and nonwhite U.S. women; increased second births for white U.S. women). From hunter-gatherers to women today, lost *time* may decrease fertility, but wealth, even women's income, tends to increase it. Education and work cost time and opportunities for women; in contrast, income has a positive impact on their fertility—but it is very difficult to separate the three.[30]

Figure 15.4. In biology we typically make two assumptions (cf. Harvey and Zammuto 1985) in testing the hypothesis that life expectancy and age at first birth are correlated: (1) that whatever population from which we have data will adequately represent the species, and (2) that variables are relatively stable and not changing rapidly. What about humans? **(A)** Fertility across 172 nations does indeed show a positive relationship between life expectancy at birth and age at first birth, but it clearly varies across nations (From Low et al. 2008). **(B)** Mortality has changed in nine major patterns across the last half century. Thus, neither assumption is met for humans, although the general hypothesis, that longer life expectancy at birth is associated with later age at first birth, is supported (From Low et al. 2008, 2013).

Figure 15.5. Age-specific fertility is lower at all ages for countries that have higher life expectancy at birth, compared to countries with lower life expectancy at birth. (Modified from Low et al. 2008).

Other dynamics (e.g., the per capita investment in children) clearly differ for wealthy professional women and poor single mothers—but both face trade-offs in terms of time, money, and fertility. Thus women who earn their own income are not likely to be comparable to middle-income families in which household income arises only from the man's work. As a result, in Western industrialized nations today, in which women are a large part of the labor force and in which divorce is prevalent, the wealth-fertility correlation for women is probably not linear.[31]

This discussion has centered on developed nations, for which we thought we understood the dynamics of fertility. We know even less about the dynamics of fertility in developing nations; so policy prescriptions of "women's work and women's education to lower fertility" may or may not be effective, depending on the relationships.

An Evolutionary Perspective: Reducing Both Fertility and Consumption Is Novel

Let us return to my third strand of interest. With economic development and increased competition, fertility is likely to decline; within populations, wealth and fertility may be positively, negatively, or not correlated. But with development, per capita consumption increases: estimates are that the resources used per child raised in the developed world are eight to twelve times the levels of the developing world.[32] Reduced fertility does not mean reduced consumption.

These are the core issues of population, consumption, and long-term sustainability. Population patterns—birth, reproduction, and death—are the sum of what individual men and women do: they mate and marry, have children, and die, consuming resources along the way. Meanwhile, policy is being made and implemented. If, as most ecologists think, we face resource and population limitations in the near future, the problems of wealth, health, and fertility for men and women are not just of interest to a few ivory-tower academics.

As I noted above, what happens to population growth depends not just on numbers, but also on age and sex ratio of populations, early versus late fertility, per capita consumption, and impacts of resource extraction and production. In 2011, the United Nations announced that the world's population had reached 7 billion; The 2013 estimates suggested the world's population will reach 11 billion people by 2100. Most growth will be in developing nations.[33]

The "carrying capacity" of the earth—the human population that can be sustained indefinitely—depends on one's assumptions. How many people can the earth support? Remember $I = PAT$; these terms will depend on how people want to live—how many resources they want to consume (what standard of living) and what technologies they use to get the resources, how many children, and how evenly resources and children are spread among families. This is clearly a problem of political planning, institutional effectiveness, and prevailing ideas and norms about social equity and economic restructuring. It doesn't take an expert to imagine that one cannot increase both numbers and per capita consumption endlessly if resources are not infinite. Many experts are increasingly uneasy about the level of realism and comprehensiveness in current approaches to the issues of population and environment in the twenty-first century. Assumptions of continued growth forever and the use of Gross Domestic Product (GDP) as a measure of a country's "health" may be far too simple.[34]

How Our Evolved Tendencies Affect Our Future

Because we are so numerous today (P), consume so much (A), and use technologies that can have serious environmental impacts (T), we not only may have trouble living sustainably, but we also face a dilemma. Today we smart and social hominids introduce rapid environmental changes, resulting in evolutionary novelty and often evolutionary mismatches (ecological and evolutionary traps).[35] Currently, much conservation effort is prescriptive ("Just Save It") and does not take into account the evolutionary history of human behavior— this is why, even when we understand intellectually what actions will help us be "green," we may have trouble convincing ourselves to perform them. Perhaps behavioral ecology can offer us guidance in shaping our conservation efforts. To what pleas will we respond?

Population	Technology	Profit	Predicted Impact

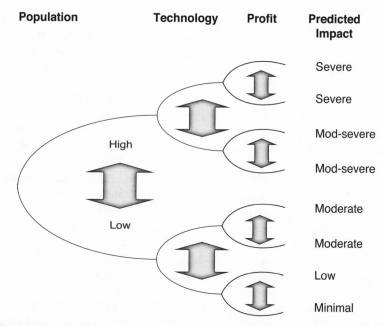

Figure 15.6. For traditional societies, it is illogical to infer a conservation ethic from the simple observation of low ecological impact by a population. All else equal, impact is a factor of population size and/or density, efficiency of resource extraction technology, and whether there is profit to be gained by harvesting the resource. The impacts shown are those we expect. One could infer a conservation ethic if one found low impact where a high impact was expected. (Modified from Low 2013b).

Ancient *Homo*, and even pre-*Homo* (other hominids) affected other species, sometimes causing extinction (see below). We know something of the evolution of primates, including humans, but mostly we rely on physical remains: seeds, skeletal remains, potsherds, arrowheads. These tell us, for example, where and when people lived, how large and fast they grew, how big and complicated their populations became. They tell us how people made a living—did they hunt and gather, fish, or depend on agriculture?

We know that ancient and traditional societies have sometimes overexploited resources to the point of causing extinction. There are documented human-caused extinctions since the Quaternary. In Pleistocene North America, we humans extirpated large mammals, some large birds, and some small mammals. Less well known, Late Pleistocene (which ended about 10,000 years ago) extinctions in South America, Australia, and New Zealand followed the same pattern. As Polynesians spread across the islands of the Pacific, they extinguished all of the giant flightless birds, a number of other flightless and flying birds, many marine mammals, frogs, lizards, and flightless insects—not to mention a number of plant species. The arrival of Polynesians on the Hawaiian Islands, for example, appears to have wiped out more than half (54%) of the endemic bird species. Humans have caused harm in several ways: unintended fellow-travelers like rats, whose competitors and predators were left behind; intentional introduction of agriculture, of animals we husbanded (goats, sheep, cattle, pigs), and of carnivorous pets like cats and dogs; weapons that made hunting efficient; and more.[36]

There are broad generalizations (fig. 15.6): people strive to get resources (and are quite clever at that), but how much impact they have on a species depends on their population size or density, the effectiveness of their extraction technology, and whether there is anything to be gained from further hunting or extraction (the marginal value of continued hunting, for example). For example, when steel axes were introduced in New Guinea, which had a market available for wood exports, serious ecological degradation followed. On the other hand, among the Ye'kwana, enhanced technology increased hunting efficiency but did not increase exploitation: meat could neither be stored nor traded regardless of the presence of a market economy.[37]

Are there general patterns? A cross-cultural study of societies in the Standard Cross-Cultural Sample found that in about a third of societies (39/122) for which data existed, severe environmental degradation existed. Degradation was positively correlated with recent population growth and recent technological improvements ($p < 0.0005$ in both cases), although neither was universally present as a driver of degradation.[38]

Our Past and Our Current Resource Use

Our short-sighted, self-centered tendencies, which have served us well in the past, can cause us difficulties today. Today we have far more technological effectiveness than the Piro or the Ache. We can harvest more resources for less human effort.

But efficient technology can sometimes mean, in hunting or fishing or animal husbandry, that we destroy too much "natural capital" (i.e., the biomass of the species we are hunting or fishing, or, in pastoralism, the grasses on which our focal species depends). When that happens recovery is likely to be neither quick nor easy—and sometimes impossible. Some of the problems we face are, like some past extinctions, inadvertent by-products of our "progress" like global climate change; in this case, because the ecological impacts are long-term, solutions will be difficult because, as you will remember, we tend to discount the future, and it might be decades before we know how effective our "fixes" have been).[39]

Here, following our path of human life history and behavioral ecology, I want to give some examples that highlight how our evolved traits interact with the life history and behavioral ecology of other species—how these interactions render some species easy to manage or conserve, and others very difficult. I'll use three fisheries: lobstering, sea urchin harvesting for roe, and cod fishing (many other groundfish tell the same story).

The Behavioral Ecology of Humans
and Other Species Interact

Recently a colleague of mine stopped his lunch to talk intensely with three local lobstermen and their friends, who had arrived in a 36-footer at his dock in Maine. "Come with us?" they asked. A shotgun was highly visible behind the windshield. It seems that many of their lobster traps, which they checked in the early mornings, were empty. One lobsterman, whom no one saw in the morning, nonetheless had plenty of lobster to take to market. When asked, he said he had just been lucky, checking his traps in the afternoon. Needless to say, the other lobstermen were suspicious. They watched him from shore for a couple of weeks until they were convinced that he was hauling some of their traps. My colleague joined the three lobstermen and cruised until they found the "afternoon" lobsterman, with whom they struck up a conversation, sitting in their boats and chatting. No threats (except the visible shotgun) were exchanged, and there was no mention of the oddities of his success and their failure. Just a chat in which the empty traps were discussed and the afternoon guy was asked to keep an eye out for whoever might be hauling their traps. After that, the lobsterman checked his traps in the morning, like everyone else, and the empty-trap problem disappeared.

Why would such a strategy, one of no direct threat but simply an acknowledgment that others knew one fisher was cheating, work? Lobstering in Maine is known for the "lobster gangs," groups of senior, experienced lobstermen whose opinions and approval count greatly in the small, stable communities of lobstermen.[40]

Conversely, why would such a strategy not work among, say, cod fishermen? The answer lies, as most conservationists and managers suspect, in the interplay between the life history and behavioral ecology of the species we hunt, husband, and fish, and our own. Sometimes, getting along with your competitors matters more than the market profit from cheating. Here, I explore when these things are true—when we can hope to find or shape communities and situations in which resource extraction can be sustainable.

Human Consuming Responds to Costs and Benefits

It takes (and always has taken) skill and training to get resources from the environment, whether hunting, gathering, fishing, or farming. Because environmental conditions vary, so will the various costs and benefits. In some ways, fish are the most difficult organism to manage and conserve. It is hard to measure with accuracy either the numbers or biomass of a species of interest. For long-lived ("slow" life history) species, the age structure of a population matters in predicting its future—but this, too, is difficult to estimate accurately.

However, not only these difficulties matter. Other factors affect how fishermen act—how they cooperate (or do not). This is a collective action problem, no different in its basics from the collective action problems of traditional societies of earlier chapters—but at a larger scale. The situations in which continued cooperation without cheating are likely involve a set of conditions that, especially in the modern world, are hard to meet.[41] Continued cooperation is successful only if others (in these cases) engage in frequent communications and develop from those communications lasting individual relationships; they form small coalitions based on those relationships, and come to share the same mental model of how things in the fishery work. This means that most successful common property regimes (in which insiders in the group all have access to the resource) tend to be small, have a very stable membership, are often comprised of related individuals, and (if embedded in a larger group) have the right to manage themselves. In the three fisheries I examine here, these factors vary—and so do the conservation outcomes.

The three fisheries, all based in Maine, are lobstering, sea urchin fishing, and ground fishing (such as cod). The ecologies of the hunted species, however, differ in ways that make cooperation differentially easy or hard for fishermen.[42]

Lobster. American lobsters (*Homarus americanus*) are found in coastal waters, typically on the coastal shelf along the Atlantic coast of North American from Newfoundland to Virginia. Most lobsters are trapped. Most lobsters, especially younger ones, are sedentary; when retrapped, they are usually caught within two miles of their release site. Lobstering success has a seasonal component to it. In late spring and summer, lobsters move into shallow water to molt; this means that lobstermen must crowd their traps into a relatively small area. Fall is the most productive time of year, when lobsters move back into deeper water (120–180 feet), and catches can be high. In winter, many fishers stop for the year. The remaining fishers catch lobsters in even deeper, muddier-bottomed areas where water temperatures are higher than in shallow water; even though the weather can be terrible, the competition is light and those fishers who remain do very well. Not only seasonal lobster movements, but informal territorial rules and formal boundary lines determine where lobstermen set their traps.

The near-shore, summer-fall lobster fishery is highly territorial, perhaps in large part because of the biology of the lobsters. To become a lobsterman, a person must be accepted by those already lobstering in one harbor. The rules are set by the established and experienced lobstermen, who are the lobster gangs. Once accepted, a lobsterman works only within this group's territory (averaging about 100 square miles); this means that a small group of men interact repeatedly, crisscrossing the same area. They come to know each other well. Outsiders are barred from trapping in the area.

Because of these factors, Maine lobstermen have had extraordinary success solving collective-action problems: agreeing to limit things such as the number of lobstermen allowed, the number of traps each can use, and the number of boats. They put strong social pressure on fishermen who stretch or break the rules, as in my colleague's experience above, and as a result can employ harvesting rules that are impossible in other fisheries; here, to be able to continue lobstering, one must follow the group's rules.

In one unusual problem that couldn't be solved locally, Maine lobstermen recognized that in the stage during which lobsters migrated, females carried eggs. Among themselves, they agreed to cut a V-notch in the tails of these lobsters and forbid the sale of V-notched lobsters in Maine, to combat several harmful practices. However, these actions could not solve another serious problem: lobstermen from New Hampshire, for example, would line up along the state line to harvest these lobsters. In this case, Maine lobstermen went to the East Coast Fisheries Board, and after negotiation, now, no one harvests or sells V-notched lobsters north of Cape Cod.

Cod and other groundfish. The biology of cod (*Gadus morhua*), is strikingly different from that of lobsters in ways that make cooperation and restraint difficult. Nonetheless, groundfish such as cod have been managed (since 1977) as if each species were a single stock that inhabited the entire Gulf of Maine. Internationally, in the nineteenth and early twentieth centuries, cod were so plentiful that no one could imagine their demise—comments about their abundance are rife.[43] But cod live in "structured metapopulations"—multiple demographically distinct stocks. This means that if a local subpopulation is fished out, neighboring fish do not move in to live and breed.[44] It also means that the value of broad-scale movements of fish, which are openly discussed among fishers, are not valued highly; they are not all that helpful. These fishermen do not interact so closely as the lobstermen. Finally, the developments in both fishing and storage techniques (the "*T*" in *IPAT*) have made it possible to fish successfully further and further from shore, with gigantic trawls or longlines. The result is that in 1992 the Canadian Department of Fisheries and Oceans banned fishing for northern cod (partially lifted in 1997) and the World Wildlife Fund placed cod on the endangered species list, although Norwegians argued that the Barents Sea cod population was healthy (this is where the Lofoten area is; see below).[45]

The area over which cod range is huge and environmentally complex with rough topography and a large tidal range. Experience in a particular locality is important, and fishers are reluctant to share expertise. Even in a new area, a fisher with good (and hard-won) experience will have a good knowledge of what he or she needs to learn and thus have an advantage, something he or she may be loathe to share.

One widely used example of a cod fishery with long-term sustainability is that of Lofoten, Norway. Because of the coastal geography, it works much like the lobster because the cod live in many small bays, and relatively small groups of fishermen work in a bay. Further, because local fishers are involved in decision making (as in Maine), the Lofoten fishery has been sustained.[46] This ecological situation is unusual for groundfish and quite different from the cod/groundfish fisheries of the eastern coast of North America, which have basically become what are known as "open access" resources; it is too difficult to monitor such a huge area and to catch and punish cheaters, so anyone who can get away with it can depredate the fishery—precisely the case on the eastern North American coast.[47]

Sea urchins. Sea urchin (*Strongylocentrotus droebachiensis*) roe is a delicacy in the Japanese market, and the Maine sea urchin roe fishery is a classic boom-bust fishery. Sea urchins are omnivores, but they are most closely associated with certain kelp, and being easy to find, there is little profit at this point for fishers to share information. Sea urchins are quite sedentary, often moving only tens of meters in months. They can detect food from several meters, and they aggregate to feed. Here is the beginning of the difficulties in managing urchins. When urchins occur in low densities, the kelp thrives; but as urchins feed and their numbers increase (both by reproduction and by immigration), the kelp disappears, and the urchins apparently turn to eating diatoms (resulting in an "urchin barren"). If divers or draggers completely remove urchins from such an area, as is common, the kelp and other seaweeds quickly regrow. This new growth provides excellent habitat for crabs that eat small urchins, leading rapidly to a local system "flip." Once a kelp bed reaches this state, there are no known cases of recovery to urchin-fishing levels. So overharvesting is as devastating as expanded cod fishing, and the result has been widespread depletion of urchins.

Management began after 1992, when an urchin-fishing license was required; then a comanagement system was devised. But the conditions of this fishery are so different from that of lobster that these practices have not been as effective as hoped. Urchin harvesters are mobile and heterogeneous—both divers and draggers. Because urchins are sedentary, one might imagine that finding them is simple—and it may have been in the early days, when little knowledge or experience was needed (so there was little need to communicate). The rapid decline of the fishery certainly led to different biological circumstances, but not to any changes in individual and social relationships among fishers. Depletion has led to very patchy distributions of urchins. A fisher may find an urchin patch which, with a little more time to grow, might provide a profitable harvest; perhaps he decides to return in a week—only to find that the urchins were harvested by someone else. The unpredictability of where and how rich the patches are makes information sharing much less profitable than it would be in a somewhat more predictable environment.

These three examples highlight two major facts: the life history and behavioral ecology of an exploited species affects the optimal strategy that humans can use, and the human strategy in turn affects the biology and biogeography of the exploited species. Both these facts have their roots in each species' evolutionary history: in our human case, of striving to be the most resourceful. Although the fish examples may be particularly complex because of the difficulty and the expense of stock assessment: for example, terrestrial mammals, birds, amphibians, and other species frequently present the same problems.

WHAT'S MISSING IN CURRENT STRATEGIES?

Clearly, we humans show remnants of our past in our resource consumption and fertility behavior, but given our great advances in technology (and the fact that we can seldom forsee the inadvertent side effects), I am far from sanguine about a number of current approaches to solving large-scale ecological problems: the contexts of different issues (air pollution and climate change, over-exploitation of other species) are very different, and none of the ideas considers fertility *and* consumption together. Let us examine briefly some current approaches, asking "what's missing?" along the way. I can think of at least five common ones: the "noble ecological savage," reduce fertility, reduce consumption (including changing norms regarding consumption), ecofeminism, and technological fixes.

The Noble Savage

The concept of the noble savage, in the sense of a morally superior human uncorrupted by civilization, was strong throughout the sixteenth to nineteenth centuries. This concept has been expanded from moral to ecological arenas, most notably by Kent Redford. A typical environmental argument runs that we have, in important ways, "lost touch" with ecological constraints as we have developed technological insulation against ecological scarcity—traditional peoples were more conserving and respectful of those resources than we and likelier to be willing to sacrifice personal benefit for the good of the group when conditions demanded it. It's all technology's fault; hunter-gatherers lived in a "balanced and harmonious" state, altering nothing.[48]

But this reasoning backward from impact to intention is fallacious: if population density is low, if extraction technologies are inefficient, and if no good profit is to be had from the resource, impact *will* be low! We can infer conservation intention *only* when the potential for high impact exists, but impact is nonetheless low (fig. 15.6). Further, humans *have*, at least since the Pleistocene, caused population declines and extinction of other species (above); the typical culprits were improved resource extraction technology, increasing population size, and a profitable market—resulting in environmental damage. Often, as Jared Diamond and others (e.g., Collier, *Bottom Billion*) have pointed out, such factors as climate change and hostile neighbors interacted with these. In ancient Greece, human habitation caused the extinction of numerous species; in North America, bison and beaver were driven to the verge of extinction.[49]

Perhaps the most cogent response to the Noble Savage view was offered by Marcos Terena, president of the Union of Indigenous Nations (UNI), in Belem, Belize, in 1990 at an international seminar on ecological problems in Amazonia:

> Why do you white people expect us Indians to agree on how to use our forests? You don't agree among yourselves about how to protect your environment. Neither do we. We are people just like you. Some of us view nature with a great sense of stewardship whereas others must perforce destroy some of it to obtain what they need to eat and pay for expensive medical treatment and legal counsel.[50]

Reduce Fertility

Reducing fertility alone, as we have seen, is an incomplete strategy, especially if falling fertility is accompanied by heightened consumption. And in any case, falling fertility only means that the rate of population growth slows down. Even if fertility rates fell dramatically today, many populations would continue to grow for some time, because their age structure is such that increasing numbers of children will move into reproductive ages for the next twenty or so years. Even if all southern African countries today—instantly—assumed only "replacement" fertility of two children per couple, because the number of ten-year-olds is greater than the number of twenty-five-year-olds, the total number of reproductive women, and population, would increase for some years.

Don't forget: "a nation's wealth" is an excellent reflection of per capita consumption (fig. 15.3B). Many successful family-planning programs around the world have clear messages that "fewer children means wealthier families." Of course, wealthier families consume more resources, and if one asks about the interplay of population and consumption, fewer-but-more-consumptive families do not lead to decreased or stabilized resource consumption.

Reduce Consumption

"Live simply, that others may simply live" reads a bumper sticker on cars around my home. Can restraint by the wealthy promote sustainability? Certainly it's true that the developing nations were right: if we in the developed world used fewer resources per capita, more resources would be available for others. And I have friends and colleagues who work hard at creating lifestyles that have ever-lower consumption.[51] But they are in the minority (for this is a "levels of selection" problem), and their actions have no direct effect on people in developing nations. If one thing is clear from earlier chapters, it is that reduced per capita resource consumption has not been, for humans or other species, a strategy that got an individual's genes into the next generation better than competitors under most conditions. Striving for what is rare and difficult to attain and showing off: these have worked, but they are consumptive.

Like many other reasonable-sounding strategies, this one must become the most common strategy and it must be continually reinvented to work; it is also highly vulnerable to cheating. It is not an evolutionary stable strategy; it is an argument that would only work if everyone acted identically, often against obvious self-interest.

Could we perhaps, using evolutionary insight, substitute nonconsumptive status competition for resource competition? For example, can we shift competition to center on hard-to-get (and thus status-enhancing) ends that are nonetheless less consumptive than many current status icons? I hope so but am not ready to bet my next paycheck on it.

There is, however, ample evidence (chapters 9, 10; the lobstermen above) that rules and norms strongly affect people's behavior. Cultural evolution can result in behaviors that have never been individually profitable in a reproductive sense. Certainly it is worth trying to promote more sustainable behavior through trying to influence norms. The difficulty I see is that human behavior, like that of other species, is sensitive to individual costs and benefits, and the further distant we perceive a problem to be (in time, across individuals, across space), the more we discount; the more difficult it seems to be to convince ourselves to pay now to avoid a future proposed crisis (like the Wimpy cartoon character: "I'd gladly pay you Tuesday for a hamburger today"). Faced with such issues, denying the problem or "manufacturing uncertainty"—claiming that because we cannot *know* the future benefits for paying now—we can argue that we should do nothing. So the conditions under which we can obtain very large-scale cooperation that involves restraint are very limited. Witness debate today on climate change.[52]

Ecofeminism

This proposition is the obverse of Professor Higgins' complaint in chapter 1: the world would fare better if only men were more like women: nurturing, caring, giving. And because intense resource consumption affected men's success more than women's in our past, it is an attractive idea: let women rule the world.

Certainly there are data that men and women differ in their valuation of the environment: men care more about what they can get out of it (including higher salaries); women care more about health in the environment. Further, women's management styles differ from men's in many circumstances; perhaps cooperation to conserve could be better obtained by women than men. And women appear to value healthy ecological conditions more than men and to be more willing to make such trades as lost income for healthier familial environment.[53]

But many of these approaches are almost stereotypic in their portrayal of sex differences. Consider: when males and females (in any species) succeed through different strategies, of course they behave differently. So elephant seal females are maternal and caring (at least to their own offspring), whereas males galumph about, sometimes squashing their own pups (whom they do not recognize) as they pursue further matings. The male loses less than the female if an infant dies. In contrast, as males' and females' successful strategies converge in modern humans (and as women's fertility is delayed and lowered), their behavior is likely to converge.

No matter that my female colleagues and I might like this idea. We look around and see that the Norwegian prime minister Gro Bruntland was a good environmentalist (and so are Michael Pollan and Bill McKibben); Margaret Thatcher was not (and neither was George Bush). This approach, in most of its forms I have encountered so far, is exhortation, not analysis, advocacy rather than logic. There are excellent feminist approaches to the problems of behavior and evolution; if these authors begin to tackle the population-environment dilemma, I would hope for some progress.[54]

Technology to the Rescue

Julian Simon was a major advocate of this position: in all of human history, there has been a correlation between an increased standard of living and increased numbers of people.[55] But what is the cause and what is the effect in this correlation? Perhaps every time we have faced an apparent "crunch," someone has invented something clever to solve it (and did pretty well by doing so)—yet technological improvements can lead to real difficulties, as in the cod case above, for example. I think our evolutionary background predisposes us to be attracted to this argument. It has never, in our entire evolutionary history, paid us to worry much about things too far in the future—we couldn't control the future. Today, we can at least predict some things further in the future than our ancestors could, but nonetheless, as a result of eons of inability to predict or control, today we still tend to use heuristic operating rules much like those of other species: do what feels good (à la Hemingway and Daniel Kahneman's System 1 of heuristic thinking) and what makes us feel successful and admired. In our past, that has led to reproductive success, and to descendants over time.[56]

Why worry? Biologists do worry; we are the Cassandras on this issue, precisely because technological advances may have caused harm as often as they have solved problems. As one of the epigraphs to this chapter notes, no other species has ever grown in numbers without limit. And while we humans are extremely clever, unless we figure out how to circumvent the Second Law of Thermodynamics or to colonize other planets or systems, we too will eventually be limited by the earth's resources.[57]

This is a longer-term view than we have evolved to hold,[58] and I wouldn't want to be foolhardy enough to predict much about the outcome or things like prices of goods along the way. But biologists, both evolutionary and ecological specialists, suspect human populations will eventually be limited: the only questions, they argue, are how pleasant the process and the result will be. Will extinction rates of other species continue to rise? Will human infant mortality rates rise as sanitation declines? Will a "sustainable" human population be able to live at the level enjoyed today in the developed nations, or will Sierra Leone set the standard?

Can New Strategies and Tactics Help?

If the approaches above strike me as incomplete or misguided, what, then, do I think are the components of a successful strategy? Can understanding the issues of earlier chapters be of any use? How can we foster strong normative conservation ethics, if we wish to? Throughout the earlier chapters I have argued that individuals (even nonhumans) act as if they could "calculate" some kinds of costs and benefits arising from their actions; those costs and benefits are current ones, not in the far future, and local, not global; and the costs and benefits are not, and need not be, monetary. Our costs and benefits as a social primate are older than the invention of barter and money, though not older than family structure and reciprocity, for example. We evolved as a highly social species, and sociality is a powerful force; the emergence of cumulative cultural transmission leading to cultural evolution is a force we may have ignored or underestimated in our attempts to encourage ourselves to act sustainably, although scale remains an issue. Information about human decision-making suggests that, faced with complex decisions (as most social and environmental decisions are), we have unconscious biases and that "knowing" overt factual information doesn't help as much as we wish.

The sum of approaches we use in the face of management and conservation problems is still rather limited, and our most-used strategies are arguably not the most effective (see below). *Information* is obviously necessary, but not sufficient. *Teaching*, or inculcation, and exhortation (e.g., to do the right thing) may sometimes be effective, but the data supporting its effectiveness tend to be sparse and anecdotal. *Social incentives*, as I have noted, may have an influence when one's reputation matters more than profit, typically in small, stable-membership groups—a situation not really common in today's Western world, at least. Nonetheless, both of these strategies play on our highly developed cultural transmission. *Economic incentives* of various sorts are the most widely used strategies today, from regulations to tradable permits to voluntary agreements. No one type of solution is applicable in all conflicts; clearly we have a large task, an uphill battle.

There are potentially useful specific examples of these approaches. Because we are such a social species, social incentives and disincentives, rather than monetary rewards, may work in small, stable-membership societies without high-profit options—as they did and do in many traditional societies, and as they do in small-scale local problems today, like household recycling.[59] It may be productive to devise policies that create real, personal incentives to conserve: the more immediate the benefit, the more successful should be the response. Nonetheless, the early dreams of finding solutions that were both effective and pleased all stakeholders have proved, as you must have suspected by now, largely unsuccessful.[60]

There is also a variety of potentially more useful approaches, which may work at various levels. One is involving stakeholders in decision making, both in persuading people to recognize that problems are *shared* problems, and in getting opposing stakeholders to come to terms and agree on small measures (the input variables), then model the outcome; because all have agreed to the inputs to the model, there is pressure to agree on the answer obtained.[61] Market-based approaches (e.g., carbon tax proposal, cap-and-trade agreements) have had varied, less-than-stellar impacts.

Cap-and-trade involves government and market forces. A central government set a limit (cap) on the pollutant emissions permitted. Then very efficient firms whose emissions are below the cap can sell the "right to pollute" to less efficient firms; this lowers overall emissions. It might seem likely at first that reducing greenhouse gas emissions would require international participation because the gases are not geographically contained; such containment has proved almost impossible.

Cap-and-trade legislation has not succeeded on a North American or national level, but there are several regions (e.g., the northeastern United States, California) in which it has been implemented successfully. This is a matter of scale: remember that the more individuals involved and the more heterogeneous they are in either interests and power, the more difficult it is to reach agreement on goals and on methods to achieve those goals. So, while cap-and-trade legislation was not possible at the federal level, some more local approaches have worked and show real promise.[62]

Other government agreements with firms include voluntary agreements to do other things (reduce energy consumption, use efficient light bulbs, etc.); these work better in European countries than in the United States (primarily because the European equivalents of our Environmental Protection Agency are more powerful in enforcing agreements).

AN EVOLUTIONARY BOTTOM LINE: CAN WE USE WHAT WE KNOW?

If, as many scholars agree, the current trajectories of population growth and resource consumption continue, few of the likely outcomes are pleasant.[63] Darwin's "hostile forces" are virtually identical to the Four Horsemen of the Apocalypse. We live in an evolutionarily novel world that we have created: can we manage it?

I wish I had an evolutionary panacea to propose; I know it is customary to make one's last chapter an exhortation: only learn or do what I propose, and life will improve. Many important proposals, like leaving the GDP behind as a measure of well-being and using measures like the United Nations' Human Development Index (which uses health, education, and a variety of measures of quality of life) are excellent—but getting national leaders to agree first, that this is a good idea, and second, even if all agree it is a good idea, *how* to do it, are almost impossible obstacles.[64]

When we are talking about large heterogeneous societies, it is not clear that any small set of strategies, evolutionarily informed or not, will be quickly or widely effective. I do think there is hope in some of the new policy proposals, which play off our concerns for relatedness and prosociality, but there is still great variation among people in different societies. These mean that we cannot simply "multiply up" strategies that have worked in small, stable-membership societies. And the Lisbon Principles are perhaps nobly phrased but rely too heavily on the ease of convincing people to think globally and in the long term.[65]

I have long suspected that there are a strong set of human biases involving the costs and benefits discussed in the earlier chapters and that the strongest strategies likely to work consistently are those that manipulate individual, familial, and reciprocal costs and benefits, using mimics of kinship and our prosociality. And it looks as though at least some policy makers are beginning to think in the same way. A thoughtful paper by Vlades Griskevicius and colleagues recently dissected five important evolved human propensities, laid out the theoretical principles involved, and gave examples of marketing strategies that match—or do *not* match—those propensities.[66] They identify the main themes covered here (table 15.1): self-interest, striving for relative status, social imitation, discounting (they discuss time discounting, but we may also have to deal with discounting of strangers, other species, and more), and "impalpable concerns"—things people cannot see or feel directly (like radon versus excrement). Their list (table 15.1) of mismatched strategies reads like a list of our most common approaches: exhortation to value the world or the distant future, emphasizing that we are well off compared to people in the rest of the world, using statistics rather than trying to evoke visceral responses, and more. I have heard Joseph Stalin quoted (see beginning of chapter 10) as saying "One death is a tragedy; a million is a statistic." We see this every day: thousands are killed in an environmental mishap like Bhopal, but the big media coverage is about a little girl named Jessica who fell down a well.

Griskevicius and colleagues suggest things like appealing to individuals' (and kin's) well-being and creating small, dense, interdependent networks (as the results Ostrom reported, imitating ancestral conditions with face-to-face communication, and like the Maine lobstermen). Encouraging competitiveness toward prosocial goals, identifying and publicly rewarding people who show (at least proximate) self-sacrifice, identifying the benefits for current conditions, and for promoting long-term strategies, and wherever possible, presenting environmental dilemmas in a locally relevant way: how these will affect *you*.

Table 15.1. Many popular current strategies are not well-matched with our evolved tendencies. Here, adapted from Griskevicius et al., are suggestions for better matches.

Ancestral Tendency	Key Theoretical Principles	Examples of Strategies Mismatched with Ancestral Tendencies	Examples of Strategies Matched with Ancestral Tendencies
Self-interest *People prioritize self-interest over group welfare*	Kin selection Reciprocal altruism Indirect reciprocity	• Imploring people to value the group above themselves • Urging self-restraint for the sake of other people or for the sake of the environment • Offering donation to green cause only if people first purchase a product	• Highlight benefits to a person's genetic self-interest • Create small, dense, and interdependent social networks resembling ancestral groups • Threaten reputations and foster group identities • Donate to green cause to create obligation and then ask people to reciprocate in return
Relative status *People are more motivated by relative rather than absolute status*	Costly signaling Competitive altruism	• Urging people to be content with their current level of status • Asking people to behave in ways that lower their status • Emphasizing that some people have more goods than people in distant places	• Encourage competitions on prosocial outcomes (e.g., the most sustainable) • Publicize lists that rank the most prosocial companies, celebrities, or ordinary citizens • Make people who engage in self-sacrificing behaviors easy to identify
Social imitation *People copy unconsciously what others are doing*	Mimicry Cultural evolution Social norms	• Depicting undesirable behaviors as regrettably frequent • Informing people of what the average person is doing	• Depict the high prevalence, or perceived prevalence, of the desired behavior • Use social approval to encourage those above average to continue their behavior
Future discounting *People value the present more than the future*	Life history theory Sexual selection Parental investment	• Calling on people to value needs of future generations as much as those of present generations • Depicting the world as fast paced and ever changing	• Emphasize consequences of nonsustainable behavior for present, not for future, generations • Highlight the stability, predictability, and safety of the world in which people live • Depict that women prefer men who engage in prosocial and sustainable behavior
Impalpable concerns *People disregard problems they cannot see or feel*	Sensory mechanisms Environmental mismatch Biophilia	• Presenting distant environmental problems by showing statistics • Depicting environmental problems by appealing to evolutionarily recent cognitive abilities	• Present local environmental problems to elicit visceral responses • Create visible links between behavior and its immediate environmental consequences

Looking to the Future

There are already small movements in many of these directions; if we can build on these, I think there is more hope for the future than otherwise. I suggested above that we have created these problems by doing what we have evolved to do altogether too well—but that gives us little guidance about what to do next. Can we design better strategies by taking our evolution into account? The most difficult resource-use problems have the following characteristics: *inadequately-known resource base, slow feedback; externalized costs; and use by many, unrelated, noninteracting individuals.* Like the groundfish and urchin examples above, and like various kinds of downstream, downwind pollution, we can't fix our negative impacts if we can't yet perceive them. This is also a problem with evolutionarily novel dangers; most of us are far more averse to mammal excrement than to something like chemical spills we cannot smell or see.[67]

Successful approaches for convincing people to shift their resource use to patterns that are more sustainable must often, I suspect, appeal to people's perceived short-term, familial, and local, interests. If the perceived benefits of conservation can be made to outweigh the perceived costs for people, then conservation strategies are likely to persist and spread. In most nations today, these are likely to involve economic or other incentives that confer immediate or short-term benefits to individuals and/or their families and potential reciprocators.

Notes

CHAPTER 1: INTRODUCTION

1. See, e.g., E. A. Smith 1992a,b, E. A. Smith and Winterhalder 1992, Ridley 1993, 1996, Cronk 1991a, Borgerhoff Mulder 1991. The evolutionary anthropologist E. A. Smith (1999) has an excellent review of three current approaches to human behavior: behavioral ecology, evolutionary psychology, and dual cultural-genetic inheritance.

2. Behavioral ecology and its intellectual relatives: human behavioral ecology is one of a number of quite similar evolutionary approaches to human behavior that have a variety of names—behavioral ecology (e.g., Davies et al. 2012, Krebs and Davies 1993), evolutionary ecology, biosocial science, human ethology, sociobiology, socioecology, evolutionary biological anthropology, and others. Depending on who is reporting, these may vary subtly, or they may be different names for the same thing. Sometimes professionals in these fields have strikingly similar models of how human behavior evolved and functions; often they differ, sometimes wildly and incompatibly. But one common thread seems to be emerging from several of these approaches: very simple rules can generate emergent complex behavior—perhaps even the complex array of human behavior. Pioneering papers on human behavioral ecology: Cronk 1991a, Borgerhoff Mulder 1991, E. A. Smith 1992a,b. Genetic influences: Hamer and Copeland 1998; historicity: Williams 1992a, b.

3. Barber 1988.

4. Heinrich 1989: 44.

5. Heinrich 1989: 302–313.

6. See Grafen 1991: 6 for the definition of the phenotypic gambit, which is common in behavioral ecology; it assumes only two things: a strategy set and a rule for determining the success of a strategy. It implies that the strategies we see are at least as successful as any nonoccurring strategy would be if it occurred in small numbers. Applying the gambit to a particular strategy set and payoff rule is a powerful way of testing the joint hypothesis that the strategy set and payoff function have been correctly identified and that the gambit is true. Endler (1986: chapter 8) has a nice discussion of how this behavioral ecological "equilibrium" approach compares to other views on the relative strength of selection, random events like mutation, and constraints.

7. E.g., Cronk 1995, Cronk and Leech 2013, Cancian 1973, Deutscher 1973, Harpending et al. 1987.

8. These two can be broken down further when it is profitable; see Holekamp and Sherman 1989. For good reviews of the issues of "proximate-ultimate" see Holekamp and Sherman 1989, Sherman and Reeve 1997, and Alcock 2009.

9. Sykes 2002.

10. E.g, Richerson and Christiansen 2013, Henrich et al. 2004, Boyd and Richerson 1985.

11. Hamilton (1971), for example, used a simple one-dimensional model of frogs on the perimeter of a circular pond to illustrate that groups can form as a way to gain individual security; Low et al. (2005) added another dimension and another independent variable (food supply) to add more realism and shed light on the structure of the groups that do form.

Models have been extremely useful in tackling the complexities of group dynamics (e.g., Richerson and Christiansen 2013, Boyd and Richerson 1985: 157–166, Richerson and Boyd 1992). As Richerson and Boyd (1992: 90) put it, "Conceiving of Darwinian models of (substantively) 'un-Darwinian' hypotheses, analyzing simplified versions of them, and considering the implications of the results clarifies complex long-disputed issues such as the possibility of group functions and the role of historical explanations."

Starting with simple assumptions one can use logic and mathematics to produce novel insights from which useful predictions follow. Specific approaches vary: the well-studied Lotka–Volterra predator-prey formal model yields rigid cyclical populations patterns, but requires unrealistically strong assumptions. Alan Rogers (1990, 1991, 1995) and Marcus Feldman and colleagues (1994), for example, use formal mathematical models to study such problems as wealth conservation and the interplay of cultural and genetic factors in sex-ratio selection, but to do so they must often make unrealistic assumptions such as haplodiploidy or single-locus determination of the trait. In contrast, E. O. Wilson, Richard Alexander, and Richard Dawkins construct less formal, verbal models to shed light on how genes affect behaviors. Both approaches are useful. When it come to adding complexity to formal and informal models—such as heterogeneities and adaptations to feedback—modelers have recently introduced "complex systems" techniques to model these complexities formally. Such techniques include cellular automata (e.g., Wolfram 2002), agent-based models (e.g., Epstein and Axtel 1996), networks (e.g., Newman 2010), and genetic algorithms (e.g., Axelrod and Hamilton 1981 and Holland 1995).

12. Mitochondrial Eve: Sykes 2002. Bull story retold by Alexander (pers. comm., 1988).

13. Selection, current utility, and historical events: e.g., Lenski and Travisano 1994, Travisano et al. 1995, and Travisano and Lenski 1996. By now, Lenski's lab has explored selection in *Escherichia coli* for more than 50,000 generations (e.g., see Elena and Lenski 2003). See also Rose and Lauder 1996. We are learning about the genetics of more human traits (including behaviors) daily, through the work of behavioral geneticists; see, e.g., Hamer and Copeland 1998 for an overview.

14. A caveat belongs here: the phenotypic gambit works best in relatively stable selective environments, when there has been time for strategies to appear, compete, and come to equilibrium—something we can not be sanguine about in human evolution. Even being born with XX or XY chromosomes does not absolutely guarantee that one will grow up to be a typical woman or man. But simply assuming that the relevant portions of the environment have changed, without testing, seems illogical.

15. What we know comes mainly from two sources: studies of twins and identification of abnormalities linked to single-locus genetic changes (e.g., Segal 1999). Mostly we describe the relevant features of an environment and can see that some strategies should work better than others in that environment. This is Alan Grafen's phenotypic gambit (see above): we examine the evolutionary basis of a character (phenotypic trait) as if a

simple genetic system controlled it, as if there were simple "payoff" rules for how many offspring each allele would generate, and as if enough time had passed and mutations had occurred to allow alternative strategies the opportunity to invade. This gambit obviously makes assumptions—that the strategies we see are the successful ones in the particular environment, and at least as successful as any nonoccurring strategy. Even this argument involves a leap of faith that can lack justification (Grafen 1991: 7).

16. E.g., Richerson and Boyd 1992, Richerson and Christiansen 2013. This is a simple, parsimonious argument, an application of Occam's razor using the phenotypic gambit: let us begin with the simplest hypothesis and add complications to it only when it no longer makes accurate predictions. We can make predictions and test them as we wait for the genetics to be sorted out.

17. The groundbreaking summaries are by George Williams 1966, 1992a, W. D. Hamilton 1964, and Robert Trivers 1972.

18. "Society" is not even a peculiarly human phenomenon. We may have more complex social rules than other species, but other organisms exhibit complex and multilayered behaviors, such as alliance formation in intragroup and intergroup conflict (e.g., Connor 2010; older classic discussions are found in Bonner 1980 and de Waal 1982), and can have reciprocal and rather complex "political" behavior. Humans have developed many such phenomena to new extremes, but it's rare to find truly unique human phenomena with no predecessors in other species. It is often hard to find a point at which to draw the line. Other species, for example, exhibit territoriality; we go further and define property rights, in which the holdings of absent owners are supported and protected by third parties. Our task is to find out when the *functional* significance changes—when does the actual selective reason for the behavior change?

19. Here is an important contribution of such approaches. If, as in traditional economics, we must "assume rational (economic) behavior" and make no other assumptions, we can get into great difficulties. The field of behavioral economics has made strides in behavioral-ecology-like understanding of "nonrational" behavior, and as Kahneman (2011) has recently summarized, our very thought processes can differ with the complexity (cost) of a task and the risk attendant on a mistake. Only if we understand the problem thoroughly, have all the currencies in mind, and people really are largely rational, will we have a hope of coming close to an answer (e.g., see Arthur 1991, 1994, E. A. Smith and Smith 1994, V. Smith 1974). Genetic algorithms and behavioral ecology may more quickly get to testable propositions.

20. See Belovsky 1987. For examples, early attempts to adopt optimal foraging models for hunter-gatherer societies were disappointing. Gary Belovsky pointed out that foraging patterns were optimal, but only if one calculated family, not individual adult, patterns of acquisition and consumption. Thus, ignoring the social complexity of providing for one's children meant that one could not figure out how the system worked. Exciting new work by such evolutionary anthropologists as Kim Hill, Magdalena Hurtado, Michael Gurven, Hillard Kaplan, Jane Lancaster, Rob Boyd, and Peter Richerson promises to enrich this insight considerably.

21. Dobzhansky 1961. Although for obvious reasons we began analyses (as in every beginning biology class) by acting as if one gene coded for one trait and no other trait of the organism mattered, that's limited for all the reasons I want to explore here. Here is a simple example: there is an allele that raises the risk of atherosclerosis (coronary artery disease). And that is the advice physicians give. But is the advice correct? Sometimes,

but not always: the "dangerous" allele, which typically raises the risk of atherosclerosis, in combination with certain other alleles, actually lowers the risk (see Sing et al. 1992, 1995, Templeton 1995). So what seemed simple at first, an apparently one gene-one response condition, actually involves epistatic interaction among genes. For complex systems in particular, I suggest that it is more profitable to leave behind the specifics of coding and concentrate on how the traits interact with the environment.

22. Dawkins 1986: 296. He credits Patrick Bateson with this metaphor. See also Bateson and Martin 2000: 16, 220–221.

23. E.g., Plomin et al. 1990: 81.

24. Plomin et al. 1990: 401.

25. Some evolutionary psychologists, for example, argue that we can only study things *known* to be adaptations: traits we can prove to have evolved specifically in response to a given pressure and for which we can identify the resulting particular structure. Typical arguments also assume adaptations formed in "the EEA," the environment of evolutionary adaptedness (usually assumed to be hunter-gatherer social systems in the Pleistocene, but without much regard for the variation that existed). Thus running away from a fast-approaching truck could not be an adaptation, for there were no trucks in the Pleistocene. See Irons 1998 for a critique. As a behavioral ecologist, I suspect humans are like other wide-ranging species in showing some basic conformity (e.g., favoring kin) but with considerable elaboration and diversity in other traits. In this view, the adaptation would be a more general rule: run away from large, fast-approaching things. See E. A. Smith 1999 for an excellent overview comparing evolutionary approaches to human behavior.

26. Lessells 1991 has a nice discussion of this problem, looking at female beetle size and egg number.

27. For good discussions of the concept of adaptation and/or current utility, see Williams 1992b, Sherman and Reeve 1997, and Seger and Stubblefield 1996.

Chapter 2: Racing the Red Queen

1. Van Valen 1973, Ridley 1993. Others (e.g., Alexander 1987) use the term "arms race" for the evolution and counterevolution of competitors, predators and prey, parasites and hosts, for example.

2. Darwin (1859) 1987, 1871. Darwin's original formulation of the theory of natural selection began with the observation that organisms seemed well suited to their habitats. Even though he wrote, arguably, before the birth of genetics, he said things repeatedly that suggest he was well aware that what we now would call "genetic self-interest" was paramount in creating the "fit" between organism traits and environmental pressures (Cronin 1991). Selfish gene: Dawkins 2006.

3. It has always amazed me that optimization approaches work as well as they do, for "success" for any allele is just producing a better-surviving or better-reproducing organism, not the best possible (see, e.g., Grafen 1991, Kirkpatrick 1996). Rolling about on fitness landscapes is not, by design, a very efficient process (one must always travel "uphill" and so can get stuck on a minor fitness foothill, whenever to get to the Mt. Everest of fitness would require first going downhill). Optimization approaches simply ask: what will work best (what can the highest peak be?) and take no note of closer, lower hills.

4. E.g., Davies et al. 2012, Dawkins 1982, 1986, 2006; Daly and Wilson 1983; Alcock 1998; Trivers 1985; Krebs and Davies 1991, 1997; and specific studies cited therein.

5. Empirical tests of selection and historical accident: Travisano et al. 1995, Elena and Lenski 2003. Lenski's lab has done amazing experiments to disarticulate the forces of adaptation for more than 50,000 generations in *E. coli* (myxo.ccs.msu.edu).

6. This is the crux of reproductive strategies in other species, and perhaps of the "demographic transition" in humans. That is, *ceteris paribus*, more resources → more offspring → more genetic persistence. But if the environment is highly competitive, more resources per offspring may be required for successful offspring. The result is likely to be fewer offspring but greater resource consumption per capita. High consumption, as it relates to persistence, is seldom important in other species, but may become an issue for humans (chapter 15).

7. E.g., Hausfater and Hrdy 1984; chapter 6 and references therein.

8. E.g., Daly and Wilson 1984. An additional complication we face in asking about human behaviors such as infanticide is that the behaviors themselves may be repugnant to some of us. Behavioral ecology has nothing whatever to say, so far as I can tell, about the moral or normative "rightness" or "wrongness" of behaviors; human individuals and societies make those determinations. So, when I observe that infanticide tends to occur in humans and in other species under specific environmental conditions, that does not mean that it is somehow "right" (the naturalistic fallacy), nor does it mean that infanticide is wrong.

9. E.g., nonreproductive helpers at the nest: Woolfenden and Fitzpatrick 1984; sterile honeybee workers: Seeley 1985. Note, however, that in several bees, worker females may lay unfertilized eggs (which become males), thus potentially getting their own genes into the lineage.

10. Darwin (1859) 1987: 236. See Cronin 1991.

11. Hamilton 1964, I and II. Actually $rb > c$ is Grafen's 1991 recasting of Hamilton's formula, which was the behavior that will evolve when $k > 1/r$, where k is the ratio of gain to loss. Grafen's reformulation is more direct and tractable.

12. Haldane 1955. See also Dunbar et al. 1995, S. Johnson and Johnson 1991, and McCullough and Barton 1991 for historical examples of how benefits, costs, and relatedness interact.

13. Notice that we have already begun to talk about different costs and benefits: energetic (calories) and reproductive (genes). Implicit in this and following arguments is the idea that organisms routinely spend calories, or take risks, to get genes. See Seger 1981, Emlen 1995.

14. Remember that this is only true when females mate only once, so all daughters have the same father.

15. Hamilton 1964, II: 19. See Alexander 1974 for a discussion of kin selection and parental manipulation in social insects; see also Grafen 1991 and Dawkins 1979 for a cogent analysis of misunderstandings of kin selection, and most recently, Gardiner et al. 2011 for a thorough debunking of mistakes in the literature about kin selection. Nowak et al. (2010) recently challenged the utility of inclusive fitness using a mathematical model that did not include genetic relatedness. E. O. Wilson, one of the co-authors, argued in a *New York Times* interview with Carl Zimmer (30 August 2010) that the existence of diplo-diploid eusocial species like termites and naked mole rats and the existence of haplo-diploid species that are not eusocial rendered the concept "tattered," although

Hamilton was clear that he was making a *general* argument, not one restricted to haplo-diploid species nor necessarily occurring in all haplodiploid species. Lively debate has followed: see especially Abbot et al. 2011, Boomsma et al. 2011, J. Strassmann et al. 2011, Ferriere and Michod 2011, Herre and Wcislo 2011 (all short communications in Nature); and online see Coyne 2010 (http://whyevolutionistrue.wordpress.com/2010/08/30/a-misguided-attack-on-kin-selection/) and Dawkins 2010 (http://old.richarddawkins.net/articles/508102-a-misguided-attack-on-kin-selection).

16. For an overview see Davies et al. 2012 and Krebs and Davies 1993.

17. As I point out in chapter 9, some authors like to treat kin selection as though it were *kin-group* selection (e.g., D. S. Wilson 1975, 1980, D. S. Wilson and Sober 1994, Reeve 1998, Dugatkin and Reeve 1994, and one might even consider Nowak et al. 2010 as belonging to this category). The most cogent response to this interpretation (as well as several other categories) is that of Gardiner et al. 2011. The functional focus is different, and even, for example, Dugatkin (1997: 44) recognized the utility of separating the approaches for most analyses. Here I agree with biologist George Williams (pers. comm.) that the appropriate interpretation of kin selection is *not* kin-group selection, but "individual selection for the adaptive use of genealogy."

18. E.g., Belding's ground squirrels: Sherman 1977, 1981; sphecid wasps: West-Eberhard 1975, 1978; wolves: Mech 1977; Florida scrub jays: Woolfenden and Fitzpatrick 1984; see Davies et al. 2012.

19. A number of studies (e.g., Knafo and Plomin 2006, Cesarini et al. 2009, Gregory et al. 2009) suggest that 30%–50% of the variation in willingness to help is genetic, and the rest due to environmental factors. Of course, calculating actual relationships can be fraught with difficulty; e.g., Dawkins 1979, Grafen 1984, 1991, Sibley and Curnow 2011. Sibley and Curnow elaborate on Hamilton's model to suggest that the relationship between the fitnesses of altruistic and selfish genotypes vary not only with individual genotype but also with the distribution of phenotypes among siblings.

20. Risk and relatives in crisis: McCullough and Barton 1991, Diamond 1992; conflict and cooperation in Icelandic (and Vikings in other countries) and English history: Dunbar et al. 1995, S. Johnson and R. Johnson 1991.

21. Cultural evolution (e.g., Boyd and Richerson 1988, 1990, 2005, Richerson and Christiansen 2013, Henrich et al. 2004, N. Henrich and Henrich 2007) is a potent force in human evolution. It will arise repeatedly as a force shaping sex differences and issues like gender equity. Although there are other species in which there is intergenerational transmission (e.g., optimality of foraging in ground squirrels) and even a crude sort of inheritance (e.g., acorn woodpeckers and Florida scrub jays), humans are extreme in the degree and influence of cultural transmission and evolution. As Kurzban and Barrett (2012) point out, it is the *cumulative* nature of our cultural transmission that allows ideas to evolve.

22. Davies et al. 2012, Krebs and Davies 1997, Dawkins 2006. For an excellent treatment of selection in a broad way, see Bell 1997a. Readable introductory material includes Alcock 2009, Davies et al. 2012, and Bell 1997b.

23. More in chapters 8 and 15. Both evolutionary psychologists and evolutionary anthropologists are interested in the question; why do we see some traits rather than others today? The psychologists typically use a forward method (Sherman and Reeve 1997), arguing that what we see now is here because it was adaptive in the Environment of Evolutionary Adaptedness (EEA, roughly some hunter-gatherers in the Pleistocene)

(e.g., Barkow et al. 1992). Evolutionary anthropologists, keenly aware of the variation in human ecologies (from the Arctic to tropical rain forests to Australian deserts), tend to use the backward method, asking, what would be useful in this environment as described? and arguing that one might be better off characterizing the adaptively/selectively relevant features of environments (the ARE, or Adaptively Relevant Environment; e.g., Irons 1998), whether current or Pleistocene.

24. Hamilton (1996: 14) does not attribute the anecdote to the bishop's wife. Whether apocryphal or true, this response to ideas linking natural selection and humans is nonetheless typical.

25. Williams (1989) called natural selection "immoral," but it has existed since the beginning of life, while moral systems came along much later; it may, in fact, give us a way to think about ethics—it has nothing whatever to say about what "ought" to be, but it might help us think analytically about systems of ethics. Thus, I wince at the cartoon in which the defendant, before the judge's bench, pleads, "Not guilty, Your Honor, by reason of genetic determinism."

26. Onerousness of invoking selection: Williams 1966.

27. The most frequent reason that animals become rare or endangered today is that their habitat is altered by humans (e.g., Myers and Worm 2003, 2005, Ehrlich and Ehrlich 1981; see also Ehrlich 1997 and chapter 15).

28. Male chimp: Goodall 1986; great tits: Krebs and Davies 1993; gull population density: Kadlec and Drury 1968, Drury 1973, Podolsky 1985; reaction to reward disparities in rhesus monkeys and chimpanzees: Brosnan 2013, Brosnan and de Waal 2003, 2004a,b,c; endowment effect: Brosnan et al. 2012, Flemming et al. 2012; effect of relatedness: Brosnan et al. 2005; corals and climate change: Palumbi 2014.

29. See Gluckman and Hanson 2006, Pollan 2008.

30. "The Pill as a Modern Wonder of the World," *The Economist*, December 25, 1993; sexual access of wealthy men: Pérusse 1993, 1994.

31. The accusation of "tautology" has been ably answered by others, both for the principles of natural selection (Maynard Smith 1969, Stebbins 1977, Alexander 1979, Dawkins 2006), and the concept of fitness (Dawkins 1982: 175–194).

32. Cronin 1991.

Chapter 3: The Ecology of Sex Differences

1. The issue of sexual versus asexual reproduction is itself an interesting and incompletely resolved question: why is sexual reproduction so widespread? For readable classic treatments, see Wuethrich 1998, Barton and Charlesworth 1998, Michod and Levin 1988, Bell 1982. For Hamilton's parasite argument, see Hamilton et al. 1981, Hamilton and Zuk 1982, Seger and Hamilton 1988.

2. Temperature-determined sex: e.g., Bull and Vogt 1979; and climate change: Mitchell and Janzen 2010. Size affecting the sexes differently: Roff 1992, Stearns 1992.

3. E.g., Warner et al. 1975.

4. Mating and parental subsets of reproductive effort: Williams 1966, Low 1978, Alexander and Borgia 1978, E. A. Smith and Winterhalder 1992; definitions of fitness: Dawkins, 1982: 175–194.

5. Questions of why and when there is sex in the first place are broad and well treated by others, e.g., Bell 1982, Ghiselin 1969, Williams 1975, Maynard Smith 1978, Michod and Levin 1988; a fun popular treatment of human sexuality is Diamond 1997.

6. This suggests that the parthenogenesis is a derived condition.

7. Some other species show similar patterns: e.g., some rotifers, aphids; see Williams 1975.

8. Parthenogenetic and hermaphroditic species are relatively rare and found in particular circumstances (see Williams 1975, Ghiselin 1969). Ryan (1998) reviews ways in which sexual selection creates sex differences.

9. This aspect of sexuality is so nearly universal that the critical arguments come from thought experiments and modeling exercises, for insufficient variation remains in the real world to do empirical tests. See, e.g., Parker et al. 1972. Arguments about the evolution of anisogamy have been expanded and elaborated by Bell 1978, 1982, Charlesworth 1978, and Maynard Smith 1978. Additionally, genomic conflict (discussed below: Hurst 1995, 1996; also Partridge and Hurst 1998) will lead to anisogamy. These two phenomena almost certainly both act; these are not true alternative hypotheses, in which only one can be true.

10. See Bell (1982: 21–23) on biologists' use of the terms "gender" and "sex." Sex is a genetic phenomenon; gender is a social or cultural trait. See chapter 4 on alternate sexualities in humans.

11. Sometimes the two sexes are physically so different that early zoologists, concerned with describing species, described males and females (of what we now know are the same species) as different species.

12. Variation in mating in dunnocks: Davies 1992. Human evolution today: we seem to be so slightly constrained by our environment that we seldom imagine that past constraints can make much difference to us; for example, a *New York Times* headline (March 14, 1995) touted, "Evolution of Humans May at Last Be Faltering" (see Stearns et al. 2012, also Pickrell et al. 2009, who discuss current hominid evolution). Our discussions of sex differences often begin with assumptions that observed differences are, by and large, "simply cultural." Here I hope to challenge that assumption.

13. E.g., see Ghiselin 1969.

14. Some species have more than two sexes—like the multi-mating-type slime molds. Laurence Hurst (1991, 1992) predicted the critical difference we find.

15. The green alga *Chlamydomonas* engages in a spectacular war of attrition between the cytoplasmic materials of its *plus* and *minus* (rather than *male* or *female*) gametes that wipes out 95% of the chloroplasts (cytoplasmic instruction carriers). And it is not an even destruction: only some of the *plus* parent's chloroplasts are destroyed, but *all* of the *minus* parent's are.

16. See Hoekstra 1987 and Hurst and Hamilton 1992. Ridley (1993: 96–97) has a lively and readable account.

17. E.g., Low 1979, 1989a.

18. Offspring-specific expenditure is true parental investment (PI): Trivers 1972; parental effort (which may be generalizable) comprises things like reusable nests, watching for predators: Low 1978.

19. Alexander and Borgia 1978.

20. Thus, any extrapolation from seasonal success to lifetime success is risky; Clutton-Brock 1988, 1991; also Clutton-Brock et al. 1986.

21. Elephant seals: Le Boeuf and Reiter 1988; butterflies: Elgar and Pierce 1988. See also Strier 1996 and Mitani et al. 2012, reviewing this problem in New World primates. Variance in male lifetimes: Clutton-Brock 1988: 474.

22. Red deer: Clutton-Brock et al. 1982, 1986; elephant seals: Le Boeuf and Reiter 1988.

23. Clutton-Brock et al. 1986.

24. Gaulin and Fitzgerald 1986.

25. Modern: e.g., Voyer et al. 2007, Ecuyer-Dub and Robert 2004, Tottenham et al. 2003, Gaulin and Hoffman 1988, Maccoby and Jacklin 1974, Moir and Jessel 1991, Silverman and Eals 1992, Silverman and Phillips 1998,; Eals and Silverman 1994, James and Kimura 1997, McBurney et al. 1997; see also Hyde 1996. Traditional: e.g., Cashdan et al. 2012, Draper and Cashdan 1988. The general pattern of sex differences suggests a history of sexual selection; because differences arise at puberty, sex hormones seem likely to be the important proximate mediators, both organizational (e.g., C. Williams and Meck 1991) and activational (e.g., Hampson and Kimura 1988, Silverman and Phillips 1998).

26. Orians 1969, Emlen and Oring 1977, Borgia 1979, Davies 1991.

27. Elephant seals: Le Boeuf and Reiter 1988; red deer: Clutton-Brock et al. 1982.

28. See, e.g., Buss 1985, 1989, 1994, Buss and Dedden 1990. It is worth noting that Buss found that among the characteristics women seek in men across at least 37 cultures today, resource holding and resource potential were always near the top of the list. This is also why parents try to bias resource control toward their own children (chapter 2), and why, especially in polygynous systems, sons are likely to be given higher inheritances than daughters (e.g., Cowlishaw and Mace 1996, Hartung 1982).

29. E.g., Betzig 1986.

30. Rattray 1923: 88–89. Laura Betzig (1986) gives other equally dramatic examples, as well as analysis.

31. In some primates, this behavior continues as the male expends some parental effort as well as mating effort; Smuts (1985) has called these more complex and lasting relationships "friendships."

32. Burley (1979) suggested that natural selection conceals ovulation to counter "a human or prehuman conscious tendency among females to avoid conception through intercourse near ovulation." Alexander (1979: 135) made a suggestion similar to Burley's about the concealment of ovulation from the female herself. Hrdy (1981) suggests that concealment allows for "paternity confusion"; in primates in which females are receptive for a period longer than the ovum is fertilizable, and in which males may give some paternal care to infants, female receptivity-without-certain-fertility might allow females to garner care for infants from more males than only the one who is the father. See also Goetz and Shakelford 2009. Alexander and Noonan (1979) argue that concealed ovulation may have enabled females to force desirable males into consort relationships long enough to preclude the male's access to other matings and simultaneously raised his confidence of paternity, making paternal investment more profitable. B. Strassmann (1991) made the further point that concealment of ovulation might favor subordinate males, who are inferior in direct physical competition for females but willing and able with regard to parental effort. More recently, it appears that pheromonal communication may mean that ovulation is not so hidden as we have thought (e.g., Grammer et al. 2005).

33. G. Brown et al. 2009, Low 2003, 2005, 2007, 2009.

34. Höglund and Alatalo 1995, Andersson 1994.

35. Result of skewed female choice: Fiske et al. 1998. Lek paradox: Taylor and Williams 1982, Williams 1992b.

36. Parasites: Hamilton and Zuk 1982, Ebert and Hamilton 1996. Gene-environment interaction: Greenfield et al. 2012; movement of smaller males: Bro-Jørgensen 2011.

37. As Daly and Wilson (1985, 1987, 1988) note, spousal and child abuse cases are in some cases appropriately viewed as violent mate-guarding techniques by males with little to offer.

38. E.g., Murdock 1967, 1981.

39. Darwin 1871, 1: 319.

40. "Expenditure" was not defined by Fisher (1958: 159), who first used the term "parental expenditure" and presumably meant caloric cost; Low 1978 defined parental "effort" as the sum of parental "investment" (Trivers 1972) over time.

41. Roff 1992, Stearns 1992.

42. E.g., S. Frank 1986. Also see the Mukogodo example in chapter 7.

43. I want to be clear here that I am not arguing that male and female humans "must" do anything or that only "genetically determined" mechanisms dictate behavior. I am talking about the relative ecological costs and benefits of different behaviors. As I will argue repeatedly, genes and environment interact, and no attempt can succeed that either ignores the genetic components or seeks to explain behavior simply as a series of genetically programmed events. We know that members of families are more alike than strangers, and the wealth of twins-raised-apart studies (e.g., reviewed by L. Wright 1997, Segal 1999) make it clear that many very subtle traits have strong genetic components. But we also know that intrafamilial correlations of traits such as cognitive ability differ among populations, and there are strong sex-by-generation and ethnic group-by-generation interactions (DeFries et al. 1982). That is, intrafamilial similarities arise both through genetic and environmental (family environment) influences and are further modified by external (cultural environmental) factors. Yet if we have a grasp of the commonalities and variations among mammals in general, we may reach a better understanding of the bias that exists. Then, if we wish to follow patterns different from our evolutionary history, we are better equipped to do so.

44. These patterns have different impacts on mating systems. Continuous two-parent care inclines to monogamy (geese); alternate "relief" care, as in sandpipers, makes multiple-clutch monogamy, and polygynous, polyandrous, or polygynandrous desertion and remating. See Davies 1992, Emlen 1982a,b, 1995, 1997.

45. Social monogamy is common; genetic monogamy is much rarer (Boesch and Reichard 2003). Biologically certain paternity is rarer than we thought (see Morrell 1998). Male confidence of paternity and male parental care: Alexander 1974, Hartung 1983, 1997, Kurland 1979, Kurland and Gaulin 1984. Mating competition as shaping male paternal investment in birds: Birkhead and Møller 1992, Davies 1992, Møller 1994; in mammals: Smuts 1985, Whitten 1987, Smuts and Gubernick 1992; male infant care as mating effort: Smuts et al. 1987, Mitani et al. 2012; among human foragers: Hawkes 1990, 1993, Hawkes et al. 1995, Hawkes, O'Connell, and Blurton Jones 1997. Frequency-dependence of trade-offs: Hawkes et al. 1995.

46. Social scientists use polygyny differently for human marriage systems; see chapter 4.

47. It is worth noting, however, that complexity in human mating/marriage systems can complicate the issue. Serial social monogamy usually reflects genetic polygyny, but as

Borgerhoff Mulder's work (2009, 2010) with the Pimbwe of western Tanzania has shown, it might in some cases work to women's competitive advantage, and even may sometimes be genetically polyandry rather than polygyny.

48. See Boesch and Reichart 2003. It is important to note that genetic monogamy is associated with equal—but *not necessarily low*—variance in reproductive success (Clutton-Brock 1983). See figure 3.3.

49. See also G. Brown et al. 2009, Low 2003, 2005, 2007, 2009.

50. Borgerhoff Mulder 2009; also Hrdy 2000, Starkweather and Hames 2012.

51. Intensity: Wade 1979, Wade and Arnold 1980; high variance in monogamy: Clutton-Brock 1983.

CHAPTER 4: SEX, STATUS, AND REPRODUCTION AMONG THE APES

1. Male care of offspring: review by Langergraber 2012.

2. Cover, food, predation: Schülke and Ostner 2012, Fichtel 2012. Bottom line: Ellis 1993, Dewsbury 1982: tables 1, 3, Silk 1987, Cowlishaw and Dunbar 1991, von Rueden et al. 2011. Tradeoffs: Ellis 1993, Packer et al. 1995. There is an excellent review by Muller and Thompson 2012.

3. Alternate male mating strategies: Alberts 2012, Smuts 1985, 1987a,b, de Waal 1986, McGrew et al. 1996.

4. The phenomenon of male mating-season mobility is well documented, but the reproductive consequences are difficult to establish for particular males, since we typically have data on mating but not on actual paternity. Male care of offspring, even those not their own, may sometimes function as mating effort: Muller and Thompson 2012.

Female choice: a caveat is important here. Since female mate choice (e.g., Andersson 1994) can be constrained by male behavior, including sexual coercion (Smuts and Smuts 1993, Clutton-Brock and Parker 1995a,b, Hrdy 1997a,b, Hooks and Green 1993), in a number of primate species (e.g., Watts 1989, 2012) mate *choice* in primates may not be the simple reflection of mate *preference* that we usually assume it to be.

5. Rank and copulations in macaques: D. Hill 1987, Paul et al. 1993; copulations not the issue: Rowell 1988, Fedigan 1983. In fact, in some species, females appear to trade copulations for a male's contribution to group defense. A male is more likely to take risks in defense of a female with whom he has copulated; females in some species actively seek copulations with subordinate males when they are not fertile (e.g., see Kano 1996, Inoue et al. 1991, Shively and Smith 1985). Hrdy (1997a,b) suggests that female solicitation of multiple copulations (either simultaneously or sequentially, depending on the breeding system) is characteristic of prehominid females, and a tactical response to male efforts to control the timing of female reproduction (see also Hooks and Green 1993). For a recent overview across the primates, see chapters in Mitani et al. 2012.

6. Alberts et al. 2003, 2012.

7. E.g., Cowlishaw and Dunbar 1991, Strier 1996, Alberts 2012. Within-group competition can affect social relationships, aggression, dominance relationships, and hierarchies, as well as kin support and likelihood of dispersal (Schülke and Ostner 2012). Studies from the wild measuring paternity and showing a positive rank-success correlation: Pope 1990, deRuiter et al. 1994. Studies of captive populations (e.g., Berard et al. 1993,

Inoue et al. 1991) sometimes fail to find a positive relationship. One potential explanation is that in these captive groups, the intensity of male-male competition is skewed;
the groups are of unusual demographic composition (proportionately more males live
in these groups than live in wild groups). Several relevant papers are found in volume 34
of the journal *Primates* and in Mitani et al. 2012.

It also is true, as will be clear shortly, that the *ways* in which males compete depends
on the social arrangements: monogamy, single-male groups, and multi-male groups.

8. De Waal 1982, 1986, Watts 2012.

9. Power and social power: Schülke and Ostner 2012. Complexity of interactions,
including female-male: Nöe et al. 1980. Male negotiations: de Waal 1986, 1996a,b. De Waal
suggested that, for chimpanzees, these sharings are essentially mechanisms to reduce
aggression, to appease, and to reassure. Males form coalitions that change over time and
show little correlation with social bonds, as measured by association. These coalitions
are reciprocal and shifting: and if a male defects, the coalition fails. These males may be
keeping track of who has helped and who has harmed them: sex, power, and politics
among the apes. No wonder de Waal titled his 1982 book *Chimpanzee Politics*.

10. Interpopulation variation: Takahata et al. 1996; females seeking furtive matings:
Gagneaux et al. 1997. More reproductive access by high-ranking males: Nishida and Hosaka 1996, Tutin 1979, Hasegawa and Hiraiwa-Hasegawa 1983, Nishida 1983; no clear
relationship between rank and reproductive success in one population: Tutin 1979. In
the Gombe population, while the dominant male had the highest reproductive success,
he did not father two infants for whom paternity is known (Morin et al. 1994). In the
Bossoi population, which has had a single male since 1985, one of four infants born
since that time was not fathered by this male (Sugiyama et al. 1993).

11. Diversity among chimpanzee populations: Kappeler 2012, Muller and Thompson 2012, Alberts 2012. Bonobo female alliances: Parish 1996, Watts 2012; rank and
mating, female choice: Kano 1996; mildness of male aggression: Nishida and Hosaka
1996, Kano 1996.

12. See reviews by Alberts 2012, Ross 1998, and Strier 1996 (who also examines the
ecology of mating systems in New World primates).

13. Reviewed by Pusey 2012, Silk 1987, Dewsbury 1982, Fedigan 1983; geladas:
Dunbar and Dunbar 1977; chimpanzee females: Pusey et al. 1997. Pusey's review has
extensive references. See also Boyd and Silk 1997, chapter 7.

14. Pusey 2012.

15. Costs: Packer et al. 1995; variation across populations: Pusey 2012.

16. The most informative review is Pusey 2012.

17. Food as an ecological force sometimes leading to female kin-groups remaining
together (as well as other influences such as predation): Chapman et al. 2012. Food
defense and female bonding: Wrangham 1980, 1987, van Schaik 1989, Barton et al.
1996; predation: Cheney and Wrangham 1987. Low et al. 2005 extended Hamilton's
model to encompass two dimensions and two currencies (safety and food, which were
traded off).

18. Mitani et al. 2012.

19. Excellent overview of all the hypotheses: Boesch and Reichard 2003. Two-parent
care: P. Wright 1984, Kurland and Gaulin 1984; male inability to monopolize multiple
females: Mitani 1984; infanticide protection: van Schaik and Dunbar 1990, van Schaik
1996, Opie et al. 2013. Davies et al. (2012) have an excellent overview. Recently, high-

lighting the difference between genetic and social monogamy, Reichard (2010) documented extra-pair copulations in a "monogamous" gibbon *Hylobates lar*).

20. Lonsdorf and Ross 2012, Terborgh and Goldizen 1985, Goldizen 1987. Mitani et al. (2012) have an excellent overview.

21. But as an interesting aside, though white-handed gibbons are low-density and widely dispersed, there is now evidence of extra-pair copulations in them; Reichard 2010.

22. Group size and structure, and male strategies: Watts 2012, Schülke et al. 2010, Schülke and Ostner 2012, Fernandez-Duque et al. 2012. Body size, testis size, and canine size comparisons: Harvey and Harcourt 1984. Note that large body size may confer additional advantages in species like gorillas and orangutans, in which males coerce females into copulations (Watts 1989, Wrangham and Peterson 1996). Davies et al. (2012) and chapters in Mitani et al. 2012 have a good review.

23. This general statement is true for other primates, but with an important difference: in nonhuman primates, resource distribution typically influences female distribution and group size, and male distributions follow. In humans, biparental care and male control of important resources appear to lead to a situation in which males comprise the primary groupings (e.g., patrilocality is common), and female distribution follows.

24. E.g., Geary 2010, Freedman 1974, Kagan 1981, Monroe and Monroe 1975.

25. For example, females with extra X chromosomes (XXX is most common, but up to XXXXX exist) have an extra Barr body for each X chromosome, and these females are at risk for retardation (Plomin et al. 1990). Males with an extra X (XXY; Kleinfelter's syndrome) are sterile and have very small testes after puberty; they represent about 1% of individuals institutionalized, about 10 times their representation in the general population. Males with an extra Y chromosome (XYY) were for some time thought to be more aggressive than others; this is not true (see review by Plomin et al. 1990), but such males are taller, tend to show delayed language development, have learning problems at school, and may have a slightly raised incidence of mild retardation. So there are a few clear observable sex differences attributable to the sex chromosomes.

26. See Skuse et al. 1997, Henn et al. 1997. The X chromosomes from mother and father are thus "imprinted." One kind of genetic imprinting, reviewed in chapter 6, arises from conflict between alleles from Dad and alleles from Mom (e.g., Haig 1992, 1993). The conflict theory does not explain this imprinting case well (Pagel 1999). Rather, a form of dosage compensation appears to operate: because females have two X chromosomes and males have only one, females can receive twice the dose of gene products found on the X. This could be disruptive to other genes shared by males and females, and so females have evolved to switch off genes on one of their X chromosomes (Iwasa 1998). There can be further complication with regard to the precise ways in which the X chromosome is unusual (e.g., Sybert and McCauley 2004, http://turners.nichd.nih.gov/genetic_fig3.html; accessed May 19, 2014).

If, as seems likely, for whatever reason, selection favors better social skills in females than in males (or, socially clumsy males are not at such a disadvantage as inept females), then that information must be contributed by the paternal X (which always goes to daughters, while a maternal X can contribute to a daughter or son). Thus we see the apparently counterintuitive result that Turner's syndrome daughters, who have only one X, from mother, are socially less adept than normal XX daughters. (see also Thornhill and Burgoyne 1993 for a related effect in mice). The dosage compensation and the conflict theories are appropriate for explaining imprinting in different evolutionary settings.

27. Rats: e.g., Pellis 2002; rhesus: Wallen 1996; humans: Berenbaum et al. 2007.

28. Betzig 1986; Turkmen: Irons 1979a,b; Kipsigis: Borgerhoff Mulder 1987, 1988a,b, 1990, 1995; Ifaluk: Turke and Betzig 1985; Yanomamö: Chagnon 1979, 1982, 1988; Ache: K. Hill and Kaplan 1988a,b, K. Hill and Hurtado 1996; Bushmen: Richard Lee 1979a,b. In another dozen societies reviewed in less detail by J. Hill 1984, resource control enhanced reproductive success in 10 out of 12. Mueller and Mazur (1996, 1998) discuss military men today.

29. Alvard 1994, 1998.

30. Alvard 1994, 1998; see also E. A. Smith and Wishnie 2000, Ruttan and Borgerhoff Mulder 1999, Hunn 1982, and Hames 1991, 2007.

31. See Hawkes 1990, 1991, 1992.

32. Concept: Fisher 1958; uses: Keyfitz 1985: 142–161.

33. Societies with bridewealth: Murdock 1981.

34. You might also say that women with high reproductive value are free to choose men with greater resources, although direct female choice is difficult to demonstrate in many societies, because marriage negotiations are typically done by fathers, uncles, and other relatives (e.g., Whyte 1978). This situation holds not only in marriage markets but also frequently in remarriage. In most societies, widows and divorcees remarry less frequently than their male counterparts, and those women who do remarry are younger (of higher reproductive value) than the women who do not (e.g., Bideau 1980, Åkerman 1981, Cabourdin 1981, Imhof 1981, Bideau and Perrenoud 1981, Corsini 1981, Knodel 1981, Wolf 1981, Glick and Lin 1986, Griffith 1980). Demographers, concerned only with economic aspects of mating markets, have been at a loss to explain this widespread pattern (e.g., Knodel and Lynch 1985), but it appears obvious to behavioral ecologists.

35. Sweden: Low 1991, also chapter 8; Norway: Drake 1969; England: Hughes 1986.

36. Women's status and reproduction: Mealey 1985, Mealey and Mackay 1990, Essock-Vitale and Maguire 1988. Women, resources and politics: Low 1992. See also chapter 12.

37. For the wives of such polygynous men in traditional societies, fertility may be uneven; in a number of societies, second and subsequent wives, for whatever reasons, have fewer children than first wives (e.g., Dorjahn 1958, J. Smith and Kunz 1976, Daly and Wilson 1983, Bean and Minneau 1986, Garenne and van de Walle 1989, Pebley and Mbugua 1989, Sichona 1993). The fertility effects are sometimes small (e.g., Timaeus and Reynar 1998) and sometimes ephemeral in terms of how a woman's fertility correlates with the number of her grandchildren (Josephson 1993). There are many covariants (e.g., women who remain unmarried until their reproductive value is low are unlikely to become first wives and also likely to have few children in their remaining lifetimes). For more on modern utility of resources, see chapter 15.

38. Uneven fertility: Daly and Wilson, 1983: 283; general female reproductive ecology: Low, 1992, 1993a,b. A. Campbell (1995) explores the conditions under which young women do engage in physical aggression, usually triggered by three issues related to reproductive fitness: management of sexual reputation, competition over access to resource-rich men, and protection of heterosexual relationships against rival women.

39. Packer et al. 1995, Hawkes, O'Connell, and Rogers 1997; Lessells and McNamara 2012 show that negotiations between parents (as a result of sexual conflicts) reduce overall care. See also chapter 7.

40. E.g., Geary 2010, Low 2013a, and chapters 1–3. Self-promotion versus derogation of competitors: M. Fisher et al. 2009, Buss and Dedden 1990.

41. Red deer: Clutton-Brock et al. 1986; elephant seals: Le Boeuf and Reiter 1988.

42. This is a database of 186 societies stratified by geographic region, and within region by language group and the specified time periods during which trained anthropologists worked with the society; the idea is to minimize Galton's Problem of autocorrelation of nearby societies. Because of the arrangement, if only a smaller sample can be used, one uses either the odd-numbered, or the even-numbered, societies. For further description, see http://eclectic.ss.uci.edu/~drwhite/worldcul/sccs.html.

43. Barry et al. 1976, Low 1989a.

44. Benenson, Antonellis et al. 2008, Benenson, Hodgson et al. 2008, Kokko and Pulkkinen 2005.

45. Within-family reproductive competition: Mace and Alvergne 2012; conception risk affecting competition and cooperation between attractive men and women: Lucas and Koff 2013; hormonal contraceptive use affecting female competition: Cobey et al. 2013, Roberts et al 2012, 2014.

46. Deaner and Smith 2013, Deaner et al. 2012.

47. VanderLaan et al. 2012.

48. E.g., Whitam 1983. Biologists track genes, social scientists, norms.

49. Studies of identical and fraternal twins: J. Bailey and Pillard 1991; metastudy of twin studies: Hirschberger 2001. Swedish twin study: Långström et al. 2010; maternal uncles and genetic linkage studies: Hamer et al. 1993, Hu et al. 1995.

50. Problems of gender orientation and the difficulty of solution: Bailey 2005, Bailey et al. 2000; intrauterine sexual differentiation: Bao and Swaab 2011; impact of number of older brothers: Blanchard 2001. Epigenetics and birth order effects: Puts et al. 2006, Blanchard and Klassen 1997, Bocklandt et al. 2006.

51. Stages of development: Blanchard et al. 2012; early marriage and delay between marriage and sexual activity: J. Whiting 2009, Kramer 2008; overview: Hames and Blanchard 2012. See also chapter 8: Nedertorneå and Gullholmen.

52. See Wickler and Seibt 1981.

53. Though see Borgerhoff Mulder 2009.

54. Such societies are what Alexander et al. 1979 termed "socially imposed monogamous" systems: socially "monogamous" but biologically polygynous. Today we distinguish genetically monogamous versus socially monogamous (Boesch and Reichart 2003). Male versus female variance in reproductive success in 18 societies is reported by Marlowe 2012; the ratio of male variance to that of females ranges from 0.8 to 4.8.

55. Usually measured as the percentage of men and women polygynously married.

56. Low 1988b, 1990b,d. White and Burton (1988; see critique by Low 1990b) suggested that polygyny is "favored by homogeneous and high-quality environments," perhaps reasoning that rich environments are easier to exploit. As this chapter and others make clear, the real relationships are difficult to determine, because very different climate zones are lumped in their analyses, as if the tropical rain forest were comparable to seasonal, high-rainfall areas. Even brief reflection suggests that it is far too simplistic ecologically to assert that there is a single measure of environmental quality, one that decreases with cold or aridity, so that dry polar regions are lowest on environmental quality and moist tropical regions highest. This not only confounds extremeness, range of variation, and predictability, but ignores, for example, the fact that pathogen stress, hardly a contributor to "high environmental quality"—but a factor promoting polygyny (see above)—is highest in moist, tropical regions, and that protein availability,

probably a requirement for "high" environmental quality, can be quite high in cold and dry regions.

57. This argument is, of course, a major focus for work on the evolution of sexual reproduction itself (see chapter 3). The difficulty lies in defining the sort of uncertainty that will favor the advantages of sex, or variable offspring, to a degree sufficient to compensate for the loss of genetic representation. W. D. Hamilton and his colleagues (Hamilton 1980, Hamilton et al. 1981, Hamilton and Zuk 1982, Ebert and Hamilton 1996) have argued cogently that pathogen stress is one of the few, perhaps the only, environmental uncertainty that will meet the criteria.

58. Polygyny threshold: Orians 1969; ecological correlates of polygyny in humans: Low 1988b, 1990b.

59. See, e.g., Cashdan and Steele 2013, Letendre et al. 2012, Thornhill et al. 2009, 2010 Fincher et al. 2008, Fincher and Thornhill 2012) .

60. E.g., Emlen and Oring 1977, Borgia 1979, Alexander and Borgia 1978. Excellent review by Davies et al. (2012).

61. This is a highly significant relationship ($p = 0.00001$); Flinn and Low 1986.

62. Borgerhoff Mulder's (e.g., 1988a,b, 1995) studies are some of the most detailed; she found that younger (higher reproductive value) women commanded higher bride price (see chapter 7).

63. Kanuri: R. Cohen 1967; Yomut Turkmen: Irons 1975; Tsembaga-Maring: Rappaport 1968. Interestingly, this phenomenon may not be immune from deceit. A. Fisher (1984: 58, 50) shows a picture of a young Dinka man wearing about 17 necklaces, each advertising his family's wealth. However, he can borrow necklaces from family members, in a way misrepresenting his own wealth.

64. E.g., Native American societies of western North American: Jorgensen 1980: 167.

65. Low 1988b, 1990d.

66. Distribution of dowry: Gaulin and Boster 1990. See also Gaulin and Boster 1997.

67. This is particularly interesting because India, like Korea, China, and Japan, has long had skewed birth sex ratios (or infanticides), resulting in far fewer girls than boys having births registered. Very recent data (CIA Factbook, 2013, https://www.cia.gov/library/publications/the-world-factbook/) suggest this may be changing in several of these countries, as potential brides become rarer.

68. Rao 1993a,b. Further, domestic violence and spousal abuse have correlated with these increases in dowry worth; although alcohol is one causative factor, "insufficient dowry" is the second (Rao 1997).

69. Aswad 1971, Barth 1956, Alexander 1979; see review by Flinn and Low 1986.

70. Alexander 1979, but see Flinn 1981.

71. Fathauer 1961, Flinn 1981, Flinn and Low 1986.

72. K. Hill and Hurtado 1996, Nordenskiold 1949; also Davies et al. 2012: 183.

73. E.g., Wade 1979, Wade and Arnold 1980. Also see Betzig 2012, Marlowe 2012, his Table 20.3. These data do not negate the fact that variance in reproductive success can be high for both sexes in genetic monogamy (Clutton-Brock 1983), e.g., with high death rates of offspring—but in monogamy the variance is similar between the sexes.

74. Fig wasps: Hamilton 1979.

75. Bullfrog reproductive success: Howard 1979; conditional strategies: Thornhill and Alcock 1983: 287.

76. Heritable class: Murdock 1967, 1981, White 1988. This can be an issue in demographic and other (e.g., ecological footprint) analyses of modern nations: wealthy graduates of Ivy League schools are not really in the same population as poor, undereducated Detroiters.

77. Cf. Bateman 1948.

78. E.g., Parker 1974.

79. Sunfish: Dominey 1980; ruffs: Rhijn 1973, Högland and Alatalo 1995.

80. Sade 1967, Mitani et al. 2012. This is true for a number of primate species, in which the youngest living daughter inherits her mother's rank upon her mother's death.

81. See also Trail 1985.

82. This measure is a direct descendant of Crow's (1958) widely used intensity of selection measure in which the variance in gain (fitness) is standardized by the square of the mean gains. Professor James Crow devised the following modification, which is simpler than our first attempts. Let x be reproductive success. Its variance is

$$V(x) = \sum_i p_i (\bar{x}_i - \bar{x})^2 + \sum_{ij} p_{ij} (x_{ij} - \bar{x}_i)^2 = V(b) + \sum_i p_i V(w_i),$$

where p_{ij} is the proportion of individuals of behavioral class j in stratum i, p_i is the proportion in stratum i, $\bar{x}_i$ is the mean success in the ith stratum, $V(b)$ is the variance among stratum (e.g., size class) means, and $V(w_i)$ is the behavioral variance within the ith stratum. Then, the index of total opportunity for reproductive gains is

$$I_t = V(b) / \bar{x}^2 + \sum_i p_i V(w_i) / \bar{x}^2$$
$$= I(b) + \sum_i p_i (\bar{x}_i^2 / \bar{x}^2)(V(w_i) / \bar{x}_i^2)$$
$$= I(b) + \sum_i p_i R_i^2 I(w_i)$$

where $I(b) = V(b) / \bar{x}^2$ (the index from stratum mean differences), $I(w_i) = V(w_i) / \bar{x}_i^2$ (the index from differences within the ith stratum), and $R_i = \bar{x}_i / \bar{x}$ (the ratio of average age within-stratum success to overall mean success). When there are only two strata, and reproductive success occurs in only one, this reduces to Crow's formula. If the trait is completely heritable, $I_t / 2$ is the proportion by which mean reproductive success increases through the sex being considered. This procedure highlights the fact that individuals in low-variance strata (e.g., young males in ungulates and primates, small "sneaker" male sunfish) should strive less than those in high-variance strata, for striving has little impact.

Steve Frank (pers. comm.) noted that this set of ideas has been little developed in the literature, although there is a widespread sense that they are logical. Frank (1996) reflected on Haldane's (1932) appendix, in which Haldane considered the conditions under which selection favors high variance rather than high mean response;, and went on to connect Haldane's discussion with models of the evolution of sex, such as Williams (1975).

83. E.g., data summarized by Marlowe 2012, White 1988; see Low 1988b, 1990b,d. Data are available on variance and heritability of resources (extent of stratification) for a number of societies, although individual status-and-success data for stratified societies are rarely published. See also Borgerhoff Mulder et al. 2009 and papers in vol. 51 of *Current Anthropology*.

84. E.g., see Hames 1996 and discussion of the Yanomamö in chapter 7.

85. Costs to women: e.g., Chisholm and Burbank 1991, Strassmann 1997, 1999. Relative fertility of first and other wives: e.g., Dorjahn 1958, J. Smith and Kunz 1976, Daly and Wilson 1983, Bean and Minneau 1986, Garenne and van de Walle 1989, Pebley and Mbugua 1989, Sichona 1993. However, in at least some societies, these differences disappear after a generation: Josephson 1993.

86. Children' survival: Dorjahn 1958; divorce: Betzig 1989, 1996, Strassman 1997, 1999.

87. Yanca and Low 2004.

88. See Borgerhoff Mulder 1992 for review of this and other influences.

89. Remember that biologists define genetic monogamy as a system in which the relative variance in reproductive success is equal between the sexes—ecologically influenced genetic monogamy. Anthropologists and sociologists define a monogamous system as one in which an individual may be married to only one spouse (at a time). If men remarry more often than women, and if they have children from second and subsequent unions more often than women, the system is socially monogamous but genetically polygynous (male success will vary more than female success).

90. Polyandry is most prevalent in Tibet, India, and nearby areas. See Peter 1963, Beall and Goldstein 1981, Tambiah 1966; Crook and Crook 1988; Hiatt 1981; on the fragility of polyandry see Levine and Silk 1997, Haddix 2001.

91. However, see Starkweather and Hames 2012; polyandry may be more common than the standard sources suggest.

92. E. A. Smith (1998b) calculates the fitness effects and analyses Tibetan polyandry as a member-joiner game; he also includes a perceptive analysis of the tangled literature on the functional significance of polyandry.

93. Gorer 1967.

94. Peter 1963: 453.

95. Tambiah 1966.

96. E.g., Beall and Goldstein 1981, Goldstein 1976, Hiatt 1981, Peter 1963, Alexander 1974, Tambiah 1966, Yalman 1967, Durham 1991. Quote: E. A. Smith 1998b.

97. Starkweather and Hames 2012.

Chapter 5: Sex, Resources, Appearance, and Mate Choice

1. Walker et al. 2011.

2. Expensive displays: Zahavi 1975 (overview in Zahavi and Zahavi 1997); genetic dissimilarity and heterozygosity: Lie et al. 2010, Brown 1997; European barn swallows: Møller 1994; grasshopper choice of good foragers: Belovsky et al. 1996; influence of social environment on female grasshoppers' choices: Rebar and Rodriguez 2013; cockroach discrimination: Clark et al. 1997; elephant seal choice of beaches: Le Boeuf and Reiter 1988; red-winged blackbird choice of marshes: Langston et al. 1997.

3. E.g., Buss 1994, 1999.

4. See D. Jones 1995, 1996, D. Jones and Hill 1993; conflicts of interest between the sexes are pervasive, both in ancient and modern environments (e.g., Kenrick et al. 1996, Smuts 1996, Malamuth 1996).

5. R. Smuts 1992.

6. Tsimane male social status: von Rueden et al. 2011.

7. It is clear that natural selection and sexual selection have shaped our mate choice (and perhaps still may) and in turn, our mate choices matter in shaping other traits; for a readable overview see Miller 2000. Intelligence: Prokosch et al. 2009; social function of intelligence: Humphrey 1976; choice of formidable men in dangerous environments: Snyder et al. 2011; masculine traits: Dixson and Brooks 2013, Mueller and Mazur 1996. Ethnographies from the 93 odd-numbered societies of the Standard Cross-Cultural Sample (Murdock and White 1969) turned up more comments made by men assessing women (76) than by women assessing men (19) (but see Pound et al. 2014, *Proc Roy Soc B: Biological Sciences* 20141639). About half of the comments (52/95) were about clear fitness signals such as clear and/or unwrinkled skin, firm breasts, symmetrical features and breasts; 20/95 concerned wealth or skills; 6 concerned traits that are profitable only in specific environments (fat or steatopygy in harsh environments); and 17 comments were recorded about largely or entirely culturally influenced preferences (light skin, long fingers, etc.; this includes the comment about disliking short hair when the practice was to cut hair during illness). In this preliminary sample, direct fitness comments were reported significantly more than other sorts of traits ($n = 95$, $\chi^2 = 49$, $p < 0.005$).

8. The dowry societies (chapter 4) reinforce this point. See Darwin 1871. Marlowe 2012, Trivers 1972, Low 1990b, and Flinn and Low 1986 review the evidence on human mating patterns; Ras (1993a,b) notes a specific dramatic example.

9. Dugatkin 1996, Godin and Dugatkin 1996; human female copying: Place et al. 2010.

10. Prevalence of polygyny: Murdock 1981. Hair and cultural inference: Herrmann and Tomasello 2012.

11. Cross-cultural study of facial attractiveness: D. Jones and Hill 1993. Iban illness and haircut: Roth 1892.

12. Changes in female preference across the cycle: Gangestad and Thornhill 2009, B. Jones et al. 2008); ; masculinity versus intelligence across the menstrual cycle: Gangestad et al. 2010, preferences in scent, Thornhill et al. 2013; however, see Peters et al. 2009. Contraceptive use effects: Roberts et al. 2014.

13. Singh 1993a,b, 1994; Singh and Luis 1995. Yu and Shepard 1998 note one exception. Low RV for women who delay childbearing: Low et al. 2002.

14. Fat and deception: Low et al. 1987, 1988; lactational difficulties: Niefert et al. 1985.

15. Fat preference: review in Low 1990d, also R. Smuts 1992, Darwin 1871, 2: 345; fat and deception: Low et al. 1987, 1988, Low 1990a.

16. Age and reproductive value: Fisher 1958: 27–30; mate choice: Buss 1994, Pérusse 1994; female status and mate choice: Gallup 1982, 1986, Strassman 1991.

17. Men prefer short-term mates more than do women, and the characteristics men seek in short-term mates are more about physical attractiveness and youth than their preferences in long-term mates: see Confer et al. 2010, B. Jones et al. 2007, and overviews by Geary 2010 and Buss and Schmitt 1993.

18. French Canadian study: Pérusse 1994; cross-cultural: Buss 1989, 1994.

19. See review by Puts 2010. Neoteny in female faces: D. Jones 1995, D. Jones and Hill 1992; hips, breasts, and buttocks: Low et al. 1987, 1990a; fluctuating asymmetry: Gangestad et al. 1994, Thornhill and Gangestad 1994, Thornhill et al. 1995.

20. Stress and preference: Lass-Hennemann et al. 2010; women's skepticism about commitment as a life-history strategy: Cyrus et al. 2011; women's preference and memory:

Allan et al. 2012; selflessness: Moore et al. 2013; target specificity of kind, selfless behavior: Lukaszewski and Roney 2010.

21. E.g., Dosmukhambetova and Manstead 2011.

22. Health and mate preferences: DeBruine et al. 2010; religious inculcation: McCullough et al. 2012.

23. Knodel et al. 1997.

24. Fisher 1958: 152.

25. Zahavi 1975, Zahavi and Zahavi 1997. See Pomiankowski 1987a,b for the formal proof.

26. Møller 1994. Not only do females choose longer-tailed males, they make up their minds about such males early: the premating delay is shorter for longer-tailed males than for others. And females prefer males with symmetrical tails, apparently because such symmetry reflects a life history of good health—freedom from disease and parasites. See also Møller 1987, 1989, Andersson 1994.

27. Bower birds: Borgia 1985.

28. E.g., Lewis 1959, 1970.

29. Brown 1997.

30. Low 1979, 1990a, e.

31. Alexander et al. 1979.

32. Whyte 1979, Low 1979.

33. Low 1979, 1990e.

34. Whyte 1978, 1979.

35. Borgerhoff Mulder 1987, 1988a,b, 1990, 1995.

36. Schlegel and Barry 1991.

Chapter 6: Sex, Resources, and Human Lifetimes

1. Jared Diamond (1992) called humans the "third chimpanzee," recognizing both that we face the same ecological problems as other primates, and that we are most closely related to common chimpanzees (*Pan troglodytes*) and pygmy chimpanzees, or bonobos (*Pongo pygmaeus*); see fig. 6.1.

Life histories—the lifelong patterns of maturation, courtship, and reproduction, and death in any species—are the outcome of competing costs and benefits of different activities at any point in the life cycle. Many life history patterns arise from allometric (size-related) relationships. For example, the timing of many life history events (e.g., early or late reproduction) is strongly driven by age-specific mortality patterns—and small organisms die earlier than large ones; they simply tend to be more at risk (e.g., Charlesworth 1980, Charnov 1991, 1993, Stearns 1992, Roff 1992, 2001). One can compare the "relative value" of a number of life history traits in the primates, including humans, controlling for these allometric effects (Harvey et al. 1987). The relative value describes the variable after the effects of size have been removed, so that we can ask: Does this trait follow the pattern we would expect from size alone? If not, what other factors are important? Gurven 2012 has an excellent overview of human life history.

2. Charnov 1993, Roff 1992, 2001, Stearns 1992.

3. See Mace 2000, Gurven 2012.

4. Maternal energy budgets and energy transfers: Kramer and Ellison 2010; resources and maternal decision-making: Krushnick 2013; intrasexual selection: Clutton-Brock 2009.

5. Altricial descriptions: e.g., Ricklefs 1983, Nice 1962, Case 1978; altricial growth rates: Ricklefs 1983; possible advantages of altriciality: Ricklefs 1983, Dienske 1986, Alexander 1990.

6. See Isler and van Schaik 2014. Older work (e.g., Humphrey 1976, Alexander 1990) have argued that large brains have been favored by the importance of intelligence in social evolution—that the uncertainties and risks of social life require more complex intellectual skills than, say, simply avoiding predation and finding food (e.g., chapter 9). Certainly, human brains are relatively expensive metabolically; in a resting human, brain tissue consumes 20% of the energy budget (compared to 9% for a resting chimpanzee or 2% for a typical marsupial).

7. Growth constraints: Martin 1983; metabolic expense: Hofman 1983, review by B. Smith 1990, 1992; postnatal growth rate: Martin 1983, Jolly 1985, Harvey et al. 1987.

8. Also see J. Jones 2009, 2011.

9. General prediction of early maturation: Roff 1992: 347; subfecundity: Lancaster 1986, Kramer 2008. See also Hawkes O'Connell, and Blurton Jones 1997, who, using Charnov's 1993 dimensionless comparative approach, predict later maturity for humans.

10. E.g., Gray 1983.

11. Hunter-gatherer interbirth intervals: Harvey et al. 1987; Sweden: Low 1991; Germany: Knodel 1988: 322; chimpanzees: Harvey et al. 1987; dental maturation: B. Smith 1991, 1992; death risk before sexual maturation: Stearns 1992, Roff 1992, Charnov 1991, Promislow and Harvey 1990; allocare and short interbirth intervals: Gurven 2012. B. Smith 1991, 1992 notes other oddities in the pattern of human developmental life history events, compared to other primates. Comparing *Pan*, *Australopithecus*, and *Homo* (from *H. habilis* to modern humans), the length of infancy to the eruption of the first permanent tooth has increased by 3 years, the time to the last permanent tooth has increased by 10 years, and life span has increased by 30 years. But we are weaned younger than would be expected from size alone.

12. Hawkes, O'Connell, and Blurton Jones 1997, Hawkes, O'Connell, et al. 1998, 1999.

13. See Richerson and Christiansen 2013, Dennett 1991, Alexander 1979, 1987, Cavalli-Svorza and Feldman 1981, Lumsden and Wilson 1981, Boyd and Richerson 1985, 1992, 1996, Richerson and Boyd 1992, Durham 1991.

14. Parent-offspring conflict: Trivers 1974; romantic conceptions: Hrdy 2009, Haig 1993.

15. E.g., Haig 1992, 2004, Moore and Haig 1991; see also Hurst and McVean 1997. Burt and Trivers 2006 have a nice summary, even identifying specific allelic effects. "Genomic imprinting," the term for this maternal-paternal conflict of interest within the fetus, may even affect brain (especially cortex) patterns in mammalian evolution (Keverne et al. 1996). There are several theories of genomic imprinting, and while they share commonalities they are not identical; see Hurst and McVean 1997.

16. E.g., Haig 1992, 1993, 2002, 2004 and Hurst (1992) argue that paternal genes in the fetus may function to manipulate maternal physiology for the fetus's benefit even if serious cost to the mother may result. Hrdy (1999, 2009) has explored maternal trade-offs clearly, and Burt and Trivers (2006) have reviewed genetic conflicts.

17. Gendered play: Hines 2010; vervets: G. Alexander and Hines 2002; personality and cognition: Mathews et al. 2009; brain/sex differences: Blum 1997, J. Becker et al. 2008.

18. Continued differentiation: De Bellis et al. 2001; boys' activity as discipline problems: e.g., Freedman 1974, 1980; Meta-analysis of gender differences in scholastic achievement: Voyer and Voyer 2014. Gender gaps: Halpern 2006, Halpern and LaMay 2000.

19. Eliminating questions on IQ tests that favor one or the other sex: Halpern and LaMay 2000; variance in boys' vocabulary scores: Halpern 2006; post-pubertal boys and mathematics: Halpern 2006, Halpern and LaMay 2000; the sexes converging in their performance: Spelke 2005.

20. Maccoby 2002, Matthews et al. 2009, Chapman et al. 2006; brain structure: Hines 2009; sexual orientation and gender identity: Hines et al. 2004, Meyer-Bahlburg et al. 2008; overviews by Hines 2011 and Chung and Auger 2013; neuropsychiatric disorders: Bao and Swaab 2011.

21. Trivers 1972.

22. Lawson and Mace 2009. See also Lawson et al. 2012 for the trade-offs between fertility and infant survival.

23. Maternal physiological responses: Peacock 1990, 1991; nutrition in traditional societies: R. C. Bailey et al. 1992; high-status women: Kasarda et al. 1986. Even in some proto-industrial societies, in which women might, for example, weave rugs at home for sale, there remains a conflict in investment in childcare versus rug weaving. See chapter 8 for more.

24. !Kung IBIs: Blurton Jones and Sibley 1978; backload model: Blurton Jones 1986, 1987a, 1989. !Kung patterns in compounds: Pennington and Harpending 1988. See also Blurton Jones 1997. Anderies (1996) extended the model; inclusion of maternal mortality further predicts optimal ages at first and last birth.

25. See Downhower and Charnov 1998, and chapter 15. See chapter 15 for more about modern ages at first birth and life expectancy.

26. E.g., see Kramer and Ellison 2010, Krushnick 2013, and the section on cooperative breeding, which may help defray costs in many cases. Kramer and Ellison examined life history traits and energy allocation (cf. our unusual life history aspects as a primate) and show that we need to understand more about energy transfers and allocation to explain life history features. Krushnick showed that among the Indonesian Batak, access to resources shaped maternal decisions.

27. Ache and Hiwi: Hurtado et al. 1985, Hurtado et al. 1992; Ye'kwana: Hames 1988a. In the Ache, an average of 1.3 adults may help in provisioning.

28. Turke 1988. There are, however, some data that suggest that peer-raised children are socialized differently than children raised primarily by their mother (Whiting et al. 1992).

29. Hrdy 1992. See also Hrdy 2009 and Lawson et al. 2012.

30. Like the high-status baboons in chapter 4, these high-status women experienced unanticipated costs. Here, the very-high-fertility women "endured a range of problems ranging from chronic anemia to prolapsed uteruses" (Hrdy 1992).

31. Harem takeovers in langurs: Hrdy 1974, 1978, 1979; in lions: Packer and Pusey 1982, 1983, 1984; in gorillas: Watts 1989 (also an overview in Watts 2012). See also Struhsaker and Leland 1987: table 8.4. Treatment of stepchildren: Daly and Wilson 1984, 1985, 1987; Anderson et al. 1999, 2007. Paying child support is correlated with reduced remarriage and lower subsequent fertility for men (Anderson 2011). Stepfathers, like langurs taking

over a harem, reduce not only their costs for a nonrelative but also reduce the potentially competitive investment by the mother; and women may be less able to prevent a spouse's abuse than a man. Good investment in adopted children: Gibson 2009. For an accurate and readable summary of these topics and more, see Gray and Anderson 2010.

32. E.g. Hughes 1988, Daly and Wilson 1988. C. Hill and Ball (1996) suggest that cultural "ill omens," as well as biology cues, may be used. Note that "adaptive" means here only "reproductively profitable in these circumstances," not "an evolved adaptation." We can measure something about the former, but the latter I think would be a difficult question.

33. Overview: Daly and Wilson 1984, 1988, Hrdy 1992. Daly and Wilson (1984) compared 60 cultures in the Human Relations Area files to identify causal circumstances; Hrdy (1992: table 1) compared infanticide rates in a set of traditional societies for which there are relatively good data. Ill omens: C. Hill and Ball 1996; maternal circumstances: Bugos and McCarthy 1984.

34. France: Fuchs 1984; Spain: Sherwood 1988; Russia: Ransel 1988.

35. Boswell 1990.

36. Abortion patterns: E. Hill and Low 1991, Torres and Forrest 1988; attitudes: Betzig and Lombardo 1991. Political shifts in the United States recently appear to have made abortions less available: Eckholm 2014. More abortion restrictions were enacted in the United States in 2011–2013 than in the entire previous decade; states are listed in Guttmacher Institute 2014: www.guttmacher.org/pubs/fb_induced_abortion.html (accessed May 21, 2014).

37. Women's attitudes: Betzig and Lombardo 1991; "issue evolution" in U.S. politics: Adams 1997. Teen rates: http://www.icontact-archive.com/s5XvpXoL5HYHElU3Q3k_wq5BevxKXDv6?w=4 (accessed 21 May 2014).

38. General model: Sibley and Calow 1986; human model: E. Hill and Low 1991.

39. Kelly and Alonzo 2009.

40. Fisher 1958.

41. E.g., Whyte 1978, 1979, Low 1990e, 1992, 1994a.

42. E.g., Chagnon 1979, 1982, 1988, 1997; Irons 1975, 1979a,b, 1980, 1983a,b. Coalitions in other male and female mammals also follow these patterns (chapters 11, 12; Mitani et al. 2012). Some of these differences arise simply from being mammals, captive by the differences in mating-effort versus parental-effort return curves. Other things being equal, male mammals achieve maximum reproductive success through expending their reproductive effort as mating, rather than parental, effort and by expending generalizable parental effort rather than true offspring-specific parental investment. Female mammals, equipped to nurse their young, do best by producing healthy, viable offspring, optimally apportioning effort to specific offspring (fig. 3.2, chapter 3).

43. Trivers and Willard 1973. See also Charnov 1982, Leimar 1996.

44. That is, young men have usually only the promise of resources, whereas young women have current reproductive value; successful older men have real current resources whereas older women have declining reproductive value.

45. Low 1989a.

46. Fisher 1958: 159.

47. It does not explain, however, why, for example, red deer females in excellent condition may have sons than daughters, as Clutton-Brock (1983, 1988) and Clutton-Brock et al. (1982, 1986) have shown, while less well-off hinds are likelier to have daughters (chapter 3). I think it is a "packaging" problem.

48. Exceptions include son-preference societies such as Korea, China, Japan, and India, in which female infanticide may occur so early that records may not be trustworthy.

49. Trivers and Willard 1973, Leimar 1996. Sons are usually more expensive for mammalian mothers to carry and rear than are females, so conception and birth sex ratios should be biased toward the more valuable sex. If local resource competition and maternal status affect resource access, we should see variation in birth sex ratios sometimes with maternal status (as in red deer and some baboons). G. Brown and Silk (2002) tested this in primates, and concluded that data are insufficient to reject the null hypothesis.

50. See Clutton-Brock 1991, Charnov 1982.

51. Leimar 1996, Low 1991, Williams 1966. Note, however, that for small clutch and litter sizes, there can be a "packaging problem"—one very large, or two smaller offspring? See Charnov 1993, Downhower and Charnov 1998. Recent work by Silk and Brown (2008) shows that local resource enhancement and local resource competition do indeed affect birth sex ratios; the authors also examined the differential cost of rearing and found, in 102 primate species, that sex ratios are biased toward the dispersing sex if they do not breed cooperatively and are biased in favor of the reproductively more valuable sex.

52. Gorillas: G. Mace 1990; Sweden: Low 1991; parental sex-preference: Knodel 1988.

53. Inheritance importance: e.g., Borgerhoff Mulder et al. 2009, Borgerhoff Mulder et al. 2010, Gurven et al. 2010, Shenk et al. 2010. Biases: e.g., J. Hill 1984, Betzig 1986, Flinn and Low 1986, Low 1991.

54. Traditional polygyny and sex bias: Hartung 1982, 1997. Cowlishaw and Mace (1996) have reanalyzed these data using a phylogenetic approach, enriching the inferences we can make. Cultural shifts to polygyny are most commonly associated with male-biased inheritance, while changes to monogamy are associated with decreased sex bias in inheritance.

55. Portuguese nobles: Boone 1986, 1988; Kipsigis: Borgerhoff Mulder 1998a; Gabbra pastoralists: Mace 1996.

56. Swedish siblings: Low 1991. Further, men who inherited land had more children than their landless brothers. See chapter 8.

57. Germany: Voland 1984; United States: Vinovskis 1972.

58. Tennessee: Abernethy and Yip 1990; United States generally: Gaulin and Robbins 1991, Mealey and Mackey 1990. This well may be a broader phenomenon than only the modern United States; see Cronk 1991a,c,d, 1993.

59. Reproductive conflicts of interests (which can reduce overall care, and thus fitness): Lessells and McNamara 2012; sex differences in bequests: Judge 1995, Hrdy and Judge 1993, Judge and Hrdy 1992. In one study (Judge and Hrdy 1992) decedents with two or more daughters treated them more equally than decedents with two or more sons. College student survey: Gaulin et al. 1997. The results were obtained through surveys about perceived care. Interestingly, bequests fit "rational" economic models rather than "altruism" models (Altonji et al. 1997).

60. Fisher 1958: 159. Laland et al. 1994 have shown that persistent sex-preferential infanticide may create selective forces that feed back and influence the birth sex ratio. See also Das Gupta and Mari Bhat 1997 and chapter 10.

61. Inuit: E. A. Smith and Smith 1994, E. A. Smith 1995; hypergynous societies: Dickemann 1979, 1981; male-biased infanticide in high-status families: Parry 1979; fitness in stratified societies: Harpending et al. 1990.

62. E.g., Barry et al. 1957, 1976, Konner 1981, Whiting and Whiting 1975, Whiting and Edwards 1973, Blurton Jones and Konner 1973, Ember 1981, Rosenblatt and Cunningham 1976. Early analyses of sex differences in child rearing have not found any clear logical patterns in the existing variation. In part, this may have arisen from a failure to use the extraordinary advances in evolutionary theory of the past three or so decades.

63. Barry et al. 1976, Low 1989a.

64. Because broad patterns in the intensity of training boys and girls generally covary, it is possible that the patterns in boys' training simply reflect general patterns in child training and are not the result of sexual selection; to eliminate this possibility, we must eliminate the possibility that the traits of interest simply covary for boys and girls. There are twelve traits that show some pattern in the training of boys or girls, with the degree or intensity of polygyny, in stratified or nonstratified societies. Seven of these are exclusive to one sex: six are exclusive to males, one to females. As intensity (maximum harem size) of polygyny increases, boys, but not girls, are trained to show fortitude, competitiveness, sexual restraint, and obedience (nonstratified societies), or industriousness (stratified societies). As the degree of polygyny (percentage of men and women polygynously married) increases, boys, but not girls, are taught to show fortitude and be aggressive and industrious (nonstratified societies). In stratified societies, as the degree of polygyny increases, the boys are taught to show less sexual restraint and more self-reliance. Girls, but not boys, are trained to be responsible in nonstratified polygynous societies. For girls, training in responsibility and industriousness increases with both intensity and degree of polygyny in both stratified and nonstratified societies. No such general pattern is evident for boys. Thus, the observed patterns do not appear to be simply due to covariance in training of boys and girls (Low 1989a).

65. Low 1989a, 1990e.

66. E.g., McElreath and Boyd 2007; for nonmathematical treatment, Richerson and Boyd 2005. Older work includes Laland et al. 1994, Kumm and Feldman 1997.

67. In this context, it is interesting that in our society children themselves quickly show sex differences in perceptions of dominance and aggression. By age three, boys play in groups and play more aggressive games than girls (e.g., Laland et al. 1994; also see Omark and Edelman 1975). By age four, boys tend to advertise themselves as toughest; by age six, they have formed dominance hierarchies, perceive them accurately, and attempt to manipulate their position—while girls find the entire question irrelevant!

68. E.g., Dweck 1975, Dweck et al. 1978, Dweck and Wortman 1982.

69. Rates of praise: Dweck and Goetz 1978, McCandless et al. 1972; type of praise: Dweck et al. 1978.

70. Human maximum lifespan = 115 years, versus 29 for macaques and 44 for chimpanzees: Stearns 1992, table 8.3. Allometric calculation: Harvey et al. 1987.

71. Harvey et al. 1987 note Sacher's 1959 finding that bigger-brained primates live longer, although they point out that (1) other measures, such as adrenal gland weight (Economos 1980), correlate even better than brain weight with longevity and that (2) correlational analyses do not suggest any appropriate causality.

72. E.g., K. Hill and Hurtado 1991, Early and Peters 1990, Howell 1979; also see Hill and Hurtado 1991 for a review.

73. Hutt 1972.

74. Death clears the way for others: Curtis 1963; selective irrelevance: Comfort 1956.

75. Williams 1957 and Hamilton 1966. Hamilton (1996: chapter 3) has precisely the right quote from James Joyce's *Portrait of the Artist as A Young Man*. Temple says: "The most profound sentence ever written is the sentence at the end of the zoology. Reproduction is the beginning of death." Hamilton had the wit most of us lack to see the profundity in it.

76. Reproductive value, you will remember, is the average number of daughters a female of a given age will have in the rest of her life, given the prevailing age-specific fertility and mortality rates.

77. Ache: K. Hill and Hurtado 1996; British men: Sheppard and Sear 2011. Children defraying costs: Kramer 2005.

78. Snopkowski and Sear 2013, Royle et al. 2012.

79. E.g., Mace 2013, Borgerhoff Mulder and Rauch 2009.

80. Fox et al. 2010.

81. Cooperative breeding among the Ache: K. Hill and Hurtado 2009; among the Aka: Meehan et al. 2013; goals of paternal direct child care: Winking et al. 2009; maternal kin versus paternal kin, and social transmission of contraception in Gambia: Mace and Colleran 2009: effects of living with kin in Thailand: Snopkowski and Sear 2013; wealth and fertility in modern wealthy nations: Schaffnit and Sear 2014; effects of kin on second births in modern Britain: Mathews and Sear 2013.

82. Alvergne et al. 2011, 2013.

83. Female schedule of reproductive decline: e.g., Dunbar 1987, Gaulin 1980; percentage function at 65: e.g., Mildvan and Strehler 1960; life expectancy at first reproduction: e.g., Nishida et al. 1990, T. Smith and Polacheck 1981: 108–110. See also Gurven 2012 for a comparative overview.

84. See K. Hill and Hurtado 1991: fig. 1, which combines data from Wood 1990 and Mildvan and Strehler 1960. "Menopause" in rhesus: Walker 1995, R. Johnson and Kapsalis 1998; the latter suggest that up to 10% of rhesus macaques experience the equivalent of menopause.

85. Environmental and life-course effects can shift the specific age at menopause in definable groups in modern societies; see Murphy et al. 2013. Interestingly, this also appears to be true for some cetaceans: see Johnstone and Cant 2010.

86. Whether this represents a shift from patterns in traditional societies is unclear. Traditional societies: Ravenholt and Chao 1984, Wood 1990, reviewed by K. Hill and Hurtado 1991; developed nations: Snowden et al. 1989.

87. Early reproductive cessation in mammals: Packer et al. 1998. Female elephants (Croze et al. 1981: 306) and perhaps horses show increasing length of interbirth interval with age. In toothed whales (Marsh and Kasuya 1986) and some strains of laboratory mice (Festing and Blackmore 1971, E. Jones 1975), female reproductive function appears to cease entirely well before the end of life, as in humans.

88. Reproductive shift: Williams 1957; other primates: Fairbanks and McGuire 1986, Cheney and Seyfarth 1987; traditional human societies: Hawkes , O'Connell, and Blurton Jones 1997, Hawkes, O'Connell et al. 1998, 1999, Hurtado and Hill 1990, Lancaster and King 1992. Postmenopausal women's resource extraction rates: Hawkes, O'Connell, and Blurton Jones 1997. Contemporary patterns of parental support: Cooney and Uhlenberg 1992, who find that parental support declines somewhat after children reach age 30 but is responsive to children's status and needs.

89. K. Hill and Hurtado 1991, 1996. See also Rogers 1993, Kaplan 1997. The conflict over grandmothers may be a twofold issue. First, as Hawkes and colleagues suggested, it may be that reproduction has not been shortened but that other systems have been extended; the gain from this extension seems supported by the wide patterns of grandparent investment and the comparative life histories of humans versus other primates. Second, there may be issues of scale of analysis: once the systems have come to equilibrium, we don't expect to see longer-than-profitable reproductive lives. So within the remaining existing variation, we would still expect to see later menopause correlated with higher total fertility. And finally, evidence from rhesus macaques (R. Johnson and Kapsalis 1998) suggests that costly female reproductive failures may be simple senescence (although this does not explain the rate problem). Model: Kachel et al. 2011.

90. Sear et al. 2000, 2003, Sear 2008, Sear and Mace 2008, Shanley et al. 2007, K. Hill and Hurtado 1991, 2009.

91. Maternal grandmothers in The Gambia: Sear et al. 2000; in-laws and fertility Sear et al. 2003; lack of such effects in Malawi: Sear 2008; conflicting results in cross-cultural studies: Sear and Mace 2008, Shanley et al. 2007, K. Hill and Hurtado 1991, 2009. Data from baboons and lions: Packer et al. 1998.

92. Kaplan et al. 2010, Mace and Alvergne 2012.

CHAPTER 7: SEX AND RESOURCE ECOLOGY IN TRADITIONAL CULTURES

1. See Ridley's nice discussion (1993) of this paradox.

2. See Nakahashi and Feldman 2014 on the evolution of division of labor within groups.

3. Murdock and Provost 1973. The Standard Cross-Cultural Sample of 186 societies, used here and elsewhere for comparisons, is chosen to represent high-quality data from all the world regions and representative language groups within regions. It stratifies societies by geographic region and within regions by language groups, to minimize Galton's Problem of cultural diffusion.

4. E.g., Murdock and Provost 1973.

5. Of course, I am not saying anything whatsoever about women's work today. I do not think that past patterns dictate current utility or individual choice.

6. Gurven et al. 2004.

7. Collective action and sharing: Hawkes 1992; similar data arise from Cashdan's 1985 studies in Africa. Overview: Winterhalder 1996a,b,c, 1997, Connor 1995a,b; tolerated theft: Blurton Jones 1987b, Bleige Bird and Bird 1997; showoff men: Hawkes 1991, 1992, 1993.

8. See Whyte 1979, Low 1990e. In societies in which women can control the fruits of men's labor and the fruits of joint labor, they are also likely to inherit property of some economic value. Societies in which women can inherit property are also the societies in which women are likely to be active in community affairs. Although the ability to hold formal leadership positions in the kin group is not associated with the ability to control the fruits of labor or to inherit property, it is positively associated with the ability to be active in community affairs. The ability of women to hold political posts outside the kin

group does not correlate with any other measure of resource control or power. Women's ability to gain political power appears to approach men's in only two societies. Women's resource control and influence showed no relationship to the risk of starvation or protein deficiency. Women's ability to control resources is not related to group size or mobility, suggesting that women do not have a greater voice in small, mobile, hunter-gatherer societies and that their ability is not related to the presence of community-wide male work groups, which appear to be organized around problems of male-male competition. Divale and Harris 1976 suggested that in areas of the world with protein insufficiency, male dominance characteristics that contribute to successful hunting, and male power, become overvalued, resulting in female powerlessness. The cross-cultural data do not support this hypothesis. Similarly, Leacock and Lee (1982) suggested that women in hunter-gatherer societies had more power, but the data do not support this hypothesis either.

There are several problems in trying to assess the actual degree of control of resources by either sex, and they are frequently not resolvable from the ethnographies. For example, when women are reported as controlling the fruits of men's labor, there are several possibilities. A woman may control only a small proportion of the resources a man garners, and only because he allows her to distribute this portion after he has disposed of all he wishes. The man's and woman's interests may overlap completely so that either may dispose of resources without conflict. The woman may distribute significant proportions of the resources garnered by the man by following the directions he gives for their distribution, and for an outsider this condition may be difficult to distinguish from true female control. Finally, she may control a significant proportion of the fruits of his labor and make decisions without regard to his desires. The meanings of each of these conditions are quite different, and they are not easily discerned from codes or many ethnographies.

9. Chagnon 1979, 1982, 1988, 1997.

10. Chagnon 1997: 7.

11. See also Marlowe 2012, Betzig 2012. Betzig has a cogent comparison of male and female variance in reproductive success among a variety of traditional societies. Among the Yanomamö, male variance is 36.64; female variance is 8.02. The mean reproductive success does not match for the two sexes, because the society is not closed: men capture women from other groups. Because of the difference in means, the coefficient of variation is more appropriate: for men, it is 7.09, for women, 1.91.

12. Female alliances: Yanca and Low 2004.

13. See Hames 1996.

14. K. Hill and Hurtado 1996 is the definitive study. Earlier ethnographic accounts include those of Clastres 1972a,b and Bertoni 1941.

15. K. Hill and Hurtado 1996: 316–317, Hurtado et al. 1985. See also Hawkes 1993, Winterhalder 1997.

16. K. Hill and Hurtado 1996: 316–317, Kaplan and Hill 1985. New work suggests that personality has heritability as well as fitness implications through mate choice (D. Bailey et al. 2013).

17. Other species: e.g., Møller 1987, 1989, 1994; Clutton-Brock et al. 1982, Le Boeuf and Reiter 1988. Human: e.g., E. A. Smith 2004, E. A. Smith and Bliege Bird 2000, 2005, E. A. Smith, Bird, and Bird 2003, Gintis et al. 2001.

18. Manners 1967.

19. Borgerhoff Mulder 1988a,b, 1990, 1995.

20. Borgerhoff Mulder 1988a, 1992, 1995 reviews the few cases reported.

21. Borgerhoff Mulder 1997. Temporal changes: Borgerhoff Mulder 1995.

22. Cronk 1989, 1991b,c,d,e, 1993, 1999.

23. Interestingly, this income does not appear to make these women's brothers, for example, able to marry more. So favoring daughters is not a matter of local resource enhancement.

24. Irons 1975, 1979a, 1980.

25. Irons 1979a.

26. J. Lee and Campbell 1997, J. Lee et al. 1993, J. Lee and Guo 1994, Wang et al. 1995.

27. Wang et al. 1995: table 1; see also Lavely and Wong 1998. The European and Chinese patterns differed: marital fertility in Europe was high, but a large proportion of adults failed to marry (see chapter 8); among the Qing, marriage was far more common, but marital fertility was lower.

28. Monogamous men who survived to age 45 had, on average, 4.5 children, and polygynous Qing men had on average 6.5 children, compared to 8–10 children for monogamous and 16–20 for polygynous European men (Wang et al. 1995 and references therein).

29. See J. Lee and Guo 1994 and Wang et al. 1995.

30. J. Lee and Campbell 1997; see also chapter 6.

31. J. Lee and Campbell 1997: 138.

32. J. Lee and Campbell 1997; chapter 9. Cf. note 82, ch. 4 (p. 269).

Chapter 8: Sex, Resources, and Fertility in Transition

1. The most general description a biologist might make is: environmental pressures shape life history and behavior (cf. Roff 1992, 2001, Stearns 1992), and when environmental pressures change, so does selection on life history patterns. Demographic transition theory rests on the argument that environmental changes (in most arguments, linked to industrialization) caused family sizes to drop, so in a way it is a subset of life history theory. Sometimes the arguments are made at the population level (e.g., Coale and Watkins 1986, Wrigley 1983a,b, Viazzo 1990), sometimes at the level of perceived value of children by their parents (Hammel et al. 1983). Simpler, more parsimonious arguments exist that require no group selection and which are a subset of the well-tested and well-supported behavioral ecological explanations of changes in reproductive patterns in other species (e.g., Hawkes and Charnov 1988) . If fertility is influenced by environment, as earlier chapters suggest, many patterns of starting, stopping, and spacing children are possible adaptive responses to environmental conditions, rather than not-optimum-because-not-maximum patterns.

2. In earlier chapters, it is apparent that the behaviors that become and remain common are those that produce reproductive profit for their performers. Even for non-human species, because the world is often far more complex than one might at first imagine, we have seen that "maximum" fertility is not always optimum. In fact, "most successful reproduction" does not necessarily mean producing the most offspring, or even the most surviving offspring (e.g., Lack 1947, 1966; Dawkins 1982, 1986, 2006; Williams 1966; Daly and Wilson 1983). No species produces offspring at its maximum physiologically defined rate; reproductive success depends not only on production, but

also on the investment required to produce viable, competitive, reproductive offspring (e.g., chapters 6, 15). Thus producing fewer, better-invested offspring, compared to the maximum physiologically possible, can be, depending on environmental conditions, reproductively more efficient. The variation in success is almost always the result of the failures—what statisticians call the "zero success" reproductive group (Falconer 1981). Costs and benefits may differ for male and female parents (chapter 3). This reiterates and extends the "quantity versus quality" dilemma first raised by Darwin in 1871.

Such a view gives a new perspective on the demographic transition, the period in the late nineteenth century during which western European and North American family sizes fell dramatically, and on today's demographic transition in developing nations around the world. The competitive environments into which humans are born can be extremely complex, in part due to the long lifetime over which individuals must respond to changing conditions that will affect themselves and their offspring. This, then, may be both a simpler and more general approach, compared to classical transition theory.

3. This is an extremely important point, to which I will return. These data, like the data on traditional societies in previous chapters, are comparable to the nonhuman animal data and unlike most modern analyses, which use "aggregate" data. Later I want to explore how different forms of data can affect the answers we get. Consider the aggregate data of a census: these are summary measures that do not track individuals over complete lifetimes; they also lack any estimates of within-class variation. There are systematic biases in just which individuals are most likely to be censored and which missed; these mean that aggregate data won't work to answer the questions I am asking here.

4. Voland 1990.

5. E.g., Voland 1990 and references therein; Low 2013a. See also chapter 15.

6. Beise and Voland 2008.

7. Thanks to Robert Ostergren.

8. Spatial and temporal variation: it is difficult to elucidate relationships that vary both temporally and spatially; hence the analyses are structured to make comparisons by decade and by parish. Doing this allows us to analyze the relationship between resources and lifetime reproduction through the changing economic and social times of the demographic transition, and in parishes of very different base fertility levels. Thus we can follow sometimes subtle relationships through the changing times of the nineteenth century.

9. Late marriage was sometimes called the "European" pattern (Hajnal 1965). Proportion never marrying: Low 1989b, 1990c; Low and Clarke 1991, 1992; proto-industrialization: Mendels 1981, see also Flandrin 1979, Tilly 1978. In Sweden it is probably related to land enclosure and inheritance changes during the nineteenth century (Jörberg 1972, 1975). From 1686 to 1810, the nobility practiced *fideicommiss*, or male primogeniture, with the constraint that the eldest son must continue the practice (Malmström 1981, Inger 1980). Until 1845, sons inherited twice as much as daughters; after that date, daughters had equal inheritance rights, although in practice sons had first choice of the land and goods that were to be their inheritance, and sons could purchase their sisters' inheritance from them (LoJohansson 1981, Inger 1980). This meant that disputes occasionally arose over the value of the exchanged inheritance items; purchasing needed land from a sibling could prove economically onerous, but also siblings sometimes complained that they did not receive fair value (not uncommon elsewhere in Europe). Even after the shift from fideicommiss, and even after the establishment of equal inheritance laws for both sons and daughters, inheritance biased by birth order was often evident (see Gaunt

1987, Low 1989b, 1990c) and a bias toward the first son was perhaps more evident in the northern areas. Legal agreements in which a father ceded his land to one (usually the eldest) of his sons before his death, typically in return for room, food, and certain other rights, were common. But as Gaunt (1987) noted, during the nineteenth century, the payments delivered to the retiring father increased in size, and receiving a farm became an economic burden. Indeed, default was common, and contemporary jokes abounded about arsenic as "retirement medicine" (Gaunt 1983, 1987). Thus, there probably existed some tension both within and between generations over resources.

10. Low and Clarke 1991.

11. Geographic descriptions and land enclosure: Gerger and Hoppe 1980; demographic and economic data: Low 1989b, Low and Clark 1993. For an excellent and detailed study of marriage, fertility, and crop prices in England (where they also covaried), see Wrigley and Schofield 1981.

12. Low 1989b, Low and Clarke 1991.

13. See Knodel 1988 regarding the usual correlation between age at marriage and peak fertility.

14. Population and demographic behavior: Low and Clarke 1991, 1993; also see Jörberg 1972; Sundin 1976; iron foundries: Ostergren 1990, Sundin and Tedebrand 1981, 1984; industrialization: Norberg and Rolén 1979. Tuna parish fertility data: Low 1990c.

15. Low et al. 1992. For a general view, see Cigno 1998.

16. Cow's milk: Brändström 1984, Low 1991.

17. Low 1990c, Low and Clarke 1991.

18. Sundin 1976.

19. Low 1989b, 1990c, Low and Clarke 1991, 1992; Low 1991; Clarke and Low 1992.

20. Geographic and temporal variation are summarized in Low and Clarke 1991, 1992. Making lifetime fertility comparisons within local populations at the same time is parallel to many nonhuman demographic analyses, but it is seldom done for humans (see chapter 15).

21. Interestingly, in both wealthier and poorer families, daughters survived better than sons, and there was no evidence of significant undervaluing of daughters among wealthier or landed families (cf. Voland et al. 1990, who did find such a pattern in the Krummhörn study).

22. Clarke and Low 1992 discuss the influence of skills and liquidity of resources on migration.

23. E.g., Low 1994b.

24. Low and Clarke 1992. See also Wrigley and Schofield's 1981 classic *Population History of England*.

25. Fahlgren 1956.

26. A *mantal's* relationship to actual land area thus varied across parishes and even within a parish across time, as technologies changed (Lext 1968). In this sample, the average landholding decreased during the study period. Inheritance laws mandated that only men could own land; widows might pay the *mantal* but were holding the land in trust for sons. If a widow remarried, her new husband did not get claim to the land.

27. For example, bönder are defined in the Swedish Demographic Database as owning land; yet some men categorized in the household records as bönder do not appear in the *mantalslängder*, the record of tax payments by landowners. Also, some men who were characterized as relatively low status but had land-associated status (*torpare*, for

example, who were basically cottars) *were* recorded in the mantalslängder, which means they paid the land tax.

28. Borgerhoff Mulder et al. 2009, Shenk et al. 2010, Gurven et al. 2010, E. A. Smith et al. 2010a,b, Bowles et al. 2010, Borgerhoff Mulder et al. 2010.

29. Low 1990e, 1991.

30. Others also have found this generally to be true (e.g., the series of papers in note 26 above; also see Røskaft et al. 1992, Voland 1990; Turke 1989, 1990). Individual patterns in such important items as age of marriage typically vary with resources (e.g., Wall 1984, Sharpe 1990, Cain 1985, McInnis 1977, Pfister 1989a,b, Thompson and Britton 1980, Hayami 1980, Schultz 1982, Simon 1974). Depending on their own resource bases (e.g., Galloway 1986, Schultz 1985), families may respond quite differently to such influences as market shifts and treat their children quite differently (e.g., Georgellis and Wall 1992, R. E. Bailey and Chambers 1998, Merrigan and St.-Pierre 1998, Mitterauer and Sieder 1982: 110); even aggregate data tend to reflect resource influences as individuals make decisions (e.g., Thomas 1941, Wrigley 1983a,b).

31. Statistically, it is those who fail to reproduce (the "zero success class") who affect the overall pattern most; these are the individuals lost, and they are not random in the population—they are poor. Then, too, census data are not designed to elucidate information about *family lineages*, but about *households*. This has two impacts. First, it increases the proportion of homeless poor who are not counted. Second, within households in most censuses, it is impossible to tell "own" children from stepchildren and from other relatives or nonrelatives living in the household. This problem of census techniques and reliable predictions remains an issue today—and not just for academic reasons. In 1998, the U.S. Congress brought suit against the U.S. Census Bureau for counts that miss individuals (mainly poor and homeless ones). However, the new requirement for total counts, rather than improved statistical sampling, will be expensive and difficult.

32. Exogenous survival: Tilly 1978; investment leading to success: MacArthur and Wilson 1967, Rogers 1990, 1991, Low 2013a.

33. This ecological approach echoes some classic demographic models centered on "individual decision" and "proximate variables" (e.g., Becker 1993, Becker and Lewis 1974, Easterlin 1978, Tilly 1978, Bongaarts 1978, 1982, Lindert 1978, Simon 1974, Lesthaeghe and Wilson 1986), as well as Mosk's 1983 "leveraging" approach to fertility and Easterlin and Crimmins's (Crimmins and Easterlin 1984, Easterlin and Crimmins 1985) models of the factors favoring a shift to conscious control of fertility. All of these models have two important characteristics: individuals are not assumed to be uniform, and there is an explicit trade-off between quantity and "quality" (probable success) of children.

34. See review by Low et al. 1992, Low 2013a.

35. More in chapter 15. Rank (1989), studying white women in Wisconsin, found that at "low" socioeconomic levels, women on welfare have fewer children (age-specific fertility is lower at all ages) than women not on welfare, and that these welfare recipients specifically cited the need for resources to invest in their existing children as the reason for avoiding further pregnancies.

36. Education: e.g., Knodel et al. 1990, Knodel and Wongsith 1991; labor markets as driving forces requiring more investment: Kaplan et al. 1995; raising the stakes: Turke 1989. Turke has also suggested that as kin networks are disrupted, fertility may decline, especially when older children and nondescendant relatives initially comprise a re-

source—nepotistic effort. When that resource declines, children pose an increased cost to their parents, a cost no longer defrayed by kin help. Thus, fertility will decline.

In sum, the investment level required to produce successful offspring may vary with environment, and specifically with the threshold level of investment required for a child's success—often a correlate of competition, and in this sense, precisely analogous to the proper use of MacArthur and Wilson's biological concept (r- and K-selection). If poorer parents cannot substantially enhance their children's success, then we might expect larger families and children leaving early. Couples at the high end of the socioeconomic "ladder" might do better by investing more per child to allow them to be competitive with their peers (e.g., education, clothing, status acquisitions). The required investment may limit the number of children they can afford. Within subgroups, however, those with more than sufficient resources may be able to support additional children and still have all be adequately invested.

37. E.g., Easterlin 1978; Becker and Barro 1988; Turke 1988, Low et al. 2008, 2013.

38. E.g., Hartung 1982, 1997, Low 1990e, Low et al. 1992.

39. Rogers 1990, 1991. One of the most consistent differentials in a child's ability to turn investment into successful reproduction, of course, is that of sex (chapter 6).

Chapter 9: Nice Guys Can Win—In Social Species, Anyway

1. See the original at www.google.com/search?q=Gary+Larson+lemming+cartoon& ie=utf-8&oe=utf-8&aq=t&rls=org.mozilla:en-US:official&client=firefox-a&channel=sb.

2. Alexander (1987: 128) wrote bluntly: "Modern society is filled with myths perpetuated because of their presumed value in regard to self-images: that scientists are humble and devoted truth-seekers (of course we are! Trust me!); that doctors dedicate their lives to alleviation of suffering; that teachers dedicate their lives to their students; that we are all basically law-abiding, kind, altruistic souls who place everyone's interests before our own; that our country (church, family) is always right and benevolent and has never done anything with malevolence as a part of the motivation; that our nation is on the side of right." Indeed, there are individuals who clearly behave in selfless ways (e.g., the teacher who died protecting children in a recent school shooting incident); the important questions are: (1) Are these behaviors widespread? (2) Are these behaviors helpful to their doers—typical high-risk-high-gain strategies?

3. True, an individual's life (and by implication, what solutions will be effective) is influenced by his or her physical and social environment, particular genetic makeup, sex, and genetic relatedness to neighbors. Nevertheless, it is a simple but central biological fact that not all individuals are identical; very few species routinely produce genetically identical offspring.

4. Following Hamilton 1964, 1970, and Alexander 1974.

5. Celibacy may, in some conditions, be genetically profitable: see chapters 2 and 10.

6. Of course, one can sometimes turn true altruism to genetic profit after the fact, as one of my students, disturbed by the depressing nature of our discussion of altruism, gave away: "But I gave to charities for a year before I told my friends!" He, of course, meant that he was a true altruist; I am afraid I got a different message! Advertising our "altruism" (whether it was originally genetic altruism or not) may convince our friends to trust us more: "Look how noble and good I am; cooperate with me." This is Robert

Frank's (1985, 1988) "commitment problem" (chapter 10). I suspect we ask our friends, in subtle ways, to make just such declarations and that we make it clear that it's more effective if the show-off altruism has some cost.

7. J. Brown 1983.

8. See especially Pusey and Packer 1997, who make this point and discuss several empirical examples. Consider grooming, often counted as reciprocity—is it really costly to the groomer? If it is directly profitable, we would call it pseudoreciprocity. Assuming that a behavior such as grooming costs the groomer is, in game theory terms, equivalent to making up the relationship between payoffs. That's fine in game explorations (what would we expect if these were the payoffs?) but circular and confusing when employed to explain observed behavior.

Several issues are easy to muddle in this complex problem. First there is the issue of whether our observations are correct: in a Walt Disney film, for example, lemmings were being driven over cliffs and thrown out of boats, not sacrificing themselves to lower population numbers and relieve nonrelatives of competition. In the real world, striking migrations of large numbers of lemmings are observed; they tend to be adolescent individuals unable to establish themselves in the crowded natal habitat. So leaving is risky but likely to be reproductively more profitable than staying and getting trounced by established adults (Chitty 1996). It is true that lemmings and their relatives do live in populations that fluctuate, but not that lemmings leap off cliffs routinely.

Next we must ask whether we are measuring all relevant currencies, and whether we're measuring them correctly. In the lion example, the females are sisters: nursing one's sisters' cubs costs energy, gains genes, and can be genetically profitable. Being relatives seems to matter in other lioness interactions, as well: when groups of (related) lionesses defend territories from intruders, some lioness must take the risky "point" position in hunting (Heinsohn and Packer 1995). Here you would expect tit for tat, sharing the risk. But in fact, some lionesses almost always lead, often looking back at laggard lionesses. There is no evidence of punishment for the laggards or of reciprocation in some other currency. In lions, as in other species, laggards can get away with a lot, if they are close relatives. And again do we really know that grooming is costly to the groomer? There is no convincing evidence that it is (Dunbar 1988, Hemelrijk 1994). Once we have eliminated such sources of error, we gain much clarity. We can distinguish situations in which cooperation pays regardless of what the other guy does ("by-product mutualism") from situations in which it pays to discriminate. We may identify true genetic altruism: there are indeed a few Mother Teresas in the world, but I do not expect us to find it the dominant practice over generations. In fact, evidence (Segal and Hershberger 1999) suggests that even monozygotic (identical) twins may not routinely cooperate, although monozygotic twins cooperate more than dizygotic twins. All are influenced by our prosociality.

9. E.g., Connor 2010.

10. As we saw in chapter 2, Nowak et al. (2010) recently challenged the utility of inclusive fitness using a mathematical model that did not include genetic relatedness. E. O. Wilson, one of the co-authors, argued in a *New York Times* interview with Carl Zimmer (30 August 2011, "Scientists square off . . .") that the existence of diplo-diploid eusocial species, like termites and naked mole rats, and the existence of haplo-diploid species that are not eusocial rendered the concept "tattered." See Gardiner et al. 2011 for a thorough debunking of such mistakes in the literature about kin selection. Lively debate has followed: see especially Abbot et al. 2011, Boomsma et al. 2011, J. Strassmann et al.

2011, Ferriere and Michod 2011, Herre and Wcislo 2011 (all short communications in *Nature*); and online see Coyne 2010 (http://whyevolutionistrue.wordpress.com/2010/08 /30/a-misguided-attack-on-kin-selection/) and Dawkins 2010 (http://old.richarddawkins .net/articles/508102-a-misguided-attack-on-kin-selection)

11. Mutualism: West-Eberhard 1989; win-win from win-lose: e.g., mitochondria may have become part of the genomes of other species from a parasitic beginning; now they are an intrinsic part of the very cells of those species. It is also of interest, I think, that the cleaner-client fish example is vulnerable to mimicry: for example, a cleaner wrasse, *Labroides dimiatus*, is mimicked by a blenny, *Aspidontus taeniatus*, which, rather than cleaning ectoparasites from its "client," nips pieces of flesh.

12. Axelrod and Hamilton 1981; also see Axelrod and Dion 1988, Axelrod 1984. See Binmore 1994 for comments regarding how initial conditions may have restricted possible outcomes.

13. Prisoner's Dilemma is one of several non-zero-sum symmetric games (see also Binmore 1992, Dugatkin 1997, 1998). There are 24 ways to order the four payoffs P, R, S, and T, but this reduces to 12 because of the games' symmetry. Eight of these possess unique optimal pure strategies for both players and are thus less interesting. We are left with four well-studied non-zero-sum games: Prisoner's Dilemma ($T > R > P > S$), Leader ($T > S > R > P$), Chicken or Hawk-Dove ($T > R > S > P$), and Battle of the Sexes. Maynard Smith 1974, 1979 and Maynard Smith and Price 1973 recognized Chicken years ago as the payoff probabilities for mate desertion in animals, and applicable in many territorial disputes, where it is called Hawk-Dove. The order of the payoffs affects the probability of cooperating: the highest payoff goes to a deserting mate if the partner stays, but in that case the faithful partner gets a better payoff that if s/he had also deserted. Here the ESS is a mixed strategy. Dugatkin 1998 reviews animal examples, and Hofbauer and Sigmund 1998 treat evolutionary games broadly.

14. Repetition affecting R: Dugatkin et al. 1994. This problem is cost sensitive, and small cheap favors may become widespread; there is some evidence that the proximate (perceived) trigger is often emotional (self-esteem, friendship) (e.g., Nesse 1990, Nesse and Borridge 1997). For example, the economist Robert Frank has noted that most of us leave tips even when traveling, in situations in which we are unlikely to see the server again. But the cost was only 15% (now about 20%) of the meal, and there is some possibility of a cost to self-esteem and of being observed (more about this in chapter 10) if one considers "defecting." A rule to expand cheap cooperation may have multiple effects. (Note, for example, that for large parties, for which the tip may be a nontrivial amount, restaurants do not typically rely on their customer's observance of practice: the tip is automatically added into the bill.)

15. E.g., Emlen 1982a,b, 1995, Reeve and Nonacs 1992. See Binmore 1994 on why Prisoner's Dilemma, even when repeated, is so resistant to cooperation. Reputation is important, as both game theorists and political scientists know. Mailath and Samuelson (2006) have treated the phenomenon of repeated games and reputation, and Ostrom (1998b) set political scientists alight with her presidential address for the American Political Science Association by focusing on reputation. Barclay (2012) discussed broad issues of the social and psychological importance of reputation.

16. Strategy dependence: e.g., Binmore 1992; Axelrod's tit for tat and variations: Axelrod 1984; see also Axelrod and Hamilton 1981, and Axelrod and Dion 1988. The idea of tit for tat (tft) is familiar, but it does not seem to be really common, at least in its

pure form (see Mesterton-Gibbons and Dugatkin 1992 and Pusey and Packer 1997). It can be beaten, not only by defection but by surprising variations. Tft is very vulnerable to mistakes or imperfect information, both rife in the real world. If I think you have defected, I respond by defecting, and a downward spiral has begun; but if I am wrong, I have cost myself. Mathematical modelers Martin Nowak and Karl Sigmund (1994) developed a probability-driven tournament in which there was no certainty, only a likelihood, that your opponent acted in a certain way. The winning strategy? "Generous" tit-for-tat (gtft), which "overlooks" single defections about a third of the time. Gtft wins over tft (which can be too harsh) and does well against a teachable opponent and in a noisy environment in which mistakes occur. It is also less vulnerable to exploitation than tit for two tats (which always forgives single defections and is resolute in punishing two defections). Now, a complication: gtft holds its own against nastier versions, but it is easily invaded by really sweet variations like "always cooperate" (ac)—but of course, as ac increases, it will lose to any defecting strategies.

So among these strategies, there is no evolutionarily stable solution (see Binmore's 1994 discussions of the complexities we usually miss about Prisoner's Dilemma). Is there no chance for stability? Add one slight improvement in mimicking reality: memory. The rule is fairly simple: remember what your opponent and you did; if the last thing you did let you win, repeat it; if you lost, change strategy. This sounds remarkably like what people do, and describe doing, in personal relationships. Not only humans, but chimpanzees, appear to use memory of past interactions to make decisions about current behavior (de Waal 1997b). This strategy goes by the inappropriate name of "Pavlov," even though it is the opposite of a reflex strategy. (A highly readable review of this is in Nowak et al. 1995). It is evolutionarily stable in these simple two-actor games. Remember, though, that all these games involve some highly restrictive assumptions: pairs of animals making simultaneous decisions without knowing anything about what the opponent is about to do (see Nowak and Sigmund 1994 for alternating decisions). This pairwise aspect is important; Boyd and Richerson (1988) have shown that reciprocity over communal resources is unlikely if there are more than two individuals. However, in groups of more than two, a common pattern I'll say more about, is: watch, and not only refuse cooperation to defectors, punish them. Experimental economists are discovering a phenomenon they call "strong reciprocity," which is characterized in part by a "taste" for punishment.

17. Trivers 1971, Connor 1986; see also Winterhalder 1996a,b,c. For optimal discrimination, see Reeve 1989.

18. Other species enforcing norms: e.g., de Waal 1996b: 89 gives an example of chimpanzees in a group punishing adolescents who were late for a feeding (the group had to wait to be fed until all individuals had arrived). Nash outcomes in untrained human experiments: See also Axelrod 1986, V. Smith 1982, McCabe et al. 1998.

19. Cohen et al. 2001, Riolo et al. 2001.

20. See especially Binmore 1992, Sigmund 1993, Maynard Smith 1982, Dugatkin and Reeve 1998, and Hofbauer and Sigmund 1998.

21. This can mean either that some proportion of the population plays each strategy, or that individuals play each strategy some proportion of the time.

22. Vampire bats: Wilkinson 1984, DeNault and McFarland 1995.

23. Pusey and Packer 1997. See also Axelrod 1986, who described the evolution of metanorms such as "punish those who fail to punish cheaters." But it also may be true

that in some potentially complex relationships (e.g., soil microorganisms), some (perhaps vague and cryptic) mutualisms will turn out to be tit for tat.

24. See Axelrod 1986.

25. Machiavellian intelligence: Humphrey 1983, Byrne and Whiten 1988, 1992, Alexander 1971, 1979, 1987, D. Wilson et al. 1996, M. Wilson et al. 1998.

26. These conditions may be simple but are not equally easy to generate in varied societies; for example, widespread cheap-to-enforce norms are most likely inside small homogeneous populations. Conditions for an evolutionary stable strategy: Maynard Smith and Price 1973. Boyd and Richerson (1992, 1996) suggest that cumulative cultural adaptation may be likely to be rare, for the conditions required to generate (as opposed to maintain) observational learning are rather stringent. But see also Richerson and Boyd 2005, Kurzban and Barrett.

27. Richerson et al. 2002. This is particularly interesting because it examines evolutionary theory as it might apply to common property regimes and institutions.

28. I should note that when, e.g., Boyd and Richerson use the term "cultural group selection," they offer a quite clear definition.

29. Wynne-Edwards (1962) collected observations on social behavior of all sorts of vertebrates in an attempt to demonstrate that anything other than maximum fertility (smaller clutch or litter size, not reproducing in certain seasons, and so on) constitutes behavior that contributes to the well-being of the group, even though it is detrimental to the individual. In his earlier work, he had argued that all other species were group selected, but humans were not—and that's why other species' populations never outpaced their resources but human populations would! Recently he argued that all species, including humans, are group selected (see Hawkes and Charnov 1988, Low 1993a, and others for reviews).

Wynne-Edwards fell prey to the problem of not assigning cost correctly; he assumed that less-than-maximum fertility imposes an individual cost, but we now know this is often not true (the winning strategy might be: do not reproduce in bad times; make fewer, raise them all; see chapters 8 and 15). If lower fertility were really a cost, the immediately obvious problem is that highly fertile "cheaters" would win.

30. Bell 1997a: 530.

31. This may not be as damning as it sounds; for those who manage or study populations, focusing on the population may seem sensible. For example, even the classic *Population History of England* (Wrigley and Schofield 1981: 462) accepted Wynne-Edwardsian "population regulation," and so did many wildlife managers. For a cogent commentary, see Hawkes and Charnov 1988. Bell (1997a: 519–532) has a nice review of how selection can affect different kinds of groups. Further progress is being made; see, for instance, Griskevicius et al. 2012.

32. See Wright 1945, Hamilton 1975, D. S. Wilson 1980, E. O. Wilson 1975.

33. Males homozygous for the *t* allele are either sterile or die before maturity: Lewontin 1970; cytoplasmic war: Hurst 1991, 1992, Hurst and Hamilton 1992; parliament of genes: Leigh 1971.

34. Lewontin 1970.

35. Leigh 1971.

36. E.g., Axelrod 1986, Boyd and Richerson 1992. Of course, some of the strongest reciprocal and coalitional relations exist to harm "outsiders" rather than to spread comfort within the group.

37. E.g., D. S. Wilson 1998a. At some level, interdemic selection, like kin selection, involves a group structure that generates genetic correlates (although these are not required in the specification of interdemic selection, it is the outcome of viscosity). For problems in partitioning the variance in outcomes of actions, this kind of hierarchical approach may be analytically useful. However, evolving "Mother Teresa" altruism in either system is empirically unknown.

38. Quote: Boyd and Richerson 1985: 240. Group norms affecting individual costs and benefits: Boyd and Richerson 1992, Alexander 1979, 1987, Axelrod 1986, S. Frank 1990. Note also that humans are not likely to have rigid heritable strategies, but rather to follow conditional ones (see, e.g., E. A. Smith 1998a).

These models are analytically interesting: they can be used to highlight the tension between individual good and group good. Richard Alexander (1987) first noted the importance of suppressing costly competition within the group, and Steve Frank (1995, 1998) has shown how, once individuals are unable to leave the group, policing and suppression can occur even in very simple models such as replicators within a cell (see chapter 1). High degrees of relatedness lead to self-restraint in such models; low degrees of relatedness lead to mutual policing, the extent and intensity of which depends also on whether the cost is borne by individual actors or the group as a whole. Even in these extremely simple systems, then, the problems of "public goods" exist.

Finally, these models are a way of partitioning variance into within-group versus between-group variance. This can be useful for some questions. But the history of this approach highlights a real danger: some (perhaps mostly nonspecialists) have made what Binmore (1992) called the "fallacy of the twins" in reference to the Prisoner's Dilemma; later (1994: 202) he demonstrated why a cooperative solution to the Prisoner's Dilemma is so difficult: that is, "However, the evidence does strongly suggest that it is unlikely that a pure equilibrium will get selected." As W. D. Hamilton (1975) noted: "But, of course, being able to point to a relevant and generally non-zero part of selective change is far from showing that group selection can override individual selection when the two are in conflict."

39. Coalitions and laws: Boyd and Richerson 1992, 2009; Alexander 1987, Irons 1991. Advertisement: R. Frank 1985, 1988, 2004. See Irons's 1996a,b reviews. Coalition-coercion models and culture-gene interaction models share much; I think the distinction is mostly an issue of the kind of approach—mathematical or verbal/heuristic.

40. Nonrandom copying: Boyd and Richerson 1990, 1996; Mukogodo daughter-son real versus claimed preference: Cronk 1991b.

41. Thus, many species of long-lived social animals behave in ways that among humans would be termed "ethical egoism," and long-term reciprocation and assistance are common, though not randomly dispensed; see de Waal 1996b. Failure to share and "tolerated theft": see Blurton Jones 1984, 1987b, Winterhalder 1996a,b,c, Hawkes, Bliege Bird, and Bird 1998.

Cosmides and Tooby 1992, 1996 suggest that humans have mental modules that predispose them to share, for example, when hunting is boom-and-bust and one makes a really rich kill. Thus the sharing is not learned, but evoked. Similarly, in the Mukogodo example of the previous paragraph, because daughters can do better than sons, better care of daughters may be easily evoked, whether or not there is someone to imitate.

42. See especially Boyd and Richerson 1985, 1990, Richerson et al. 2003, Fehr and Fischbacher 2004.

43. E.g., Boyd et al. 2011a, b, Mathew and Boyd 2011.

44. For this reason, Boyd et al. (2011b) suggest that cultural adaptation can facilitate the evolution of large-scale cooperation.

45. E.g., Kurzban and Barrett 2012.

46. E.g., Boyd et al 2011a, b, Boyd and Richerson 1985, 1990, Durham 1991, Laland et al. 1994.

47. Even other primates show this tendency; Brosnan and de Waal 2004 a,b,c; Brosnan 2013.

48. Henrich et al. 2004.

49. Gurven 2004.

50. See Henrich et al. 2004. These papers compare how people in a variety of traditional societies around the world play Ultimatum and other games; there is considerably more variation that one would have supposed from previous work, all in developed nations.

51. Hadza: Marlowe 2004; Tsimane: Gurven 2004.

52. Men offering more when they knew player 2 was a woman: Saad and Gill 2001; sex differences in acceptance as player 2: Solnick 2001.

53. Eckel and Grossman1988, 2008.

54. E.g., see Dawkins 1982. Recently D. S. Wilson (1998b), Sober and Wilson (1998), and Wilson and Sober (1994) in lumping almost all interactions as "group selection," conflate kin selection, clade selection, interdemic selection, and "cooperator selection," including reciprocity and coalitions. Nowak et al. (2010) chose to interpret kin selection as a subset of group selection; but see Gardiner et al. 2011 for a reply. Indeed, it may be useful to partition the within-group and between-group variances, but such lumping, I think, obscures the *functional* aspects of genetic costs and benefits and yields approximately the same analytic power as the *New York Times* crossword puzzle. See E. A. Smith 1998a, Hawkes et al. 1998, Harpending 1998, and Palmer et al. 1997.

55. http://www.timesunion.com/local/article/One-of-last-Shakers-talks-about-the -life-4298710.php. Accessed February 16, 2014.

56. We need caution here; we may not recognize all the currencies involved. For example, when lionesses hunt, male lion are free riders, but they take a share of the kill. Unless we understand that a strong male coalition can prevent infanticide by invading males, this behavior would be inexplicable.

57. Even more vulnerable to free-riding exploitation than common-pool resources (which anyone in the specified group can access) are open-access resources, which cannot be protected from outsiders at all. In the first case, free riders inside the group are problematic, and coalitions and mores may be of help; in open-access resources, the free riders are strangers. A recent and highly readable work on the social and evolutionary roots of cooperation, and how resource distribution affects rules (Is the resource subtractable? Not? Can others be excluded?) is Cronk and Leech 2013.

58. Hawkes 1992, 1993 and E. A. Smith and Boyd 1990 have nice discussions of this problem in traditional societies. Ostrom 1990, 1998a,b and Ostrom et al. 1994 have comparative and experimental data on common-pool resources. Atran 1993 has examples of how relatedness and village stability affect use practices. David Sloan Wilson (1998b) has proposed commons issues as another example of group selection; commentaries (e.g., Hawkes, Bliege Bird, and Bird 1998, E. A. Smith 1998a, Harpending 1998) make it clear that truly costly behaviors are theoretically possible but not shown to exist

in any examples or applications. Richerson et al. (2002) link evolutionary theory to problems of common property regimes.

Chapter 10: Conflicts, Culture, and Natural Selection

1. Karlsen 1987: 51.

2. Wealthy women (estates valued at greater than £500): Karlsen 1987: 79; poor women and women with no male protectors: Karlsen 1987: 73.

3. Except in the Salem outbreak (which was somewhat anomalous in many regards), fewer than 35% of witchcraft accusations were directed at women under age 40 (Karlsen 1987: 65, 86). And if one considers only married women, those from poorer families (with estates less than £200) are overrepresented in convictions.

4. The different return curves of mating and parental effort suggest some biases in patterns of sex differences (e.g., see M. Wilson et al. 1996 on sex differences in Machiavellianism). Add human cultural and historical particulars, and (even though human culture, like other behavior, has basic operating rules) there is no way we can predict a priori all possible outcomes (see especially Boyd and Richerson 1992, but also McElreath and Boyd 2007, whose modeling approaches are very helpful). But even knowledge of the basic rules may cast a new light on complex examples—may help us see a pattern when it was obscure before.

The fact that in our evolutionary past a man's value as a mate was largely based on his ability to control resources, while a woman's value derived primarily from her ability to produce children and not the resources she brought to the marriage, seems perhaps at once simple and obvious and irrelevant to our lives today, when we have so few children and many women work. Yet a number of social patterns, both cross-cultural and historical, seemingly unconnected to resource and reproductive value, make sense in this context. Similarly, we can make some testable predictions about the strength and enforceability of norms in small stable-membership groups versus large heterogeneous and mobile societies. Alexander 1987 described moral systems as emergent phenomena resulting from these conflicts of interest. Irons 1991 and Cronk 1995 have made the point about manipulation of systems eloquently. Flinn 1997 has reviewed the important hypotheses about the role of social learning as well as the difficulty of testing them.

5. This conflict between "my genes, my lineage" and "all our fates," sometimes called the "levels of selection" problem, is crucial to understanding human resource use. Chapter 9 has suggested why it is so hard to convince ourselves to make really costly sacrifices for some unrelated group or individual. More subtly, why are we eager to make the sacrifice of, say, a benefit dinner or concert, especially if our friends are aware of our sacrifice? Why is "nepotism" a dirty word in politics and administration? Why are some patterns widespread, repeated in culture after culture even if the cultures are otherwise quite different? These become central issues as we face large-scale complex problems such as population growth and individual consumption, the temptation to externalize costs and let them be borne by others who do not share the profits, and so on.

6. Psychological factors and prosociality: Jaeggi et al. 2010, Chudek and Heinrich. 2011, Caprara et al. 2012; importance of culture as vehicle: A. Bell et al. 2009; utility of modeling: Gintis 2003; review and summary: Boyd and Richerson 2009. Alvard 2012 is a nice up-to-date overview.

7. Many authors think that "moral thinking" may be literally an adaptation (constraints on harm of others within the group, solidarity for intergroup competition) that allows the formation of larger groups than would otherwise be possible, (e.g., Cronk and Leech 2013, Henrich et al. 2004, Alexander1987 [also Alexander 1974, 1977, 1979, 1988]; Campbell 1975, 1979, 1983; Gibbard 1992, Irons 1991, 1996a,b, Ruse 1982, 1986, Cronk 1994, 1995. The most readable recent synthesis is Cronk and Leech 2013.

8. See especially Cronk and Leech 2013, Richerson and Boyd 2005, and Alexander 1987.

9. Kohlberg 1981, 1984. See Cronk 1994, and Henrich et al. 2004.

10. E.g., Cronk 1991b,c,d,e, 1999, Alexander 1987, Hamilton 1975, R. Frank 1988, Irons 1996a,b.

11. Subtlety: Benenson, Antonellis, et al. 2008; Benenson, Hodgson et al. 2008.

12. Perhaps this was especially important in the southern United States, where I grew up. Rules included: if you are smart, never show it to a male; never contradict a male; be deferential and subordinate in all conditions to males and your elders; never dress sloppily or as if you were going on a field trip; be demure—there were even rules about the material of shoes—heaven forfend they should reflect up your skirt (of course, you had to wear a skirt).

13. Child training: Barry et al. 1976; Low, 1989a. Tsimane female competition through the lifespan: Rucas et al. 2012.

14. Recent review: Mace 2013a. Divorce and conflict among co-wives: Betzig 1989; Decrease of female competitiveness in a patriarchal society: Andersen et al. 2012; indirect aggression most common: Osterman et al. 1998; contexts of female physical aggression: Levy 2012, Lin 2011, Sikes 1997.

15. E.g., Lessels and McNamara 2012.

16. Traditional societies (in this case Tsimane): Steiglitz et al. 2011; modern disinvestment by stepfathers: Anderson 2005, Anderson et al. 1999; power asymmetries: J. Jones and Ferguson 2009, risk of not living with both genetic parents: Daly and Wilson 1985, 1987, 1988.

17. Overall "rising tide" of gender equity: Inglehart and Norris 2003; variations across countries: Low 2012. Human development index, gender ratio, and gender empowerment measures from the UNDP: http://hdr.undp.org/en/statistics/hdi. Principal religions, and mixes of religions, of countries: CIA Factbook 2012 (2013 update is at https://www.cia.gov/library/publications/the-world-factbook/).

18. E.g., Cronk 1994, 1995, Irons 1996a,b. Dawkins 2006. See also R. Frank 2004 on the role of factors promoting cooperation in competitive environments, and Binmore 1994: 160–161 on the difficulties and misleading assumptions often involved in most of our views on commitment.

19. We might begin by asking waiters how much the tip percentage changes, depending on whether there is more than one person at the table.

20. See Cavalli-Sforza and Feldman 1981 for the classic treatment.

21. Richard Dawkins coined the term in the first edition of *The Selfish Gene*, in the 1970s. See Dawkins 2006, Binmore 1994: 66, Dennett 1995, and Henrich et al. 2004.

22. Cultural transmission in simpler species: Ritchie 1990, 1991; cumulative culture: Kurzban and Barrett 2012.

23. Great variation may reside in very local cultural responses to environmental pressures. In the high desert regions of Peru, for example, Quechua mothers swaddle

their infants tightly. This swaddling alters the microenvironment for the infant, making the temperature higher and more stable and the humidity higher (Tronick et al. 1994). Such practices may, over time, select for infant characteristics that do well under the cultural practices.

24. Pulliam and Dunford 1980: 66.

25. Boyd and Richerson 1985, 1990, 1996, Richerson and Boyd 1992, McElreath and Boyd 2007, Henrich et al. 2004, Borgerhoff Mulder et al. 2009. As a result Ernest Hemmingway put it so well in *Death in the Afternoon*, "So far, about morals, I know only that what is moral is what you feel good after and what is immoral is what you feel bad after."

26. Richerson and Boyd 1992 and references above. Boyd and Richerson (1996) noted that the conditions for generating observational learning (as opposed to maintaining it) are stringent, and this is "why cultural transmission is common, but cultural evolution is rare."

27. Kinds of biased transmission: sign language: Aoki and Feldman 1994; hunter-gatherer skills: Hewlett and Cavalli-Sforza 1986; Stanford undergraduates: Cavalli-Sforza et al. 1982. Cultural and reproductive success: Irons 1983a, Alexander 1979, 1987; cultural manipulation: Cronk 1995, Cronk and Leech 2013; parent-offspring political correlations and genetically-influenced personality traits: Hibbing et al. 2013, Tuschman 2013.

28. Cultural and reproductive success: Irons 1983a, Alexander 1979, 1987; cultural manipulation Cronk 1995, Cronk and Leech 2013, Boyd and Richerson 2009.

29. Durham 1991; see also Swallow 2003, Vesa et al. 2000.

30. Guttentag and Secord 1983 review these data.

31. Peng 1991.

32. Chinese policy and birth sex ratios: Hull 1990, Johansson and Nygren 1991, Yi et al. 1993, Li and Choe 1997.

33. Two-province, all-births sample: Wen 1993; China-wide one percent census: Yi et al. 1993. Formal models confirm that a culture-gene interaction can be powerful, and indeed, the relative value of women in China, at least as brides, appears to be increasing (Kumm and Feldman 1997).

34. Ting 1994.

35. Durham (1991: 194ff.), Boyd and Richerson (1985), Richerson and Boyd (1992), and McElreath and Boyd (2007) model the conditions and limits under which cultural variations that do not enhance inclusive fitness can be maintained or spread. Daniel Nettle and colleagues (2011) and David Sloan Wilson and colleagues (2009) both showed how diverse cultural attitudes could be even within a city. Nettle and colleagues also found great differences in the Dictator game across neighborhoods within the city. See also Logan and Qirko (1996) on the role of psychological mechanisms. "Gene-culture coevolution theory" (Laland et al. 1994), derived from Cavalli-Sforza and Feldman (1981) is the most open to nonadaptive interpretations, for, as Laland et al. note (1994: 151), it treats culture as purely ideational; culture is a "pool of knowledge which is stored in the nervous systems of individuals."

36. Of course, religions are not immune from hierarchical treatment of members or risk taking among young men.

37. See, e.g., Palmer et al. 1997.

38. Kibbutzim: Sosis and Ruffle 2004, Ruffle and Sosis 2007; unreliability of self-reports: Preston et al. 2010, Norenzayan and Shariff 2008.

39. In AD 304, the Council of Elvira (Cannon 33) ruled that all bishops, presbyters, and deacons were to abstain completely from sex and from having children. Practices about celibacy changed, but in 1022, Pope Benedict VIII declared all clerical offspring as serfs—clearly an attempt to deprive clerical offspring of resources that otherwise could then likely revert to the Church; he prohibited the children of priests from inheriting property (thus gaining property for the Church). A few decades later, Pope Gregory VI issued a decree against clerical marriage. This prohibition in many cases was not obeyed. In 1059, bishops tried to persuade priests to make their marriages "discreet," and there were often monetary penalties for priestly marriage, which nonetheless continued to increase. Today, a major reason seminarians give for leaving the religious life is the requirement of celibacy. (Heinemann 1990, Sipe 1990, Rice 1990, Sweeney 1992, Saunders 2004)

40. The anthropologist William Irons (pers. comm.) calls this the "spare tire" (cf. bet-hedging) theory of parental investment: when resource richness is uncertain, starting more offspring than can be made independent can be a reasonable strategy. Among raptors and several other avian groups, a simple version of this exists: parents start incubating the first egg before the second is laid, and continue. The resulting clutch has chicks of different ages. When food is plentiful, all chicks are reared, but when resources are constrained, the smaller chicks die. In pre-contraception eras and areas, later sons often were left to fend for themselves as they reached adulthood.

41. Borgias: Mallett 1987: 109–110, 112.

42. Low 2001, fig. 1.

43. Here is an example ripe for comparing predictions from two slightly different approaches. Conditions can change, and thus change the utility of any strategy, as in this example. This important phenomenon is a topic of dispute between (at least some) evolutionary anthropologists and evolutionary psychologists. Evolutionary anthropologists, focusing on the diversity of human behavior, are likely to ask about the "adaptively relevant environment" (Irons 1998): what are the costs and benefits of this behavior in this environment? This leads to the sorts of comparisons biologists and behavioral ecologists are likely to make. Evolutionary psychologists, in contrast, think about the Environment of Evolutionary Adaptedness (EAA; Barkow et al. 1992) in the Pleistocene, which leads to consideration of what behaviors might be currently maladaptive. Although both approaches are valuable for particular questions, I think that the EEA approach is currently more limited, for it imagines "an" environment in the Pleistocene to which human brains became adapted, remaining relatively unchanged since. From what we know about the diversity of habitats in recent geological time, and about the genetic variability and cultural complexity of humans, this seems to me an overly rigid view.

44. R. Smuts 1992; see also Irons 1998, Cronk 1994, 1995, Cronk and Leech 2013, Boyd and Richerson 1992, Flinn 1997, Kurzban and Barrett 2012, Kurzban 2011.

45. See Irons 1998, Cronk 1994, 1995, Boyd and Richerson 1992, Flinn 1997.

46. Social competition fostering intelligence; Cronk and Leech 2013, Henrich et al. 2004, N. Henrich and Henrich 2007, Humphrey 1976, 1983, Jolly 1966, Alexander 1979, 1987; costs of large brains: e.g., Byrne 1995: 213. See also Mitani et al. 2012.

47. Machiavellian intelligence: Byrne and Whiten 1988.

48. Social complexity and neocortical enlargements: Schülke and Ostner 2012, Dunbar 1992; tactical deception: Byrne and Whitten 1992, Byrne 1993; communications as manipulation: Dawkins and Krebs 1978, Cronk 1994, Alexander 1979, 1987; knowledge of social relations: Seyfarth and Cheney 2012; overview: D. S. Wilson et al. 1996.

49. Cross and Guyer 1980, Cross 1983, Costanza 1987.

50. Using Kahneman's 2011 relatively cheap System 1 heuristic thinking.

51. Kurzban 2011.

52. Quote: Dawes 1988: 103; group biases: Plous 1993: chapters 17, 18. Status has a self-reinforcing pattern: we use interactions to make previously irrelevant traits into status traits, and dominant individuals use status in interactions to pressure subordinates into accepting beliefs that may be disadvantageous to them: Webster and Hyson 1998, Ridgeway et al. 1998. Social traps: Cross and Guyer 1980, Cross 1983, Costanza 1987.

53. See, e.g., Kahneman 2011, Kahneman and Tversky 1982, Nisbett 1980, and Dawes 1988, who give examples. Cf. Cheng et al. 1986, Cosmides and Tooby 1992.

54. Kahneman 2011, Kahneman and Tversky 1982 and the references within each; Kahneman and Tversky are the premier experimenters in how people think. For a fun exercise in checking your own biases in thinking, take the introductory test in Plous's 1993 introductory text *The Psychology of Judgment and Decision Making*.

55. Larrick et al. 1993, Cosmides and Tooby 1992, Cheng et al. 1986, Nisbett et al. 1987.

56. For a readable overview, see Blum 1997, J. Becker et al. 2008.

57. E.g., Cosgrove et al. 2007, and De Bellis et al. 2001.

CHAPTER 11: SEX AND COMPLEX COALITIONS

1. There is evidence that males do in fact coerce females and physically keep them from escaping by biting them or doing a full-body slam, and that it is the pressures of controlling access to females that leads to temporary supercoalitions among males (Connor et al. 1992).

2. Single males as reproductive targets: Young et al. 2013; varied tactics of coalitions: Widdig et al. 2000.

3. De Waal 1989 (especially his diagram 4), 1996a. Later, coalitions reformed, and in a bloody conflict, the male Luit was killed. Matt Ridley (1996: 159) draws a wonderful parallel with the War of the Roses: Margaret of Anjou as Luit, Edward VI as Nikkie, and Warwick the Kingmaker as Yeroen. The roles do fit.

4. This is what Pandit and van Schaik 2003 might call a "leveling coalition," although they model a broader set of issues.

5. Tiger and Fox 1971.

6. Co-wives may not be related (as in non-sororal polygyny), but in some ways they have little power to leave the coalition in most polygynous societies.

7. E.g., Low, 1989a, 1990b,e 1992, 1993b, 1994a,b. Men, cooperating in the community sphere, often with nonrelatives, can in many systems gain substantial direct reproductive benefits—resources to buy another wife, for example. Because women, like other female mammals, make the most substantial direct reproductive gains not by maximizing mates but by maximizing the survival of their children, they may historically have had more to gain by seeking parental investment from males and by cooperating with other women in the familial, rather than the community, sphere. This will obviously be influenced by the distribution and predictability of resources and by the reproductive consequences of cooperation versus defection (chapter 3, Emlen and Oring 1977, Daly and Wilson 1983, Krebs and Davies 1991, Low 1990e, 1992).

8. Chapter 4; see also B. Smuts 1996: 236, Hooks and Green 1993.

9. Baboons: B. Smuts 1985; fiddler crabs: Milner et al. 2010.

10. In Battle of the Sexes, there are two Nash equilibria: baseball-baseball, and Stravinsky-Stravinsky, and the payoffs are symmetrical. This is a puzzling game: in most presentations, the payoffs for John: Stravinsky, Mary: baseball are exactly the same as for John: baseball, Mary: Stravinsky. The former seems to me worse for each player compared to the latter. See O. Henry's *The Gift of the Magi*.

11. See Maynard Smith 1982.

12. Beissinger 1987.

13. Common examples are polygynous species with internal fertilization resulting in female-only care; see Clutton-Brock 1991 for additional influences determining which parent does the care. Examples of monitoring and desertion: Trivers 1972. In almost all species one sex or the other (not either sex) can desert. In Beissinger's 1987 example of the snail kite (in the United States, which resides in the Florida Everglades and is endangered) *either parent* may desert.

14. Pusey and Packer 1987, Greenwood 1980, Mitani et al. 2012. Remember that in humans females usually move away from the natal home; males remain (philopatry); I suspect this is because groups of males can be so effective in getting and controlling very useful resources, and coalitions of male kin are relatively stable, which lowers the cost to members and renders them more effective.

15. This very allocation of effort to harassing subordinates and maintaining dominance can itself have a cost, as Packer et al. 1995 have shown. Resisting male sexual coercion: rhesus, Bernstein and Ehardt 1985; general patterns: Smuts 1987a, 1996, Smuts and Smuts 1993.

16. Redtails and patas: Cords 1986; gorillas: Fossey 1983; geladas: Dunbar 1983. In multimale groups, both canines and testes are likely to be larger (chapter 4). See Mitani et al. 2012 and Smuts et al. 1987 for more.

17. Nishida and Hiraiwa-Hasegawa 1987, Nishida et al. 1985. See also McGrew et al. 1996 and Wrangham and Peterson 1996.

18. Struhsaker and Leland 1987, Smuts 1987a,b. For the most recent updates, see Sterck 2012.

19. Importance of reproductive resources: de Waal 1982, 1984, 1986, 1996a,b. De Waal 1996a contrasts the more egalitarian nature of chimpanzee coalitions with the more hierarchical coalitions of, e.g., macaques.

20. Quote: de Waal 1989: 53. Lethal coalitions: de Waal 1989: 61–69; chimpanzees and human intergroup aggression: Wrangham and Peterson 1996: 24. De Waal (1996a,b) notes that chimpanzees show a more "egalitarian" dominance style than many other primates, with interventions on behalf of the subordinate more often than, for example, in rhesus macaques (which he defines as having a despotic dominance style) or stumptail macaques (tolerant).

21. Gilby et al. 2013.

22. Newton-Fisher 2006.

23. E.g., Watts 2012; an excellent review, including discussion of bonobos, is that of Aureli et al. 2012. See also Parish 1996. As in chimpanzees, females are smaller than males (female bonobos are four-fifths the size of males); their control comes from coalitions, not size. G-G rubbing: Parish 1996; low level of male-male aggression over copulations: Kano 1996.

24. Silk et al. 2004.

25. Prevention of others' success: B. Smuts 1985, 1987a; harassment: Wasser 1983. See also Hrdy 1976, 1978, Packer et al. 1995 and Mitani et al. 2012.

26. Wasser 1983.

27. Bonnet macaques: Silk and Boyd 1983; reproductive costs for dominants: Packer et al. 1995.

28. E.g., examples in Mitani et al. 2012. Also Hrdy 1978, Wasser 1983, Gouzoules et al. 1982. Pattern of female variance: If the most successful female in a year has one offspring and the least successful has none, statistically the variance in reproduction is less than among a group of polygynous males, in which the (rare or unique) top male(s) may have ten or more offspring, while more than half the males have none. Statisticians recognize that the "zero success class" has the most influence on the measurement of variance. Further, in at least some primates, dominance striving itself carries reproductive costs (Packer et al. 1995).

29. De Waal 1982, 1984.

30. Bygott 1974, Bygott et al. 1979, Goodall 1986, Clutton-Brock and Harvey 1978.

31. Wrangham 1979, Halperin 1979, Nishida 1979, Goodall 1986.

32. Wrangham 1979, Nishida 1979.

33. E.g., Watts 1996. Coalitions against nonkin: Stewart and Harcourt 1986; coalitions for maternal kin: Silk 1987; sex of participants in aggressive encounters: Smuts 1985, 1987a.

34. Wasser 1983, Chapais 1983.

35. Benenson et al. 2009.

36. Fallacy of the Twins: Binmore 1992: 310ff. And remember (chapter 2): even siblings are likely to kill one another if the stakes are high enough.

37. Interestingly, this works out to be a group-selection argument of the Wynne-Edwards sort (chapter 9) if we all just . . . ." It shows up occasionally also in popular antiwar and proenvironment movements. In Chicken, or Hawk-Dove, the order of the payoffs is $T > R > S > P$; the evolutionary stable strategy depends on the relative value of winning versus the cost of repair or getting well: it will comprise some proportion of cooperators and defectors.

38. Ostrom et al. 1994, discussed below; see especially their table 9.1. Perhaps communication is a proximate trigger that signals a repeated interactor; here is a problem for evolutionary psychologists. This phenomenon, that it pays more to cooperate when interactions are repeated and one doesn't know when interactions will end, underlies many observations: that cooperation and self-sacrifice are more common in small, related, stable-membership groups and that successful management of "commons" resources is easier in just such groups—for these are the group conditions that make the payoffs work for cooperation. See also Richerson et al. 2002.

39. Cook et al. 2007, Ostrom and Walker 2005.

40. Ostrom et al. 1994. They tested both relatively low-stakes (low-endowment) and high-endowment commons situations; no-communication versus communication options; and fine (sanction) versus no-sanction environments. They then calculated the average "yield" (production minus fees and fines).

41. The biologist Garrett Hardin in 1968 described the classic English grazing commons: a village green on which the sheep of all villagers grazed. Obviously, any pastoralist who could add a sheep or two to his own herd got all the profit of an extra sheep. The

whole village shared the cost of an extra sheep—so there was great incentive for each individual to graze more sheep on the commons than was good for the village as a whole—exactly the sort of "levels of selection" problem introduced in chapters 9 and 10. Modern commons problems include oceanic whaling, ocean fisheries, and atmospheric problems like acid rain, and global climate change. In each of these cases, individuals make more profit if they behave in ways that cost the group as a whole (e.g., taking more than the sustainable amount of fish; cheap-but-polluting manufacturing). Not surprisingly, such cases are notoriously hard to solve for the group's long-term good. See, e.g., Ostrom 1990, 1998a,b, Ledyard 1995.

42. This, of course, is precisely what one would expect after reviewing the "levels of selection" issues of chapter 9.

43. Interestingly, experimental economists are finding a taste for punishment in enforcing reciprocity (perhaps related to the issues of chapters 9 and 10: not only punish cheaters, but also those who fail to punish). See, e.g., Fehr et al. 1997, Fehr and Gächter 1998.

44. See especially Ostrom 1998a,b.

45. Ostrom et al. 1994, Ostrom seminar to the Complex Systems Program, University of Michigan, February 1997.

46. Rewarded competition: Axelrod and Hamilton 1981, Nowak and Sigmund 1994. See also Pusey and Packer 1997, de Waal 1996a. Nowak et al. 1995 is a readable review.

47. Thus game theory converges with behavioral ecological theory. Behavioral ecological theory predicts that cooperation will occur under specific conditions: first, among relatives who live together; second, among nonrelatives in long-lived social species who are capable of recognizing individuals and likely to have repeated interactions.

48. Hill and Hurtado 1996.

49. Traditional societies again provide an interesting example. Tolerated theft is common (e.g., see Winterhalder 1996a,b,c). D. S. Wilson (1998b) argued a group-level selective pressure for this phenomenon, but Hawkes, Bliege Bird, and Bird (1998) and E. A. Smith (1998a) clarify the confusion.

50. Low 1989a, 1996a; see also chapter 6.

51. Nomadic societies, however, tend to teach children to be generous to a greater degree than expected, a phenomenon perhaps related to the uncertainly of resource acquisition in nomadic life (Low 1989a). Information sharing and turn-taking with respect to food gathering may lower the variance (lower risk of failure), and for poorer hunters, increase the average return (e.g., Kaplan and Hill 1985). It may well be that many societies treat unpredictable variables not as uncertainties about which little is known, but as risks (i.e., assigning heuristically some probability of failure).

52. Pathogen stress shows a number of strong relationships with social factors (chapter 4); it has even been proposed as a source of kin assistance (Lewis 1998). It appears to change the costs and benefits of a variety of social relationships. Importantly, as the risk of infection by serious pathogens increases, fewer men are suitable, healthy husbands, and monogamy disappears (and, thus, the proportion of unmarried men and the degree of male-male competition). See also Cashdan and Steele 2013, Letendre et al. 2012, Thornhill et al. 2009, 2010, Fincher et al. 2008.

53. It is probably important that animal husbandry and the frequency of internal warfare (which also is associated with low trust) are positively related—that is, wars are frequent over resources, such as cattle, that are used for brideprice. Like polygyny, these

suggest heightened male-male competition. See Ember and Ember 1992 for cross-cultural studies, and Nisbett and Cohen's 1996 book on the associations between pastoralism and male violence in the United States.

54. Today we face growing problems related to these issues, since reciprocity and cooperation within a group are easier to promote when groups are small, have many kin, and have stable membership—all more typical of small traditional societies than most communities in modern nations (see chapter 15 and Ostrom 1990, 1998a,b).

55. E.g., see Keohane and Ostrom 1995.

56. Flinn and Low 1986, Alexander 1977, 1979.

57. Or brother-sister in societies with low male confidence of paternity; Alexander 1979.

58. Note that this is quite similar to Alexander's argument.

59. Henrich et al. 2012.

60. However, sometimes female-female coalitions among co-wives can become strong enough to compensate: Yanca and Low 2004.

61. E.g., Chagnon 1982, 1988.

62. Flinn and Low 1986, M. Ross 1983. Note, also, that patrilocality probably increases the strength of male-male coalitions and typically makes strong female-female coalitions less probable.

63. Divorce and co-wife conflict: Betzig 1989, White 1988; female-female coalitions: e.g., Irons 1983b, Yanca and Low 2004.

64. E.g., Beals 1961, Blood 1960.

65. See, e.g., Daly and Wilson 1984, 1985, 1987 on child abuse patterns.

66. An interesting exception to this general pattern exists among the South African Herrero (Harpending, pers. comm.): Men and women act and form groups to influence events in different spheres. In most societies, there is little ritual and cultural conceptualization of female roles. Collier and Rosaldo 1981 suggested that "marriage organizes obligations, and . . . such obligations shape political life." Their discussion of the centrality of marriage to men's lives highlights the importance of male-male coalitions, in the community sphere, as mating effort. Male-male and female-female coalition differences, which are related to reproductive gain, are broader than the various asymmetries created by different marriage arrangements.

67. Freedman 1980, Draper 1976, Whiting and Edwards 1973, 1988.

68. Draper 1980.

69. Szal 1972.

70. Piaget 1932: 32.

71. Freedman 1980, Lever 1976, 1978.

72. Lever 1976: 482.

Chapter 12: Politics and Reproductive Competition

1. De Waal 1982: 212; quote: de Waal 1982: 19.

2. The "glass ceiling" effect; for example, in 2000 only 8% of law partners in the United States were women, according to a survey by the American Bar Association; in 2013 19.9% were.

3. See Hart and Pilling 1960, Goodale 1971; review by Irons 1983b.

4. Goodale 1971: 65. The most recent U.S. census data put the difference at about two years, largely as a result of women marrying later.

5. The *Oxford English Dictionary* gives as a primary definition of politics: "the science and art of government" and defines government as "the action of ruling; continuous exercise of authority over the actions of subjects or inferiors." Political anthropologists (e.g., Swartz et al. 1966, Lewellen 1983) focus on group processes: a typical definition is "the study of the *processes* involved in determining and implementing public goals and in the differential achievement and power by the members of the groups concerned with these goals" (Swartz et al., 1966:7).

6. Acknowledging that politics is about resource acquisition and power sometimes leads to expressions of moral censure. It may not surprise anyone that Ambrose Bierce defined a politician as "an eel in the slime upon which the superstructure of organized society is reared," but even Abraham Lincoln was curt about politicians: "a set of men who have interests aside from the interests of the people and who . . . are, taken as a mass, at least one step removed from honest men." Such sly witticisms reflect that there is individual gain to be had from politics. Yet, aside from Machiavelli, few authors treating politics formally, including political anthropologists, have considered the possibility that there may be individual gains in political activity.

7. Whyte (1978, 1979), who used the odd-numbered societies (n = 93, around the world) of the Standard Cross-Cultural Sample.

8. Information is missing in published studies for a fifth of the societies, which I suspect means that women's power is slightly overrepresented here, since a major reason for not reporting relative status is the nonexistence of women as leaders. Thus I suspect a number of "no information" societies are actually male power societies.

9. Whyte 1978, 1979. Women's greater say in kin leadership—Africa: Ibo, Ashanti; Pacific Islands: Truk; Manchuria: Manchu; North America: Haida; South America: Cubeo. Equal influence:—Africa: Nubians; Asia: Semang; Pacific Islands: Iban, Marquesans.

10. Remember that here "monogamy" is the anthropological definition: social monogamy is usually mild polygyny.

11. Convergence of reproductive interest: Flinn and Low 1986, Low 1988a; women's political office: Low 1992; recent data on women in parliaments: Inter-parliamentary Union: http://www.ipu.org/wmn-e/classif.htm (accessed May 22, 2014).

12. Low 1992.

13. Rumors about women's status: Nama, Montagnai, Mbundu.

14. Whyte 1978.

15. Schapera 1930.

16. Schapera 1930: 150.

17. Schapera 1930: 332.

18. Childs 1949, McCulloch 1952.

19. Childs 1949: 178.

20. Rattray 1923.

21. Rattray 1923: 81.

22. Rattray 1923: 83.

23. Rattray 1923: 82.

24. E.g., Charlton 1984.

25. Richards 1951.

26. This is not common but occurs elsewhere in central Africa as well.

27. Richards 1951: 174. Although descent was matrilineal, rank cut across descent lines. Descent was counted back on the mother's side, but the father's name was added as a sort of surname to both sons' and daughters' names, and the father's as well as the mother's ancestral spirit was honored. Not surprisingly, men were likely to recount their father's lineage if it outranked their mother's.

28. Women's literacy, etc.: Carcopino 1941: 84; married women's emancipation: Pellison 1897: 45.

29. Friedlander 1907: 239.

30. Friedlander 1907: 240.

31. E.g., Gardner 1986, Hallett 1984.

32. Linton 1939; 155 reported 2.6/1; Otterbein 1963, in a careful review of earlier studies (1963), reported that it never exceeded 6/5

33. Linton 1939: ways of polyandry arising, Social class: Otterbein 1963.

34. Hershkovits and Hershkovits 1934.

35. Hershkovits and Hershkovits 1934: 192.

36. Speck 1931, Leacock 1955, Lips 1947.

37. Leacock 1955, Strong 1929.

38. Lips 1947.

39. Lips 1947. Note that the importance of male kin is similar to that in the Yanomamö (chapter 7).

40. Lips 1947: 400.

41. Swanton 1928b: 337.

42. Chief's power: Swanton 1928a: 696; women as chiefs: ibid. 696, 700; elections: Swanton 1928b: 331.

43. Bossu cited in Swanton 1928b: 312.

44. Mead 1935; Gewertz 1981, 1983; Gewertz and Errington 1991.

45. O'Brien 1977.

46. Shilluk: Farran 1963; Nyoro: Roscoe 1923, Beattie 1960, 1963.

47. E.g., Trivers 1985, Clutton-Brock et al. 1982, 1986.

48. In no sense do any of these results suggest genetic determinism; there is no assumption or implication that particular alleles render males or females more fit for particular kinds of coalitions or more likely to respond in particular, rigid ways. Rather, the question is: for different kinds of individuals (males, females, high/low status, etc.) in particular environments (social as well as physical), what are the costs and benefits of different strategies? This argument does not hinge on the fact that men are generally bigger and stronger than women. In primates and in human societies, the social complexities so far outweigh the impact of physical size that size alone is a poor predictor of success in getting power or influence. Similarly, this does not reduce to an assertion that women are bound by the constraints of pregnancy, nursing, and child care (e.g., because women who become politically active tend to do so later in life than men). If that were true, sterile women as well as postmenopausal women might broadly be expected to have and wield more public political power. In fact, most men do not achieve significant power when they are young, and even though women may spend more hours per day working in child care while men are "idle" (perhaps doing politics?), there are hours remaining. If there were net reproductive profit to political activities, women's political networks, or bisexual politics, should be more broadly seen.

49. E.g., H. Walker 2014, also note 2 in this chapter, on the percentage of women who become partners in a law firm.

50. Betzig 1986, Betzig and Weber 1993.

CHAPTER 13: SEX, RESOURCES, AND EARLY WARFARE

1. E.g., Keeley 1996; see also Pinker 2011.

2. Adult male red deer mortality from fights: Clutton-Brock et al. 1982.

3. Mathew and Boyd 2011.

4. E.g., Ferrill 1985, Keeley 1996. Quote: Keeley 1996: 175; proportion of men: Keeley 1996: 34; death rates: Keeley 1996: 64. See also Pinker 2011. Perhaps this represents not just a decline in the percentage of warriors' deaths, but a separation of those sent to fight and those not affected by distant wars.

5. Huntingford and Turner 1987.

6. In humans, too, within-population lethal individual encounters are a male phenomenon—and driven largely by reproductive conflicts, though we don't usually think of homicide that way (Daly and Wilson 1988).

7. Reproductive roots of human homicide: Daly and Wilson 1988; assessment of combatants' strengths: e.g., Parker 1974, 1984, Maynard Smith and Parker 1976. Quote: Parker 1974.

8. *The Ax Fight* analysis: Alvard 2009; Lamaleran whale hunters: Alvard 2003.

9. See review by Huntingford and Turner 1987. There is a limit, of course, to the effectiveness of "cheap talk" bluffs.

10. E.g., Manson and Wrangham 1991, Low 1992, 1994a,b.

11. Perhaps the closest I can think of is Barbara Fritchie, who was a childhood icon for me, for hoisting the new American flag under occupation in the American Revolutionary War. The rhyme went: "Shoot if you will, this old gray head, but spare my country's flag, she said."

12. This list is current as of 2014; it may change.

13. Darwin 1871. See Andersson 1994.

14. Aggressiveness cross-culturally: Ember 1981, Barry et al. 1976, Low 1989b; homicides: Daly and Wilson 1988; scope of male versus female politics: Ross 1983, Low 1990e, 1992; women's relative gain: Low 1990e, 1992; gender differences in aggression: Hyde 1996.

15. Quote: Holmes 1985: 102. Historical: Keegan 1987.

16. Even in ungulate species like red deer, in which status and resource control are mediated through physical combat and there is no evidence of reciprocal "political" alliances, size is not the only determinant of status (e.g., Clutton-Brock et al. 1982).

17. Note that under these conditions, women face a conflict of interest with their husbands, for their husbands may be making war upon their fathers and brothers; Adams 1983.

18. The Handicap Principle: Zahavi 1975, Pomiankowski 1987a,b, Zahavi and Zahavi 1997.

19. This is the Sexy Son hypothesis. See review by Andersson 1994: 44–45.

20. Runaway sexual selection: Fisher 1958, quote p. 152.

21. Chagnon 1988. See also Chagnon 1997 for the way ecological factors affect possible payoff in warfare.

22. Lorenz 1966; variations of this argument are found in Eibl-Eibesfeldt 1979 and Ardrey 1966. The term "instinctive" has typically been used for traits that seem to appear without learning or gradual development, though often as we discover more about development, we uncover the process. Thus "instinctive" behaviors are not qualitatively different from "learned" behaviors, simply more likely to occur when relatively fixed and simple responses are appropriate. Alexander (1979) uses the probably more correct but infelicitous term "cryptontogenetic" (hidden ontogeny).

23. Refer back to chapter 1 for the reasons a behavioral ecologist would reject this notion.

24. International Society for Research on Aggression (http://www.israsociety.com/), 1986 Seville statement. For full text (from which quote is taken) see http://eprints.ucm .es/13354/1/XXXIII_CICA_book_of_abstracts.pdf, p. 19.

25. E.g., Harris et al. 1998. Once again evidence from a pathological condition gives some information. In the early 1990s, a Dutch family was identified in which several men had a point mutation resulting in a selective deficiency of enzymatic activity of monoamine oxidase A; these men showed borderline mental retardation and considerable impulsive aggression, arson, attempted rape, and exhibitionism (Brunner et al. 1993). A "gene for aggression"? Genetic influence, certainly, but the data suggest that, here too, genes, environmental conditions, and developmental circumstances interact (Brunner 1996; see also Miles and Carey 1997, Cairns 1996, Cadoret et al. 1997).

26. Tooby and Cosmides 1988.

27. Further, although Cosmides and Tooby (1992) discuss allocation of rewards to participants (that they may be unequal), they fail to address the fact that these payoffs must be compared to payoffs from other strategies (coercion and the dilemma of disenfranchised males).

28. Ember and Ember 1992.

29. Chagnon 1997. This is an excellent study of the interplay of evolutionary trends influenced by ecological conditions.

30. Sex ratio and aggression: Mesquida and Wiener 1996. Aggression in young males: Daly and Wilson 1988. The United Nations' *World Population Prospects*, 1996 revision, suggests that in developing nations, young males will increase as a proportion of population, increasing by 25 million; in the developed world, young males are expected to decrease by 4 million. In the 2012 revision, male life expectancy at birth (1950–1955 cohort) is under 45 in most of Africa and under 50 throughout much of Asia; this phenomenon also leads to a very young male population; the number of men 15–24 has increased by just under 300,000—almost all in the developing world. Compare this to the United States, where that male life expectancy is about 75. As the proportion of young males increases in resource-limited environments, the potential for violent solutions in both local and larger conflicts seems likely to increase.

31. Wolves: Mech 1977; hyenas: Kruuk 1972; lions: Packer 1986; vervet territorial disputes: Cheney and Seyfarth 1992.

32. See Byrne 1995, especially chapters 9 and 14.

33. Gorillas: Watts 1996, 2012, Harcourt 1978; intergroup aggression in primates: Cheney 1987; chimpanzee raids: Manson and Wrangham 1991, Goodall 1986, Goodall et al. 1979, Nishida et al. 1985; chimpanzee female distribution: Goodall 1986, Nishida

1979. See also chapters 4, 10 and references therein. The best recent synthesis, which captures much of the complexity we are continually discovering, is Mitani et al. 2012.

34. E.g., Pusey and Packer 1987, Goodall 1986, Manson and Wrangham 1991. Chimpanzee fusion-fusion groupings vary in the strength of male-male bonds (Boesch 1996), and we might profit from more inquiry into variation in chimpanzees.

35. E.g., see Chagnon 1988.

36. Cheney and Seyfarth 1987, 1992, Packer et al. 1995.

37. E.g., Hrdy 1976, Wasser 1983, Silk and Boyd 1983. Strong social bonds are important in many ways, including longevity (Silk et al. 2010). Further, subordinates who cannot form strong coalitions may be harassed and suffer infanticide, with little risk to the aggressors.

38. See de Waal 1989.

39. Several anthropologists (Durham 1976, Harris 1979, Divale and Harris 1976) have suggested that in the past, wars were fought to secure scarce animal protein from the hunting grounds accruing to the winning side. Yet there is a strong argument that it was not the means of production but the means of reproduction that led to such serious escalation of competition (Chagnon and Hames 1979, Chagnon 1979; see also Ember and Ember 1992). Frequency of war: Ross 1983, Ember 1978, Ember and Ember 1992.

40. Manson and Wrangham 1991.

41. Jivaro: e.g., Karsten 1923, Keeley 1996; Pinker 2011.

42. Blackfoot: e.g., Denig 1961, Ewers 1958. It seems to be currently fashionable among some anthropologists to argue that cultural disruption (e.g., the introduction of horses) upset some balance and generated warfare. Yet the Blackfoot case is functionally identical to the Meru case (and many other pastoral societies) in which no such "disruption" could be postulated.

43. Fadiman 1982; cf. Flinn and Low 1986.

44. Manson and Wrangham 1991.

45. Land: Ember and Ember 1992; adultery and wife-stealing: Divale 1973; Capture of women: White 1988; Maori: e.g., Vayda 1960.

46. Hartung 1992.

47. Chagnon 1997: 191.

Chapter 14: Societal Complexity and the Ecology of War

1. Keegan 1987.

2. Relatedness among the Companions: Keegan 1987: 34; Keegan 1987: 35, 40–41.

3. Stiller 1973.

4. Warfare and state formation: Carniero 1970; political complexity and war: Strate 1985, Otterbein 1970.

5. Thus, you can see that war might not lead to statehood, but perhaps civil wars, as in a number of African and middle-eastern nations today.

6. For example, an early debate (e.g. Carniero [1970] vs. S. Wright [1977], Webster [1975]) centered on whether population growth is demonstrated to precede warfare. However, if competition can be driven to lethal levels by reproductive conflicts, this is no longer a relevant question. Similarly, the military historian's distinction between

"pre-heroic" and "heroic" leadership (e.g., Keegan 1987) may be a matter of scale rather than function. See also Bueno de Mesquita 1981.

7. Axelrod and Dion (1988) have shown formally that increasing the number of actors makes cooperation more difficult; and "noise" (misperceptions, incorrect information) can invite exploitation. In fact, no strategy is evolutionarily stable if the "shadow of the future" is too short. See also chapter 9 on the tensions between individual and group interests.

8. Portuguese nobles: Boone 1986, 1988; disenfranchised males generally: Dickemann 1979; Cheyenne: Moore 1990. In a spectacular misunderstanding of the principles of behavioral ecology (chapter 1), Moore used the lack of reproduction in war chiefs to argue that there is "no gene for warfare" and posited it as a response to Chagnon's work (which did not propose such a gene).

9. Proportion of young males: United Nations *World Population Prospects*, 1996 revision. See also next endnote. Recent Middle Eastern suicide missions are novel.

10. Sex ratio and aggression: Mesquida and Wiener 1996. Aggression in young males: Daly and Wilson 1988; increase/decrease in young males: United Nations *World Population Prospects*, 1996, 2012 revisions. The 2012 revision shows that the number of young men (15–24) in the world has indeed increased by just under 300,000 since the first publication of this book; almost all of that increase has been in less-developed nations.

11. Kin selection: Shaw and Wong 1989; local low-intensity warfare: Brogan 1990, Keegan and Wheatcroft 1986, Dunnigan and Bay 1986.

12. E.g., Hanson (1989: xv) argued that classical Greek warfare has "left us with what is now a burdensome legacy in the West: a presumption that battle under any guise other than a no-nonsense, head-to-head confrontation between sober enemies is or should be unpalatable." Political consequences: McNeill 1963; seasonal constraints: Hanson 1989.

13. General's importance to troop morale: Hanson 1989; generals moving to the rear: Keegan 1987.

14. Violent housekeeping and quote: Hale 1985: 13; knights fighting individually: Hackett 1983: 28; birth order and risk: Boone 1988; socially acceptable property acquisition: Hale 1985: 22; insignias representing social status: Vale 1981: 148.

15. E.g., see McNeill 1982.

16. Hale 1985: 75, Miller 1975, Elton 1975.

17. Savorgnan quote: Hale 1985: 109; Thomas Wilson quote: Hale 1985: 125.

18. Chamberlin 1965: 151.

19. These patterns are parallel to the training of boys in nonstratified polygynous societies in which male kin groups are important (chapter 6). See also G. Johnson 1986, G. Johnson et al. 1987.

20. E.g., Holmes 1985, Dixon 1976. In this context it is interesting to return to the sex differences in child training noted in earlier chapters: highly polygynous groups with no stratification train sons to be obedient—and these are the societies in which small-scale warfare is common: men fighting alongside male kin over women.

21. Hackworth 1989: 633–635.

22. Cited in Gies 1984: 157.

23. *Detroit News and Free Press*, February 10, 1991.

24. E.g., Watson 1978, Hackworth 1989, Schein 1957. Crowfoot and Wrangham 2010 offer an approach that links primate and human aggression in a clear way.

25. UK training: Leigh 2004; U.S. interrogation methods: Frater 2007.

26. E.g. Wilkinson 1986; see also Jacobson and Zimmerman 1993, and the examples above of U.S. involvement in armed conflict since 2001.

27. Low-intensity warfare reviews: Gutteridge 1986, Klare and Kornbluh 1988; genetic interests in warfare: Shaw and Wong 1989; paternalistic terrorist leaders: e.g., Aston 1986; ethnic components in large-scale war: Aston 1986, Brogan 1990, Keegan and Wheatcroft 1986, Dunnigan and Bay 1986, Goose 1988. See also Maechling 1988, Barnet 1988. U.S. Special Forces funding quadrupling from 2001- 2011: Waller 2011.

28. Deception in other primates: Byrne 1995; faulty information: Parker 1974, Alexander 1987: 239. Google "faulty information" "war" for more.

29. E.g., Putnam 1993a.

30. E.g., Alexander 1979.

31. David Hackworth (1989: 634) called the failure to mimic the familial structure of preindustrial warfare (which resulted in high casualties and an increasing need for replacements, followed by a lowering of induction requirements) "the most blatant example of the use and misuse of the poor and disadvantaged in America's wars." In the Vietnam War, these men, the Project 100,000 soldiers, were those who had (or would have) previously failed the armed services' physical or mental requirements; they proceeded through the Army "as they proceed through life, walking wounded in the center of a monstrous joke, forced to struggle with basic training as they are forced to struggle with everything else" (Just 1970: 62). The Vietnam example may be simply a dramatic recent case (Watson 1978: 34).

32. Cf. Holmes 1985: 93.

33. E.g., http://en.wikipedia.org/wiki/Women_in_combat (accessed February 20, 2014).

34. E.g., the Comanche: Cabello y Robles 1961: 178; the Natchez: Swanton 1911: 104. See also Betzig 1986.

35. *New York Times*, October 3, 1992.

36. Congo: Storr 2011; Sudan: United Nations News Centre 2009, which provides continuous updates on rape frequencies around the world. For an overview with relatively complete references and a longer time period, see http://en.wikipedia.org/wiki /War_rape#Recent_occurrences.

37. Need for and beginnings of qualitative data for analysis: Gleditsch et al. 2014; differential impacts of domestic and international terrorism: Enders et al. 2011; indirect adverse effects (e.g. trust) of both local and international terrorism: Gaibulloev and Sandler 2011, Blomberg et al. 2011; influence of inequality on local terrorism: Piazza 2011, Hultquist 2013; why starting negotiations is so hard in insurgencies: Bapat 2005.

38. Overviews: e.g., Singer 1980, 1989, Huth 1988; conflict causes: Richardson 1960, Stoessinger 1982; pacifying influence of shared ethnicity: Shaw and Wong 1989, Richardson 1960; common government and recent alliance: Richardson 1960; military strength: Huth 1988. Leader's personality: Stoessinger 1982, Winter 1989. A clear well-informed approach is Gat 2010.

39. Brogan 1990; other atlases: Dunnigan and Bay 1986, Keegan and Wheatcroft 1986, Black 2005, and http://en.wikipedia.org/wiki/List_of_ongoing_armed_conflicts (accessed February 20, 2014).

40. Cf. Stoessinger 1982.

41. Shaw and Wong 1989: 208.

42. E.g., Keohane 1984, Oye 1986.

43. Alexander 1987: 240.

44. E.g., Goldstein 1989, Keohane 1984, Oye 1986, Groebel and Hinde 1989.

Chapter 15: Wealth, Fertility, and the Environment in Future Tense

1. E.g., see Pickrell et al. 2009 about current selection on human genes.

2. Interestingly, as will become clear later in this chapter, these relationships are largely reversed in modern developed nations for women: it is highly educated, high-status career women who tend to have fewer children than others and are more likely to be childless (e.g., Anderson and Low 2003, Barthold et al. 2012, Mace 2013).

3. Even though caveats are important here, modern hunter-gatherers are still the best mirror we have for many traits (e.g., Hawkes, O'Connell, and Rogers 1997, K. Hill et al. 2009).

4. The "garbage" quote is from Lehrer's song "Pollution." Tom Lehrer is a singer/songwriter best known for witty and satirical songs written in the 1950s and 1960s; he also had a varied career in mathematics, statistics. He taught at MIT and from 1972 to 2001 at the University of California, Santa Cruz. He made many records and CDs, including *Tom Lehrer Revisited* (Reprise). Lehrer quote: "Pollution" from *That Was the Year That Was* (Reprise).

5. This is a particular problem in some common-pool resources; these can be Stiglerian, like Hardin's 1968 example of the English grazing commons, in which only a few people get benefits but everyone shares the costs, or the reverse, Olsonian situations such as hazardous-waste siting, in which costs are concentrated on a few but the benefits are widely dispersed.. A number of current environmental problems are common-pool (really open-access) resources: global temperature, the ozone layer, open-ocean fish and whale populations (e.g., see Myers and Worm 2003, 2005, and discussions that followed about the decline of fisheries worldwide). The demographer Ronald Lee argued in 1990 that, in a resource-limited world, fertility itself is a common-pool resource issue: the children you have are children I can no longer afford to have.

Neither the welfare of strangers nor the far-distant future has been of much concern in human evolution. Common-pool resources and collective action have been problems for traditional societies as well as modern ones, and getting individuals to cooperate for the group good has typically required some individual benefit for a stable solution. Hawkes 1993, in examining why men hunt and share their kills, showed that some individuals supplied meat as "public goods" because of individual incentives, including direct reproductive benefits for men who demonstrated hunting skill. In light of these issues, it is of interest that even extremely simple models (e.g., the origins of cellular cooperation) find that cooperation for the group good (e.g., energy spent for cellular regulation) is more frequent when r (the degree of relatedness) is high among cellular components, and/or the costs are paid from the group resources rather than individual components' resources (S. Frank 1998).

6. E. A. Smith et al. 2010a,b, Borgerhoff Mulder et al. 2009, 2010, Gurven et al. 2010, Shenk et al. 2010. Among agricultural societies, nineteenth-century Sweden appeared to be the single exception, but (chapter 8) this was an artifact of the need for all involved in the project to use precisely the same statistical analysis. In Sweden, the

importance of land was paramount, creating essentially landed and nonlanded classes (90% of the former married; about a third of the latter married: Low 1990c, Low and Clarke 1991); further, issues of primogeniture, land enclosure, industrial development, and the demographic transition all combined to obscure the very real impact of land ownership and inheritance of land on reproduction (chapter 8).

7. E.g., Borgerhoff Mulder et al. 2009, Shenk et al. 2010, E. A. Smith et al. 2010a,b, Gurven et al. 2010.

8. Lawson and Mace 2010, 2011. Demographers (e.g., Wolf et al. 2011) are converging on similar points.

9. Women's paid labor: http://www.oecd-ilibrary.org/employment/employment-rate -of-women-2013–1_emp-fe-table-2013–1-en.

10. MacArthur and Wilson 1967.

11. Island crowding and shift in selection ("r-K theory"): MacArthur and Wilson 1967: 149; human explorations: Barkow 1977, Beauchamp 1994, Irons 1983a, Low 1993a, 1999, Low et al. 1992, Low and Clarke 1991, Voland 1984, 1989, 1990, Voland and Dunbar 1995. See also Wood 1998, Cronk 1991a, Borgerhoff Mulder 1991, 1992, 1998b, and the references in note 4 above. Becker (1993) considered the costs of children explicitly, and Freedman and Thornton (1982) discussed heuristic rules like those proposed in chapter 9 (e.g., how am I doing compared to my peers?) as they affect fertility decisions. Parents strongly favoring investment in their own children: Borgerhoff Mulder et al. 2009, Shenk et al. 2010, Gurven et al. 2010, E. A. Smith et al. 2010a, Borgerhoff Mulder et al. 2010, E. A. Smith et al. 2010b; Thai schooling: Knodel and Wongsith 1991; market training: Kaplan et al. 1995; relative investment affected by degree of relatedness: Anderson 2005, Anderson et al. 1999. Note: we actually do not have enough data to tell whether human demographic transitions are really an example of the biological "r- and K-selection"; we have tantalizing clues like Knodel's Thai data and Kaplan's market analysis, and Mace's 1998 study of the coevolution of fertility and wealth inheritance strategies. Mace, blending empirical data and dynamic simulation, found that increasing costs of children decreased the optimum fertility and increased optimum inheritance: classic r-K selection.

12. Stepparent investment: e.g., Anderson 2005, Anderson 2011, Anderson et al. 2011; cost of schooling and fertility in Thailand: Knodel and Wongsith 1991, Knodel et al. 1990.

13. Fishery declines: Myers and Worm 2003, Heithaus et al. 2008 and papers cited therein; extreme weather events: overviews at http://www.climate.gov/ and http://en .wikipedia.org/wiki/2013_extreme_weather_events (both accessed May 13, 2014).

14. See also M. Wilson et al. 1998: 501–523.

15. "Externalization" is the economic term for keeping the benefits and sending away the costs of one's actions (e.g., shipping one's garbage or toxic wastes to poor communities). A classic example exists of a memo, for example, in which Larry Summers, then a World Bank official, argued in an internal memo that the United States should export toxic wastes to poor African countries: after all, the population density was low (so few would be harmed), their life expectancy was low and unlikely to be greatly lowered by our actions, and besides, they were poor and powerless if any ill effects followed (http://www.whirledbank.org/ourwords/summers.html).

16. I write as if we all agreed that there *is* a dilemma. I realize that we don't all agree, but my own area of research convinces me. Despite the variety of opinion, daily, in the

popular press, politics, and academe, we read and hear about environmental consequences of the combined effects of population growth and per capita consumption. The *Limits to Growth* study of 1972 argued that current growth trends would reach serious limits, but that it was possible to alter patterns in order to reach ecological and economic stability that is sustainable into the future. Since then, a number of environmentalists and political leaders have adopted sustainable development as a goal (e.g., see edited reader of Mazur 1994, works such as Edwards and McKibben 2010, the Worldwatch Institute 2013, and Gluckman and Hanson 2006). The Brundtland Report in 1987 (*Our Common Future*, chapter 2, p. 1, http://www.un-documents.net/ocf-02.htm) defined sustainable development as "development that seeks to meet needs and aspirations of the present without compromising the ability to meet those of the future." For additional examples, see Ehrlich 1997, MacKellar 1996, 1997. But the fact remains: if fertility is lowered, for example, from 8 to about 2, but per capita consumption is raised fourfold or more, sustainability is not at all enhanced (see data at http://www.footprintnetwork.org/en/index.php/GFN/page/footprint_for_nations/).

17. *IPAT*: Ehrlich and Holdren 1971. Not only the per capita rate of consumption differs between developed and the least-developed nations, but so do the kinds of side effects (e.g., type of pollution from activities).

18. Consider Nancy Birdsall's (1980) study on population growth in developing nations. She found a slight negative trend between GNP per capita and the Total Fertility Rate in developing countries (all of which had higher fertility rates than Europe and North America). Such data are widely used, for example by policy makers, to argue that high income correlates with lower fertility; thus the prescription to industrialize and to raise per capita GNP, and all will be well. But as fig. 15.3 shows, things are not that simple.

Though studies like Birdsall's come to mind when people speak about wealth, fertility, and consumption, they are inappropriate. They only tell us that the resource ecology of developed and developing countries is vastly different. The resource-fertility question can only be reasonably asked about individuals within the same population: Do wealthier families have more or fewer children than poorer families? Do they consume more or less? Birdsall's *within-society* comparisons showed a linear positive relationship between wealth and fertility in India, and a curvilinear relationship in the other three countries she examined. In *no* case did the wealthiest fifth of families have fewer children than the poorest. More to the point, the per-child wealth—a reflection of likely per capita consumption—in each country was vastly greater for the wealthiest families compared to others. The possible implications of these patterns are interesting in terms of population policies around the world. The required investment may limit the number of children a couple can afford. Within subgroups, however, those with more than sufficient resources may be able to support additional children and still have all be adequately invested (e.g., Hughes 1986, Low et al. 1992).

19. Three difficulties are most common with aggregate data (e.g., Wrong 1980, 1985): (1) fertility itself is not measured; data are "household" and one cannot tell, e.g., step- and foster children from genetic children; (2) sometimes fertility is measured—but those who never marry, or those hard to census, are missed in the analysis; (3) only averages are calculated, with no idea of variability, so it is hard to tell if differences are meaningful. So, we have a dilemma: it is likely that fertility often declines with income, but we seldom have the measurements to tell.

Aggregate data tell us nothing about variation. For example, if fertility is 1.8 children per couple in much of western Europe, and 2.1 in the United States, is that a "real," significant difference? There is no way to tell unless we have a description of the distribution. Aggregate data for different occupational groups within countries have varied patterns—but without knowing the variance within groups and whether wealth in each group is normally distributed or skewed, we cannot make any inference.

Further, aggregate data typically undersample a very important group in any population: the poor and the homeless. These are probably the least likely to have children, and statistically the biggest predictor of variance is the size of the zero-success class. For example, U.S. census data are taken from households. Suppose poorer households have more children in them than wealthier households. Does that mean wealth and fertility are inversely related in the United States today? By no means. Relationships are not given; income is not apportioned to individuals: so there is no way to know father's earnings versus mother's work effort, with its complex effects.

Finally, the poorest of the poor have both very low fertility and very low income—and seldom appear in census data; if the homeless compose a sufficient percentage of the unsampled population, the statistical analysis is compromised. Unless we ask about them as well, we have no reliable information about the overall pattern of wealth and fertility in the United States. These issues are currently being debated about earlier U.S. censuses, in which it is almost certain that serious undercounts of exactly the sort I am describing have occurred.

Evolutionary anthropologists considering these issues include: e.g., Kaplan 1994, Kaplan et al. 1995, Kaplan and Lancaster 1999; see also Betzig 1996, Low 1993a, Low et al. 1992. The investment required to produce successful offspring varies with environment, and specifically with the threshold level of investment required for a child's success, and fertility varies across natural fertility societies (Bentley et al. 1992). How harsh is the competition that offspring face? In this sense, environmentally influenced differences in successful investment in children is precisely analogous to the proper use of MacArthur and Wilson's (1967) concept of r- and K-selection (see reviews by Low et al. 1992, Low 1993a, 1999, also Mason 1997 and Borgerhoff Mulder 1998b. For empirical descriptions of developing world transitions and the failure of demographic transition theory, see Schofield and Coleman 1986, Robey et al. 1993). Do relatively wealthy women who have late and low fertility nonetheless have greater net lineage increase? I suspect not, but we have few data. In postdemographic transition societies, in which it appears that successful children cost more per capita in parental investment and fertility is relatively low, women might actually profit reproductively at the lineage level by spending more effort bringing in resources rather than filling traditional roles of child production.

20. Positive empirical correlations: e.g., Irons 1983a, Chagnon 1988, Mueller, 1993, Mueller and Mazur 1996, 1998, Low and Clarke 1992, Low 1990c, 1991, and chapter 8, Essock-Vitale 1984, J. Hill 1984, Essock-Vitale and Maguire 1988. Lineage success through fewer, better-invested children: Mace 1998, Low et al. 1992. Disappearing correlations: Kaplan et al. 1995, Kaplan and Lancaster 1999, Low 1989b. See also Rogers 1990 and Harpending et al. 1990.

21. Wealth accumulation, inequality, and success in premodern societies: E. A. Smith et al. 2010a, Bowles et al. 2010, Borgerhoff Mulder et al. 2009. Small family size and ability to accumulate wealth: Havanon et al. 1992. We are still unsure whether constraint of

fertility in competitive environments is a general picture in modern societies. It seems likely to be true for a family with some wealth; in contrast, a family with no resources may do best to maximize fertility and let children's own initiative be the deciding factor—yet really poor women in the United States today have low total fertility.

22. Models on the value of children: Becker 1993, 1993; see also Merrigan and St.-Pierre 1998, Easterlin, 1978; children's economic value: Caldwell 1976, 1982, 1983. Note that, while the only good empirical data we have suggest that children in today's hunter-gatherer societies never produce true net upward wealth flows (e.g., Hawkes, O'Connell et al. 1998, 1999; Turke 1992), children's value as *genetic* currencies means that even when they have a net economic cost, they have a total net benefit, and this answers the quote (Schoen et al. 1997: 334) asking why "Americans still want children."

23. Support for Easterlin's hypothesis: review by Macunovich 1998; effects of parity and marital status: e.g., Schoen et al 1997; effects of government policies: e.g., Gauthier and Hatzius 1997.

24. Women's workforce participation: e.g., Kasarda et al. 1986, Taşiran 1995. Low et al. 2002, 2003, using sociological data for the United States, found that when women delayed age at first birth beyond about age 30, their total fertility rate was much reduced. Delayed fertility is the strongest correlate of slowed population growth, even when it does not mean lower total lifetime fertility.

25. Harvey and Zammuto 1985. Stearns 1992 has an excellent graph of the relationship. Like many other useful generalizations, this one makes simplifying assumptions, and it is important not to overgeneralize, for examining variance often proves useful. Even in traditional societies, there is considerable variation in longevity. Here the assumptions are (1) that any population for which data exist will adequately represent the species and (2) that conditions are not changing rapidly. See Low et al. 2008, 2013 for transnational patterns—but be aware that there can be significant within-nation variation as well.

26. E.g., Daly and Wilson 1987. Wilson and Daly (1997) who analyzed 300 Chicago neighborhoods.

27. Consider the question of welfare payments and fertility. For nonwhite women in the United States today, it appears that the Aid to Families with Dependent program increases the likelihood of having a first child (Taşiran 1995, his table 30; see also Luker 1996). However, it has *no effect* on the probability of second and subsequent births; thus the AFDC program is unlikely to lead to high total fertility among poor women. In fact, contrary to widespread beliefs about the fertility of welfare mothers, today in the United States poor women have early but *low* total fertility. Women on welfare in Wisconsin have 29.4% fewer children than women across the United States, and 32.8% fewer children than women in Wisconsin; such women tended to be unmarried, black, and less educated than others (Rank 1989). More recent data from the United States, Canada, and Great Britain Summarized in Brewer et al. 2009 suggest that sufficiently high levels of welfare can increase fertility. In Great Britain, for example, 1999 child-contingent welfare reforms did increase fertility, primarily for women in couples, and in the low-education group.

28. Poor women's fertility in the United States tends to be low but early, which is adaptive for the reasons explored above (citations in notes 18 and 27). Consider poor urban black women in the United States today: as they age, they suffer excess mortality and a variety of sublethal health problems that make it not only more difficult to conceive but to raise children successfully. High mortality among black men means that the

probability of having a male partner declines with a woman's age. Even female assistance (mothers, aunts, etc.) declines. In the poor black populations of Harlem, central Detroit, and Watts today, a teenage mother had about a 75% chance that her mother would be alive and able to help when her child is 5 years old; for women who postpone childbearing until age 20, that figure was only 40%. Not surprisingly, some scholars suggest that early childbearing may mitigate the threat to family economies and care-taking systems (and reproductive success) imposed by the heavy mortality and disease burdens of these women (e.g., Geronimus et al. 1996, 1997, Geronimus 1996a,b. Lancaster 1986 made similar suggestions for more traditional societies). Further, there is evidence that these disadvantages affect the life chances of such women's children, exacerbating disparities over generations (e.g., Duncan et al. 1998).

These findings fit the predictions from life history theory remarkably well: the most important variable in setting the timing of fertility is adult mortality (chapter 6, fig. 15.4a; see also Coale and Trussell 1974, 1978, Ellison 1991). Early fertility is common among poorer women who face uncertain futures, and teenage fertility is strongly associated with low educational achievement for women—in contrast to the age-specific fertility of American women in general, which is shifting to later in life. From 1976 to 1988, overall fertility peaked at ages 20–24, but compared to 1976, pregnancy rates in 1988 were substantially higher for women aged 30–39 (Rank 1989, Luker 1996).

The most puzzling fertility pattern is that of relatively wealthy women who earn a high proportion of their household income and who routinely delay fertility past their midtwenties. My own suspicion is that two things interact. First, when very high offspring-specific true parental investment is required to produce competitive and successful offspring—that is, when environments are highly competitive, women's mate value shifts from largely-to-entirely reproductive value, to relatively more resource value. Women, like their husbands, may concentrate on getting resources for offspring rather than on producing more offspring; such production of fewer-but-better-invested offspring seems likely to be more profitable under these circumstances. Obviously, any family that can produce both more and better-invested children would profit, but such families would be a very small percentage of the population, and a linear pattern of high wealth–low fertility is not likely. Under these circumstances, we predict declining family sizes, and greater participation of women not only in the job market but in the higher-paying professional market. But I suspect a second factor operates as well: the proximate cues (chapter 2) of "doing well" are wealth, health, possessions; when both partners are highly educated, a likely result is that it will prove enticing to continue to strive and accumulate— and this will delay and decrease fertility. So we may see a very mixed bag: some small but highly successful families who concentrate resources in a few successful children, some families that attempt to pattern and fail, and some highly consumptive ("DINKs") couples who may have no children.

29. Taşiran 1995. Taşiran compares empirical studies and models for Sweden (pre-1935 to 1965 cohorts), and white and nonwhite American women (pre-1945 to 1966 cohorts), using macro-, micro-, and integrated data. Women's education, another "lost time" trade-off, has a negative effect on fertility: the longer women stay in school after high school, the fewer children they are likely to have (see also Low et al. 2002). This effect is also complex. For Swedish women, more education means they are less likely to have children, and if they have children they are likely to stop at two (Taşiran 1995: table 30). More-educated white American women are also less likely to begin having families

(see also fig. 15.2). For nonwhite American women, 5–9 years of education makes having a first child less likely; 10 or more years of education makes having a first child more likely, though later in life (which decreases lineage growth).

30. Some of these differences appear to be simple cultural preferences; others seem to reflect cultural influences on the costs and benefits of having a child. In Sweden, whether a woman is married or cohabiting (versus living alone) has no effect on her likelihood of having one or more children; significant social services exist for parental benefits. A man's wages in Sweden have a positive effect on the likelihood of having a first and second child. Interestingly, in the United States, a man's income has a negative relationship to a woman's probability of having a first, second, or third child (Taşiran 1995). We have no information (e.g., lineage persistence) to help us understand whether this is a sensible strategy shift (higher per capita investment per child, fewer children) or simply a reproductively costly cultural preference. If highly educated and wealthy men tend to marry highly educated women, in the United States, then we might see a spillover effect . . . what an interesting question!

31. Taşiran 1995, Rank 1989, Kasarda et al. 1986, Low 1993a, Low and Clarke 1992, Low et al. 2002, 2003.

32. Global Footprint Network: http://www.footprintnetwork.org/en/index.php/GFN /page/footprint_for_nations/.

33. United Nations' *World Population Prospects*, released continuously: https://www .google.com/search?q=UN+Population+Projections&ie=utf-8&oe=utf-8&aq=t&rls=org .mozilla:en-US:official&client=firefox-a&channel=sb.

34. Cohen 1995 and Lutz 1994 are excellent nonpolemic, analytic treatments. See also MacKellar 1996, 1997, Portney and Weyant 1999, and Cohen 1998 for explicit treatments of the role of equity in sustainability. For data estimates online, see http://www .footprintnetwork.org/en/index.php/GFN/, the Global Footprint Network, which is revised continually. GDP insufficient: c.f. Costanza et al. 2014. The authors suggest a number of alternative measures; curiously, they leave out the United Nations Human Development Index, which generates (in its most recent report) data for 187 countries, using a composite index based not only on Gross National Income per capita, but also life expectancy, education, and a set of defined nonmonetary influences.

35. E.g., Cross and Guyer 1980, Costanza 1987.

36. Ancient and traditional society overexploitation: Bodley 1990, Clay 1988; Quaternary extinctions: Martin and Klein 1984; late Pleistocene extinctions: A. Anderson 1984, Cassels 1984, Trotter and McCullough 1984, Crosby 1986.

37. Low 2013b. Steel axes: Salisbury 1962; Ye'kwana: Hames 1979.

38. Low 1996a.

39. Fish depletion worldwide: e.g. Myers and Worm 2003, 2005; extremes in climate change: http://www.whitehouse.gov/energy/climate-change (accessed May 12, 2014); discount rates: e.g., Nordhaus 2008: 9.

40. Acheson 1988.

41. E.g., Ostrom 1998b, 2000.

42. Empirical data: J. Wilson, Acheson, and Johnson 2013; modeling of how ecology can affect self-organization: J. Wilson et al. 2013.

43. Kurlansky (1998: 8) quotes fisher Sam Lee of Petty Harbor, Newfoundland, in commenting about an early ban on the number of fishers (125) who could work in that bay: "Nowadays everyone tries to say it was for conservation. There was no such thing as

conservation. For God's sake, there were enough fish to walk on. It was because there wasn't enough room [in the bay]."

44. These conditions have led to near-shore and then further-out extinction of local subpopulations, leading to the conditions depicted in *The Perfect Storm* (Junger 1997), later made into a movie.

45. WWF: http://worldwildlife.org/press-releases/no-more-cod-in-15-years-wwf-report-warns; Norwegian response: http://web.archive.org/web/20080629194826/http://www.aftenposten.no/english/local/article31. Both retrieved May 13, 2014.

46. Hannesson et al. 2008; see also Holm et al. 2000. The important points are that small, localized fisheries, with a stable group of fishers, can be managed relatively sustainably (Hanneson et al. 2008); Holm et al. make the point that under these circumstances, managing space (who can fish where) is also effective.

47. For a readable summary of the problem, see Kurlansky 1998.

48. Dryden's was the first recorded use of the term, in *The Conquest of Granada* (1672): "I am as free as nature first made man, / Ere the base laws of servitude began, / When wild in woods the noble savage ran." On the noble ecological savage: Redford 1991.

Rousseau, of course, used the concept effectively to anathematize civilization—and in fact, even today, there is a strong streak of romantic distaste for human company and human machinations in various environmental movements. A peculiar conjunction of historical events may have exacerbated possible misinterpretations. Considerable data were collected during the nineteenth-century explorations, at the height of the Romantic approach to ethnography and the growth of manifest destiny in the United States; there are many assumptions, but few data, about the precontact practices of traditional peoples among the noble savage proponents. A mythology developed, probably promoted by the Romantic view, of a pristine America barely peopled and populated by societies virtually without impact (Denevan 1992). American primitivist writers like Cooper and Thoreau probably fueled this image. Threatened by constraint, even extinction, it may have paid indigenous peoples to acquiesce in an overstatement of the globally conserving long-term nature of their policies.

Today we have remnants of this history (e.g., Krech 1999, Edgerton 1992, Oelschlanger 1991). For example, the words of Chief Seattle regarding an impending ecological crisis, reminding us that "the earth does not belong to man, man belongs to the earth" are widely quoted—yet it appears that those words were written by a scriptwriter in the early 1970s, and there is no empirical evidence suggesting that Chief Seattle had any such attitude (e.g., see Budiansky 1995: 32–34). And, as anyone who remembers George Bush's desire to be remembered as the Environmental President can attest, stated attitudes and actual practice may remain far apart (cf., Tuan 1968, 1970 for Chinese examples similar to the United States). In both the United States and China, protested areas were easy to set up *if* they were far from centers of population and perceived to be of little value.

Oelschlanger (1991) reviews romantic history views in which reverence is required. At its extreme, the Romantic view surfaces as the Gaia concept, which argues that the earth is a homeostatic, self-regulating unitary entity especially suited for human life (Lovelock and Margulis 1974, Lovelock 1988). See Williams 1992a for a cogent debunking.

Such romantic misconceptions might not matter, except that they generate normative prescriptions rather than understanding. Several current strategies of environmental and conservation education reflect our faith in this wisdom (see review by Budiansky 1995), but without any logical or empirical evidence. Some muddle the pragmatic and

the normative: if only we could recapture the reverence and cooperativeness of traditional societies, and expand them, we could solve our problems. And this "wisdom" is
simply not true. Empirical tests of the Noble Savage hypothesis suggest that individuals
strive to get sufficient resources efficiently and in competition with other (also striving)
individuals (e.g., Hames 1979, 1988a,b, 1989, 1991, Alvard 1994, 1998; see also chapter
4). "Excess" resources, over and above what's needed for health, are sought if they are
reproductively useful. The cross-cultural data are about what an evolutionary cynic
would predict, but not what a Noble Savage advocate would hope (Low 1996a).

Traditional peoples often do express concerns about the long-term "health" of the
ecosystems on which they depend. They often have unique knowledge of sometimes
subtle ecological relationships because such knowledge has had survival value (e.g.,
Atran 1993). And they often, particularly when the interests of kin or local community
are vested in sustainable use, can manage resources sustainably, sometimes even when
new, highly efficient technologies make overexploitation easy (White 1988). But these
are possible outcomes, not driving forces. People take resources from the environment
to feed themselves and their families, to garner favorable attention (especially for men in
sexual selection; chapter 7), and use reciprocity to "even out" variation in success. This
kind of reciprocity is typically biased by kinship relations, with closer kin receiving more
aid (e.g., Hames 1989). Because empirical studies suggest that people take as much as
they need (or sometimes as much as they can), novel technological changes do not give
them any reason to consider the potential impacts of maximum use—it has never mattered before. This, I suspect, is the source of the pattern in the cross-cultural data that
degradation frequently follows technological change.

For traditional peoples as well as modern, impact is simply a function of the population density, the relative impact of the technology used, and the consumption level (the
biggest influence here is the profit from taking the resource, if any, above subsistence).
Several combinations of population, profit, and technology will yield low environmental
impact but do not imply conservation: low population density plus inefficient technology (e.g., stone axes) plus no market, for example. To infer that low impact must imply
conservation ethic is simply not logical; we can infer deliberate conservation only when
impact is low even though there is high profit to be made, the technology is sufficient,
and/or the population density is moderate to high (fig. 15.6).

49. Invalidity of backward inference: Low 2013b. See also Hunn 1982, in which
Hunn made the point long ago that one cannot call low impact "conservation" unless
one can show intent (if one cannot, he called the result "epiphenomenal conservation").
Ruttan and Borgerhoff Mulder (1999) helped discriminate among outcomes according
to their driving forces. These are implicit in Alvard 1994, 1998. Jared Diamond's *Collapse*
(2004) reviews several major societal failures involving environmental damage, as do
Martin and Klein (1984).

50. I am indebted to Professor Emilio Moran, Indiana University, for this quote.

51. See especially Princen 2005.

52. E.g., Michaels and Monforton 2005, and the conflict over the White House Report by scientists: http://www.whitehouse.gov/energy/climate-change.

53. M. Wilson et al. 1996, Low 1996a,b. Gender styles of management: American
Psychological Association 2006, Davidson and Burke 2011, Burke and Collins 2001.

54. See, for example, Lancaster 1991, Hrdy 1997b, and Gowaty 1997, an edited volume of broad scope.

55. E.g., Simon 1996. Note, however, that the lesson we draw from this correlation is very different if we think, as did Simon, that more people lead to more comfort, versus the interpretation that inventions increasing the standard of living tend to increase fertility, and thus population growth (see, e.g., Ahlburg 1998). The long-standing argument between Julian Simon and the Ehrlichs was really about different scales of analysis. Simon noted that as world population increased, so did technology and standard of living; Paul and Ann Ehrlich noted that no species ever has been able to grow without limits.

56. Using simple heuristics this way (see Gigerenzer et al. 2000) actually combines the cultural evolution approach of Rob Boyd and Pete Richerson, and Lee Cronk's insights—and also Daniel Kahneman's work in *Thinking Fast and Slow*.

57. This is in contrast to the classical economic approach of continued growth; see Costanza et al. 2001a, 2014.

58. The role of environmental uncertainty (e.g., Low 1990b) meant that saving for the distant future never paid. Would one expect conservative management for one's great-grandchildren? Probably not, under most conditions. Remember that, in an outbred population, the relatedness of an individual to her/his direct descendants decreases by 0.5^k over k generations. So ego would share only about 12% of genes with a great-grandchild; by five generations, the shared genes are 3%! So the far-distant future of one's descendants was probably of little concern, even if our ancestors had been able to control their environments. This emphasis in our evolutionary past on short-term optimization may be a serious stumbling block in our ability to ask ourselves for any significant sacrifices now for the sake of future protection.

59. For example, De Young 1993 has worked successfully with a series of relatively local conservation issues and found that in settled areas, social motivation can be more effective than, say, giving 10 cents for each returned bottle. He reviews five dimensions of criteria on which to judge success.

60. E.g., Mancebo 2013.

61. Shared problems and integration of benefits: Redpath et al. 2013; stakeholder participation in modeling impacts: Costanza et al. 2001a, 2014.

62. See Rabe 2013 for northeastern and California regions; Outka and Feiock 2012 on refining definitions to achieve better impact of local and regional plans; Krakoff 2012 on the importance of local involvement even for planetary-level problems like greenhouse gases; Millard-Ball 2012 on what drives local governments and people to become involved, and the impact of that involvement on success.

63. Of course, natural selection has never been "pleasant"—extinction is the most probable outcome for any species. It is certainly true that extinction rates have increased greatly in the last hundred years, due to human activity; it is also true that 95% of all species ever in existence are extinct today. In the case of our own species, many of us have more than an academic interest in whether we can avoid, or delay, our extinction.

64. See Costanza et al. 2014, Costanza et al. 2001b, Mancebo 2013.

65. Policy working with evolved predilections: Griskevicius et al. 2012; prosociality: Henrich et al. 2004; inability to multiply up strategies from simple small stable groups to large heterogeneous societies: e.g., Ostrom 1990, Costanza et al. 2001; Lisbon Principles: Costanza et al. 1998.

66. Griskevicius et al. 2012.

67. E.g., Penn 2003, Low 1993a, 1996a,b, 2001, 2013b, Low and Heinen 1993, Heinen and Low 1992.

Glossary

adaptation—a structure, or physiological or behavioral phenomenon, shaped by natural selection, that renders the possessor more likely to survive and reproduce than its competitors. Also, the evolutionary process leading to such a trait. The concept is onerous; no phenomenon should be labeled an adaptation without effort to determine that the benefit seen is in fact a true function (q.v.) of the trait (Williams 1966).

aggregation—a temporary grouping of animals, usually around some limited and localized resource. Animals in aggregations, in contrast to animals in longer-term social groups, generally show no, or limited, social interactions.

allele—a particular form of a gene, distinguishable from other forms or alleles of the same gene; thus at a locus for eye color, there may be competition between alleles for different eye color to occupy the locus.

allopatric—referring to populations or species occupying different geographical areas. (Cf. *sympatric*).

alpha—referring to the highest-ranked individual in a dominance hierarchy.

altricial—pertaining to young animals which are helpless for a significant period after birth or hatching (Cf. *precocial*).

altruism—in common parlance, any helpful behavior. Here I restrict its use to behavior that genetically costs the performer and benefits some other unrelated and nonreciprocating individual. Most behavior labeled as altruism should more properly be specified as phenotypic altruism—behavior that appears altruistic but may well be genetically selfish (as parental behavior, reciprocity)—to distinguish it from the usage I employ here, genetic altruism.

anisogamy—condition in which the gametes of two reproducing individuals are of unequal size. In general, we call the large gamete an egg and its producer a female; and the small gamete a sperm and its producer a male.

anthropomorphism—interpretation of what is not human or personal in terms of human or personal characteristics.

aposematic—equipped with conspicuous "warning" coloration, usually associated with poisonous or distasteful characteristics (e.g., monarch butterflies).

avunculocal residence—a pattern of residence in which a married couple lives with or near the husband's maternal uncle. (Cf. *matrilocal residence, patrilocal [virilocal] residence, neolocal residence*.)

Barr body—a condensed, inactive X chromosome found in interphase nuclei of XX individuals in mammalian species.

breeding, cooperative—the condition in which a new mother receives help (e.g., food) from others: often her mother, but also from her older children, if any, and nonreproducing members of the group.

bridewealth—a substantial gift of goods, money or service (bride service) given by the groom or his family to the bride's family at or before marriage. (See dowry.)

brood—n.: litter or clutch; v.: to incubate eggs by sitting on them.

carrying capacity—the largest number of individuals of a particular species the environment can support, theoretically for an indefinite period of time; usually symbolized by *K*. Although treated as a constant, it is clear that the carrying capacity of all but the most stable environments must change, because environmental conditions change.

clone—a population of individuals all derived asexually from a single parent.

clutch—the number of eggs laid by a female at one time. (See *litter.*)

coefficient of relatedness—the fraction of genes identical by descent in two individuals; symbolized by *r*. (See *inclusive fitness*, chapter 2.)

commons—a term loosely used to mean either (1) common-pool resources or (preferably) common-property regime, both of which are controlled and used by a well-defined group in common, while excluding outsiders; or (2) (loosely) open-access resources (for which there are no access rules), which are typically soon depleted as people pursue their individual interests at the cost of group resources. The ability to exclude outsiders and to monitor and punish cheaters within the group, are crucial to successful maintenance of commons.

conjugation—joining of cells via a cytoplasmic bridge, followed by transfer or exchange of DNA to accomplish reproduction.

constancy—the extent to which an environment is predictable because conditions remain the same with regard to the parameter measured; symbolized by *C*. (Cf. *contingency.*)

contingency—the extent to which an environment is predictable because some parameter is correlated with (contingent on) some other parameter; e.g., the seasonality of rainfall; symbolized by *M*. (Cf. *constancy.*)

cross cousins—children of siblings of the opposite sex. One's cross-cousins are father's sister's children (FZD) and mother's brother's children (MBD). (See parallel cousins.)

culture—socially transmitted information, including behaviors, beliefs, attitudes, customs, and ideas.

deme—a small, reproductively isolated population, within which reproduction is generally assumed to be random. (See chapter 9.)

dimorphism—having characteristically two (di-) "morphs," or physical types. For example, sexually dimorphic species are those in which the two sexes look different.

diploid—having a chromosome complement consisting of two copies of each chromosome. In mammals, for example, most body cells are diploid. (Cf. *haploid.*)

directional selection—selection that favors individuals with one extreme of a characteristic and disfavors individuals with the other extreme; thus the presence of coyotes as predators for rabbits is likely to favor fast-running rabbits and disfavor slow rabbits. (Cf. *disruptive selection, stabilizing selection*.)

disruptive selection—selection that operates against the middle of the range of variation in any characteristic, thus favoring the two extremes and tending to split the population. It is often proposed as the mechanism for sympatric speciation, for example, in allochronic speciation, and for the evolution of two sexes with divergent characteristics. (Cf. *stabilizing selection, directional selection*.)

dominance hierarchy—the behavioral domination of some members of a group by other members in relatively long-lasting patterns.

dowry—goods paid by the bride's family to a daughter on her marriage (rather than parents' death). The husband's family typically gains. (See *bridewealth*, chapter 6.)

drift, genetic—random changes in gene frequency, usually in small populations as some genes disappear because particular individuals do not reproduce.

effect—any result of a characteristic not produced by natural selection, as opposed to functional adaptations produced by natural selection. The function of an apple is reproduction; an effect was the derivation of Newtonian physics (Williams 1966: 9).

epideictic display—displays in which members of a population show themselves and "allow others to assess the population density" (Wynne-Edwards 1962). Wynne-Edwards proposed that the function (q.v.) of such displays was the regulation of population numbers. (See chapter 9).

epistasis—the condition is which one gene's expression depends on the presence of another "modifier" gene. The impact can be additive if the two genes do not interact (e.g., use different metabolic pathways).

equilibrium frequency—symbolized by $\hat{q}$; the frequency of any specified gene after the population has reached genetic equilibrium.

estrus—the point in a cycle at which a female mammal has ovulated (and thus can conceive) and is sexually receptive.

evolution—any change in relative frequencies of genes over time, as a result of natural selection (q.v.), genetic drift, mutation (q.v.), and recombination (q.v.). Sometimes the terms "organic evolution" and "biotic evolution" are used to specify changes across time of the kind of individuals seen and the kind of groups seen, respectively.

evolutionarily stable strategy (ESS)—a strategy that once established cannot be beaten or invaded by another.

exogamy—cultural rule specifying marriage to a person outside one's own kin or community group.

fecundity—denotes the theoretical reproductive potential of an individual; the potential to reproduce.

fertility—a term reflecting the actual, or realized, reproduction.

fitness—as originally used by Darwin, the term "fitness" meant roughly the ability to survive and reproduce. There are at least five definitions including this one (Dawkins 1982). Two of the most commonly used are W, the fitness of the genotype (the contribution to the next generation of one genotype relative to the contributions of other genotypes), and "classical fitness," an individual measure (an individual's relative lifetime reproductive success). The remaining two definitions are Hamilton's inclusive fitness (q.v.) and what Hamilton called "neighbor-modulated fitness," thought to be too unwieldy to calculate (an individual's relatives' effect on its reproduction).

function—any characteristic produced by natural selection as a result of adaptation; the purpose of the characteristic. (Cf. *effect.*)

gamete—haploid cell specialized to fuse with another haploid cell (fertilization) in a sexual life cycle (e.g., egg or sperm). Bacteria and a number of other unicellular organisms have a different mechanism for attaining genetic variation in individuals: conjugation (two cells align, form connecting bridges, and exchange DNA).

game theory—mathematical analysis of optimal choice of strategies when one's payoffs are affected by other actors. Rules define *who* the players are, *what actions are possible for each*, and *what determines success*. Actors in a game have a *strategy*, and a *payoff*, In evolutionary game theory, the payoffs are genetic.

gene—a segment of DNA encoding a polypeptide or one of the RNAs involved in translation.

genomic imprinting—defined by David Haig (1992). An epigenetic phenomenon in which conflicts of reproductive interests among the mother, father, and fetus are mediated. Certain genes from one parent may be suppressed by the corresponding gene in the other parent (monoallelic gene expression). It has been documented in fungi as well as plants and animals.

genotype—the genetic constitution of an individual organism; may be used in reference to a single trait, a set of traits, or the entire genetic constitution (genome) of an organism. (Cf. *phenotype.*)

group selection—several quite distinct arguments are all called by this term: (1) selection favoring or disfavoring whole groups rather than genes or individuals (Wynne-Edwards); (2) the effects of group structure on gene frequency (Sewall-Wright); (3) the effects of coalitions on the interests of others within a group; (4) the effects of cultural transmission on gene frequency. (See chapter 9.)

Hamilton's rule—predicts that helping behavior among relatives will be favored by natural selection whenever $rb - c > 0$, where r = degree of relatedness, c = cost to the helper, and b = benefit to the recipient. In sum, a helping behavior should evolve when there is a net genetic gain.

haplodiploidy—the condition seen in many social insects, in which the males are derived from unfertilized eggs (and are thus haploid), while females are derived from fertilized eggs (and are thus diploid, having the full complement of genes). In this condition, sons are exactly half like their mother—but because males have only one set of chromosomes, queens' daughters are 75% like each other (identical through father), unless the queen has mated multiple times.

haploid—having a chromosome complement consisting of just one copy of each chromosome. (*Cf. diploid, haplodiploidy.*)

harem—a social group, usually consisting of one mature male, several females, and sometimes immature males and females. Typically, harem-holding males attempt to keep females from leaving the group and other mature males from joining.

herbivore—an animal that eats principally vegetable material. Organisms may be more specifically categorized as frugivores (fruit eaters), granivores (grain eaters), folivores (leaf eaters), etc.

heterogametic—having different sex chromosomes, as XY in male mammals, or XZ in bird females. (See *homogametic*.)

heterozygous—having, at any specified genetic locus, two dissimilar alleles.

homogametic—having two sex chromosomes of the same type, as XX in female mammals. (See *heterogametic*.)

homozygous—having, at any specified genetic locus, two identical alleles.

inbreeding—the mating of closely related individuals.

inclusive fitness—the sum of an individual's contribution to the next generation through the effects of the individual's actions on genes identical by descent (IBD) in other individuals. (*See* Grafen 1991, chapter 1.)

interdemic selection—when groups of reproducing individuals are small, isolated, and breed largely within their group; this structuring of the population can affect gene frequency. The conditions for effectiveness of interdemic selection are stringent and seldom met. (Cf. *group* selection, chapter 9.)

intrinsic rate of increase—the maximum growth rate of a population under specified (close to optimal) conditions; symbolized by r, this parameter is usually wrongly assumed for analytic purposes to be a constant for any species.

iteroparous—reproducing repeatedly. (Cf. *semelparous*.)

kin selection—selection on genes resulting from those genes' presence in relatives other than descendant relatives; e.g., individuals may increase their inclusive fitness (q.v.) not only through offspring, but also through helping nondescendant relatives. What counts and what doesn't is a bit tricky and has often been miscalculated (Grafen 1991). (See *inclusive fitness*.)

K-selection—selection imposed by consistently high density of conspecific competitors. MacArthur and Wilson (1967) noted that a high density of competitors favored parents who were efficient in converting resources into offspring. In nonhuman species, a high density of competitors is the strongest pressure favoring production of fewer but better-invested offspring. In humans, other factors than density can result in intense competition. (See *r-selection*, chapter 15.)

lek—an area traditionally used for communal sexual or courtship displays. Typically, males display only their ability to get and hold a territory; no resources useful to females, such as food or nest sites, are involved. Males on particular (usually central, hard-to-defend) territories attract the most mates.

levirate—cultural custom in which a man marries his brother's widow.

life history—an organism's entire ontogeny, including time to maturity, degree of iteroparity, degree of sexual dimorphism, breeding system, clutch or litter size, degree and kind of parental care. Life history strategy analysis investigates the costs and benefits of each life history parameter.

lineage—a kinship group. In anthropology, the group is further defined by tracing the relationship either through males (patrilineage) or females (matrilineage).

litter—all offspring born to a female at the same time. (See *clutch*.)

locus—the location of a gene on a chromosome. (Plural: *loci*.)

marsupial—mammal (e.g., kangaroo, koala, wombat) that lacks a placenta and gives birth to young after a short period (e.g., 30 days). The young typically complete their development in the marsupium, a pouch on the female's abdomen.

mating effort—that portion of reproductive effort (q.v.) devoted to gaining a mate. Typically assumed to include both caloric and risk expenditure. Includes territorial behavior, mating displays, dominance fights, etc. (Cf. *parental effort*.)

matri—of or through the mother, such as "matrilocal," living with mother's kin; or "matrilineal," tracing descent through the female line.

matrilocal residence—a pattern of residence in which a married couple lives with or near the wife's parents. (See *patrilocal* [*virilocal*] *residence, avunculocal residence, neolocal residence*.)

monogamy—the condition in which equal numbers of males and females contribute to the next generation. Typically, a single male and female will cooperate to raise at least one brood. (Cf. *polygyny, polyandry*.)

mutation—a chance alteration in the code of a gene, often as the result of environmental insult.

Nash equilibrium—in game theory, some combination of players' strategies (some solution) from which neither player would willingly change unilaterally.

natural selection—Darwin observed that (1) in any environment, not all individuals will survive and reproduce equally well, and (2) variation that is heritable means that, over time, individuals that survive and reproduce better will come to comprise a larger and larger proportion of the population. We call this process natural selection: the selective, or filtering influence of environmental conditions. Natural selection is the only one of the contributing forces leading to evolution for which there is any predictive ability.

neolocal residence—a pattern of residence in which a married couple lives separately, and usually at some distance, from the kin of both spouses. (Cf. *matrilocal residence, patrilocal* [*virilocal*] *residence, avunculocal residence*.)

neoteny—attaining sexual maturity while still in larval form, as in *Necturus*, the mud puppy.

omnivore—an animal that eats both animal and vegetable material.

paradigm—in general, an outstandingly clear or typical pattern. In research, a logical structure or pattern used to solve problems.

parallel cousins—children of siblings of the same sex. One's parallel cousins are one's father's brother's children (FBD) and mother's sister's children (MZD). (See *cross cousins*.)

parental effort (PE)—that portion of reproductive effort (q.v.) devoted to the production of offspring; typically, egg and sperm production, lactation, etc., are included (Low 1978). Some activities (e.g., nest construction) may function as either parental effort, if specifically done for the use of raising offspring, or as mating effort (q.v.), if used to attract a mate.

parental investment (PI)—that portion of parental effort (q.v.) received by any specific offspring; what is invested in one offspring cannot be invested in another (Trivers 1972). $\Sigma PI = PE$ for any period. There is thus a conflict of interests among siblings, and between each sibling and its parent(s).

parental manipulation—adjustment or manipulation of parental investment by parents to reduce the reproduction of some offspring in the interests of maximizing the parent's inclusive fitness. Examples might be sex ratio adjustment in social insects, or the enforced sterility of mature wolf offspring in a pack.

parsimony—the requirement that no theory be more complicated than necessary to explain the phenomena observed.

parturition—the act or process of giving birth.

patri—of or through the father, as in "patrilocal" (living with husband's parents) or "patrilineal" (tracing descent through father).

patrilocal (virilocal) residence—a pattern of residence in which a married couple lives with or near the husband's father's group. (Cf. *matrilocal residence, avunculocal residence, neolocal residence.*)

phenotype—the appearance of a trait.

placental—a mammal (e.g., deer, horses, humans) that has a vascular organ (the placenta) uniting the fetus and its mother. Nutrients are carried through the placenta. (Cf. *marsupial.*)

pleiotropy—the condition in which a single gene has multiple effects (e.g., as in senescence; see chapter 6). Phenylketonuria is an example: one gene locus affects several systems.

polyandry—a biological mating system in which fewer females than males contribute to the next generation, e.g., jacanas, Arctic sanderlings. Sexual competition between females is intensified, and female variance in reproductive success is greater than male variance. Socially, polyandry is a marriage system in which a woman may have more than one husband at a time. (Cf. *monogamy, polygyny.*)

polygamy—a general term describing the possession during a lifetime of more than one mate. (See *polygyny, polyandry, polygynandry.*)

polygynandry—a biological mating system in which both males and females may have multiple mates, as in Dunnocks or the Ache; note that the Ache marriage system is serial (social) monogamy. Male variance in reproductive success probably exceeds female variance.

polygyny—a biological mating system in which fewer males than females contribute to the next generation (i.e., a few males do most of the breeding, most fail). Male sexual competition is intensified. In anthropology, polygyny is a marriage system in which at least some men are permitted more than one wife simultaneously; it is further distinguished by whether the wives are sisters (sororal polygyny) or not (nonsororal polygyny). (Cf. *polyandry, monogamy, polygynandry.*)

precocial—referring to young animals that can move about and forage very quickly after birth or hatching. (Cf. *altricial.*)

predator—an organism that kills other animals for food.

predictability—the extent to which forecasts can be made about the condition of an environment with regard to important parameters. Symbolized by *P*, predictability is mathematically defined (Colwell 1974), and is the sum of constancy, *C* (q.v.), and contingency, *M* (q.v.).

proximate mechanism—environmental (including social) conditions triggering any response, such as changes in day length for many bird migrations. (See *ultimate cause*.)

reciprocity—mutual exchange of benefits, often at different times. Sometimes called reciprocal altruism.

recombination—also called crossing over, the process of exchange of genetic material between homologous chromosomes during meiosis.

Red Queen—in biology, the concept that all competitive progress is relative, because individuals exert reciprocal selective pressure on one another. Taken from the statement by the Red Queen (whom Alice meets in *Through the Looking-Glass*) that here you must run as fast as you can just to keep up, because everything else is moving also.

reproductive effort—any energy or risk devoted to achieving genetic representation in the next generation; consists of mating effort (q.v.) and parental effort (q.v.). (See *somatic effort*.)

reproductive success—the number of surviving (and presumably reproductive) offspring of an individual.

reproductive value—the number of female offspring remaining to be born to a female at any point in life given the prevailing age-specific fertility and mortality schedules; symbolized by v_x, where x is the age of the female under consideration.

r-selection—selection imposed by environment in which density of conspecific competitors is usually low. Typically low density of competitors favors high production of offspring, as opposed to fewer, better-invested offspring.

runaway sexual selection—if sexual selection involves intersexual selection, this is the condition in which there may be no "brakes" on the extremeness the trait may reach (as there are in intrasexual selections: weapons may become too elaborated to be useful). If, for whatever reason, females begin to prefer gaudy and risky displays, only males with extremes of those will be favored. In the next generation, sons will have more extreme displays than their father(s), and daughters are likely to prefer ever more extreme displays.

rut—the breeding season of ungulates (hoofed mammals).

selective pressure—any feature of the environment that "filters" which genes enter the next generation; this results in natural selection. Usually reducible to Darwin's "hostile forces"—e.g., shortage of resources (including mates), risk of disease, and parasites.

semelparous—reproducing only once in the organism's lifetime. (Cf. *iteroparous*.)

sexual dimorphism—any consistent difference between males and females beyond the basic functional anatomy of the reproductive organs; may include appearance and behavioral differences.

sexual selection—a "special case," a subset, of natural selection that acts on an individual's ability to obtain mates. It has two parts: intrasexual selection (e.g., within-sex competition, weapons) and intersexual selection (traits favored, for whatever reason, by the opposite sex: see runaway sexual selection). Darwin originally separated it from natural selection because although he saw clearly how selection could favor survival, under sexual selection risky and potentially lethal behaviors could be favored. He was uncertain how selection could favor behaviors that might get the individual killed.

social group—a group of individuals of the same species that interacts frequently in competitive and cooperative ways and often remains together much of the time; for example, a pack of wolves, a school of fish.

somatic effort—that portion of an individual's lifetime risk and energy budget not directly devoted to reproduction but rather to self-maintenance. (Cf. *reproductive effort.*)

stabilizing selection—selection that favors individuals having characteristics near the mean of some array, and disfavoring individuals with extremes. This is typical of moderate and unchanging environments.

sustainability—combinations of fertility and resource consumption that result in persistence of the population through time. When a population exceeds K, the carrying capacity, it will become unsustainable.

sympatric—occupying the same geographical area. (Cf. *allopatric.*)

sympatric speciation—the formation of two or more reproductively isolated species from a single parental species without a period of geographic separation (e.g., perhaps *Gryllus pennsylvanicus* and *G. veletus*).

trophic—pertaining to food.

Turner's syndrome—a human female condition characterized by XO genotype; results in sterility and disturbances of secondary sex characteristics.

ultimate cause—the cost-benefit ratio, or selective conditions, leading to any response. The ultimate cause of migratory behavior in a species is always differential reproductive success (of migrating individuals versus those that stay), while the proximate mechanism (q.v.) or proximate cause of migration may be a change in day length, temperature, or hormonal level that initiates migratory behavior.

ungulate—a hoofed mammal (e.g., deer, goats, cattle, horses). There are two subgroups: artiodactylids (even-toed ungulates, such as deer), and perisodactylids (odd-toed ungulates, such as horses).

viscosity—in population genetics, a low rate of gene flow due to dispersal by individuals; a viscous population has low dispersal and low gene flow.

WEIRD—a term coined by Henrich et al. (2010) to reflect the fact that many data for human behavior had been collected in evolutionarily novel environments—modern populations that were Western, Educated, Industrialized, Rich, and Democratic.

zygote—diploid cell resulting from the union (fertilization) of two haploid gametes.

References

Abbot, P., and 141 co-authors. 2011. Inclusive fitness theory and eusociality. *Nature* 471: E1–E4.

Abernethey, V., and R. Yip. 1990. Parent characteristics and infant mortality: The case in Tennessee. *Human Biology* 62(2): 279–290.

Acheson, J. 1988. *The Lobster Gangs of Maine*. Reprinted 2012. Lebanon, N.H.: University Press of New England.

Adams, D. B. 1983. Why there are so few women warriors. *Cross-Cultural Research* 18(3): 196–212.

Adams, G. D. 1997. Abortion: Evidence of an issue evolution. *American Journal of Political Science* 41(3): 718–737.

Ahlburg, D. 1998. Julian Simon and the population growth debate. *Population and Development Review* 24: 317–327.

Åkerman, S. 1981. The importance of remarriage in the seventeenth and eighteenth centuries. In J. Dupâquier, E. Hélin, P. Laslett, M. Livi-Bacci, and S. Sogner (eds.), *Marriage and Remarriage in Populations of the Past*, 163–175. New York: Academic Press.

Alberts, S. 2012. Magnitude and sources of variation in male reproductive performance. In J. Mitani, J. Call, P. Kappeler, R. A. Palombit, and J. B. Silk (eds.), *The Evolution of Primate Societies*, 420–431. Chicago: University of Chicago Press.

Alberts, S., H. Watts, and J. Altmann. 2003. Queuing and queue-jumping: Long-term patterns of reproductive skew in male savannah baboons, *Papio cyanocephalus*. *Animal Behaviour* 65: 821–840.

Alcock, J. 1998. Unpunctuated equilibrium in the *Natural History* essays of Stephen Jay Gould. *Evolution and Human Behavior* 19: 321–336.

———. 2009. *Animal Behavior*. 9th ed. Sunderland, Mass.: Sinauer.

Alexander, G., and M. Hines. 2002. Sex differences in response to children's toys in non-human primates (*Cercopithecus aethops sabeus*). *Evolution and Human Behavior* 23: 467–479.

Alexander, R. D. 1971. The search for an evolutionary philosophy of man. *Proceedings of the Royal Society of Victoria* (Melbourne) 84: 99–120.

———. 1974. The evolution of social behavior. *Annual Review of Ecology and Systematics* 5: 325–383.

———. 1977. Natural selection and the analysis of human sociality . In C. E. Goulden (ed.), *Changing Scenes in the Natural Sciences: 1776–1976*, 283–337. Special Publication 12. Philadelphia: Academy of Natural Sciences of Philadelphia.

———. 1979. *Darwinism and Human Affairs*. Seattle: University of Washington Press.

———. 1987. *The Biology of Moral Systems*. New York: Aldine de Gruyter.

———. 1988. Evolutionary approaches to human behavior: What does the future hold? In L. Betzig, M. Borgerhoff Mulder, and P. Turke (eds.), *Human Reproductive Behaviour: A Darwinian Perspective*, 317–341. New York: Cambridge University Press.

Alexander, R. D. 1990. How did humans evolve? Reflections on the uniquely unique species. Museum of Zoology, Special Publication 1. Ann Arbor: University of Michigan.

Alexander, R. D., and G. Borgia. 1978. On the origin and basis of the male-female phenomenon. In M. F. Blum and N. Blum (eds.), *Sexual Selection and Reproductive Competition in Insects*, 417–440. New York: Academic Press.

Alexander, R. D., and K. Noonan. 1979. Concealment of ovulation, parental care, and human social evolution. In N. A. Chagnon and W. Irons (eds.), *Evolutionary Biology and Human Social Behavior: An Anthropological Perspective*, 436–453. North Scituate, Mass.: Duxbury Press.

Alexander, R. D., John L. Hoogland, R. D. Howard, K. M. Noonan, and P. W. Sherman. 1979. Sexual dimorphism and breeding systems in pinnipeds, ungulates, primates, and humans. In N. A. Chagnon and W. Irons (eds.), *Evolutionary Biology and Human Social Behavior: An Anthropological Perspective*, 402–435. North Scituate, Mass.: Duxbury Press.

Allan, K., B. C. Jones, L. DeBruine, and D. Smith. 2012. Evidence of adaptation for mate choice with women's memory. *Evolution and Human Behavior* 33: 193–199.

Altonji, J. G., F. Hayashi, and L. J. Kotlikoff. 1997. Parental altruism and intervivo transfers: Theory and evidence. *Journal of Political Economy* 105: 1121–1166.

Alvard, M. 1994. Conservation by native peoples. *Human Nature* 5: 127–154.

———. 1998. Evolutionary ecology and resource conservation. *Evolutionary Anthropology* 7: 62–74.

———. 2003. Kinship, lineage, and an evolutionary perspective on cooperative hunting groups in Indonesia. *Human Nature* 14: 129–163.

———. 2009. Kinship and cooperation: The axe fight revisited. *Human Nature* 20: 394–416.

———. 2012. Human sociality. In J. Mitani, J. Call, P. Kappeler, R. A. Palombit, and J. B. Silk (eds.), *The Evolution of Primate Societies, 585–603. Chicago: University of Chicago Press.*

Alvergne, A., M. Gibson, E. Gurmu, and R. Mace. 2011. Social transmission and the spread of modern contraception in rural Ethiopia. *PLoS ONE* 6(7): e22515. doi: 10.1371/journal.pone.0022515.

Alvergne, A., D. Lawson, Parry Clarke, E. Gurmy, and R. Mace. 2013. Fertility, parental investment, and early adoption of modern contraception in rural Ethiopia. *American Journal of Human Biology* 25: 107–115.

American Psychological Association. 2006. When the boss is a woman. http://www.apa.org/research/action/boss.aspx (accessed January 10, 2014).

Anderies, J. M. 1996. An adaptive model for predicting !Kung reproductive performance: A stochastic dynamic programming approach. *Ethology and Sociobiology* 17: 221–245.

Andersen, S., S. Ertac, U. Gneezy, J. List, and S. Maximiano. 2012. Gender, competitiveness and socialization at a young age: Evidence from a matrilineal and a patriarchal society. *Review of Economics and Statistics*. doi: 10.1162/REST_a_00312.

Anderson, A. 1984. The extinction of the moa in southern New Zealand. In P. S. Martin and R. G. Klein (eds.), *Quaternary Extinctions: A Prehistoric Revolution*, 728–740. Tucson: University of Arizona Press.

Anderson, K. G. 2005. Relatedness and investment in children in South Africa. *Human Nature* 16(1): 1–31.

———. 2011. Does paying child support reduce men's subsequent marriage and fertility? *Evolution and Human Behavior* 32: 90–96.

Anderson, K. G., and B. S. Low. 2003. Non-marital births and women' life histories. In J. Rogers and H-P. Kohler (eds.), *The Biodemography of Human Fertility and Reproduction*, 57–86. Boston: Kluwer.

Anderson, K. G., H. Kaplan, J. B. Lancaster. 1999. Paternal care by genetic fathers and stepfathers. *Evolution and Human Behavior* 20: 405–431.

———. 2000. Confidence of paternity, divorce and investment in children by Albuquerque men. *Evolution and Human Behavior* 28(1): 1–10.

———. 2007. Confidence of paternity, divorce, and investment in children by Albuquerque men. *Evolution and Human Behavior* 28(1):1–10.

Andersson, M. 1994. *Sexual Selection*. Princeton, N.J.: Princeton University Press.

Aoki, K., and M. W. Feldman. 1994. Cultural transmission of a sign language when deafness is caused by recessive alleles at two independent loci. *Theoretical Population Biology* 45: 101–120.

Ardrey, R. 1966. *The Territorial Imperative*. New York: Atheneum.

Arthur, W. B. 1991. Designing economic agents that act like human agents: A behavioral approach to bounded rationality. *AEA Papers and Proceedings* 81(2): 353–359.

———. 1994. Inductive reasoning and bounded rationality. *AEA Papers and Proceedings* 84(2): 406–411.

Aston, C. C. 1986. Political hostage-taking in western Europe. In William Gutteridge (ed.), *Contemporary Terrorism*, 57–83. New York: Facts on File.

Aswad, B. 1971. *Property Control and Social Strategies in Settlers in a Middle Eastern Plain*. Museum of Anthropology, Anthropological Papers 44. Ann Arbor: University of Michigan.

Atran, S. 1993. Itza Maya tropical agro-forestry. *Current Anthropology* 34: 633–689.

Aureli, F., O. Fraser, C. Schaffner, and G. Schino. 2012. The regulation of social relationships. In J. Mitani, J. Call, P. Kappeler, R. A. Palombit, and J. B. Silk (eds.), *The Evolution of Primate Societies*, 531–551. Chicago: University of Chicago Press.

Axelrod, R. 1984. *The Evolution of Cooperation*. New York: Basic Books.

———. 1986. An evolutionary approach to norms. *American Political Science Review* 80(4): 1095–1111.

Axelrod, R., and D. Dion. 1988. The further evolution of cooperation. *Science* 242: 1385–1390.

Axelrod, R., and W. D. Hamilton. 1981. The evolution of cooperation. *Science* 211: 1390–1396.

Bailey, D., R. Walker, G. Blomquist, K. Hill, M. Hurtado, D. Geary. 2013. Heritability and fitness correlates of personality in the Ache, a natural-fertility population in Paraguay. *PLoS ONE* 8(3): e59325. doi: 10.1371/journal.pone.0059325.

Bailey, J. M. 2005. Sexual orientation: Genetics. *Encyclopedia of Life Sciences* 2005: 1–3.

Bailey, J. M., and R. Pillard. 1991. A genetic study of male sexual orientation. *Archives of General Psychiatry* 48(12): 1089–1096.

Bailey, J. M., M. Dunne, N. Martin. 2000. Genetic and environmental influences on sexual orientation and its correlates in an Australian twin sample. *Journal of Personality and Social Psychology* 78(3): 524–536.

Bailey, R. C., M. R. Jenike, P. T. Ellison, G. R. Bentley, A. M. Harrigan, and N. R. Peacock. 1992. The ecology of birth seasonality among agriculturalists in central Africa. *Journal of Biosocial Science* 24: 393–412.

Bailey, R. E., and M. J. Chambers. 1998. The impact of real wage and mortality fluctuations on fertility and nuptiality in precensus England. *Journal of Population Economics* 11: 413–434.

Ball, S., and C. C. Eckel. 1998. The economic value of status. *Journal of Socio-Economics* 27: 495–514.

Bao, A-M., and D. Swaab. 2011. Sexual differentiation of the human brain: Relation to gender identity, sexual orientation and neuropsychiatric disorders. *Frontiers in Neuroendocrinology* 32: 214–226.

Bapat, N. 2005. Insurgency and the opening of peace processes. *Journal of Peace Research* 42(6): 699–717.

Barber, P. 1988. *Vampires, Burial, and Death: Folklore and Reality.* New Haven, Conn.: Yale University Press.

Barclay, P. 2012. The evolution of charitable behavior and the power of reputation. In C. Roberts (ed.), *Applied Evolutionary Ecology*, 142–172. Oxford: Oxford University Press.

———. 2012. Harnessing the power of reputation: strengths and limits for promoting cooperative behaviours. *Evolutionary Psychology* 10(5): 868–883.

Barkow, J. H. 1977. Conformity to ethos and reproductive success in two Hausa communities: An empirical evaluation. *Ethos* 5: 409–425.

Barkow, J. H., L. Cosmides, and J. Tooby. 1992. *The Adapted Mind.* New York: Oxford University Press.

Barnet, R. J. 1988. The costs and perils of intervention. In M. T. Klare and P. Kornbluh (eds.), *Low-Intensity Warfare: Counterinsurgency, Proinsurgency, and Anti-terrorism in the Eighties*, 207–221. New York: Pantheon.

Barry, H., III, M. K. Bacon, and I. L. Child. 1957. A cross-cultural study of some sex differences in socialization. *Journal of Abnormal and Social Psychology* 55: 327–332.

Barry, H., III, L. Josephson, E. Lauer, and C. Marshall. 1976. Traits inculcated in childhood. 5. Cross-cultural codes. *Ethnology* 15: 83–114.

Barth, F. 1956. *Models of Social Organization.* Occasional Paper 23. London: Royal Anthropological Institute.

Barthold, J., M. Myrskyla, and O. R. Jones. 2012. Childlessness drives the sex difference in the association between income and reproductive success of modern Europeans. *Evolution and Human Behavior* 33: 628–638.

Barton, N., and B. Charlesworth. 1998. Why sex and recombination? *Science* 281: 1986–1990.

Barton, R. A., R. W. Byrne, and A. Whiten. 1996. Ecology, feeding competition, and social structure in baboons. *Behavioral Ecology and Sociobiology* 38: 3211–329.

Bateman, A. J. 1948. Intrasexual selection in *Drosophila. Heredity* 2: 349–368.

Bateson, P., and P. Martin. 2000. *Design for a Life: How Biology and Psychology Shape Human Behavior.* New York: Simon and Schuster.

Beall, C. M., and M. C. Goldstein. 1981. Tibetan fraternal polyandry: A test of sociobiological theory. *American Anthropologist* 83: 5–12.

Beals, A. R. 1961. Cleavage and internal conflict: An example from India. *Conflict Resolution* 5(1): 27–34.

Bean, L. L., and G. P. Minneau. 1986. The polygyny-fertility hypothesis: A reevaluation. *Population Studies* 40: 67–81.

Beattie, J.H.M. 1960. *Bunyoro: An African Kingdom.* New York: Holt, Rinehart and Winston.

———. 1963. Aspects of Nyoro symbolism. *Africa* 38: 413–442.

Beauchamp, G. 1994. The functional analysis of human fertility decisions. *Ethology and Sociobiology* 15: 31–53.

Becker, G. 1993. *A Treatise on the Family*. Enlarged ed. Cambridge, Mass.: Harvard University Press.

Becker, G., and R. J. Barro. 1988. Reformulating the economic theory of fertility. *Quarterly Journal of Economics* 103: 1–25.

Becker, G., and H. G. Lewis. 1974. Interaction between quantity and quality of children. In T. W. Schultz (ed.), *Economics of the Family: Marriage, Children and Human Capital*, 81–90. Chicago: University of Chicago Press.

Becker, J., K. Berkley, N. Geary, E. Hampson, J. Herman, and E. Young. (eds). 2008. *Sex Differences in the Brain: From Genes to Behavior*. New York: Oxford University Press.

Beise, J., and E. Voland. 2008. Intrafamilial resource competition and mate competition shaped social-group-specific natal dispersal in the 18th and 19th century Krummhörn population. *American Journal of Human Biology* 20: 325–336.

Beissinger, S. R. 1987. Mate desertion and reproductive effort in the snail kite. *Animal Behaviour* 35: 1504–1519.

Bell, A. V., P. J. Richerson, and R. McElreath. 2009. Culture rather than genes provides greater scope for the evolution of large-scale human prosociality. *Proceedings of the National Academy of Sciences USA* 106(42):17671–17674.

Bell, G. 1978. The evolution of anisogamy. *Journal of Theoretical Biology* 73: 247–270.

———. 1982. *The Masterpiece of Nature: The Evolution and Genetics of Sexuality*. Berkeley: University of California Press.

———. 1997a. *Selection: The Mechanism of Evolution*. New York: Chapman and Hall.

———. 1997b. *The Basics of Selection*. New York: Chapman and Hall.

Belovsky, G. 1987. Hunter-gatherer foraging: A linear programming approach. *Journal of Anthropological Archaeology* 6: 29–76.

Belovsky, G. E., J. B. Slade, and J. M. Chase. 1996. Mating strategies based on foraging ability: An experiment with grasshoppers. *Behavioral Ecology* 7: 438–444.

Benenson, J., T. Antonellis, B. Cotton, K. Noddin, and Campbell, K. 2008. Sex differences in children's formation of exclusionary alliances under scarce resource conditions. *Animal Behaviour* 76: 497–505.

Benenson, J., L. Hodgson, S. Heath, and P. Welch. 2008. Human sexual differences in the use of social ostracism as a competitive tactic. *International Journal of Primatology*, 29: 1019–1035.

Benenson, J., H. Markovits, M. Thompson, and R. Wrangham. 2009. Strength determines coalitional strategies in humans. *Proceedings of the Royal Society B* 276: 2589–2595.

Bentley, G. R., T. Goldberg, and G. Jasienska. 1992. The fertility of agricultural and nonagricultural traditional societies. Working Paper 1992-02, Population Issues Research Center, Pennsylvania State University, University Park.

Berard, J. D., P. Nurnberg, J. T. Epplen, and J. Schmidtke. 1993. Male rank, reproductive behavior, and reproductive success in free-ranging rhesus macaques. *Primates* 34: 481–489.

Berenbaum, S., C. Martin, L. Hanish, P. Briggs, and R. Fabes. 2007. Sex differences in children's play. In J. Becker, K. Berkeley, N. Geary, and E. Hampson (eds.). *Sex Differences in the Brain: From Genes to Behavior*, 273–290. Oxford: Oxford University Press.

Bernstein, I., and C. Ehardt. 1985. Agonistic aiding: Kinship, rank, age, and sex influences. *American Journal of Primatology* 8: 37–52.

Bertoni, M. 1941. *Los Guayakies*. Asunción: Revista de la Sociedad Cientifica del Paraguay.

Betzig, L. 1986. *Despotism and Differential Reproduction: A Darwinian View of History*. New York: Aldine de Gruyter.

———. 1989. Causes of conjugal dissolution: A cross-cultural study. *Current Anthropology* 30(5): 654–676.

———. 1996. Not whether to count babies but which. In C. Crawford and D. Krebs (eds.), *Evolution and Human Behavior: Issues, Ideas, and Applications*, 265–273. Hillsdale, N.J.: Erlbaum.

———. 2012. Means, variances and ranges in reproductive success: Comparative evidence. *Human Behavior and Evolution* 34: 309–317.

Betzig, L., and L. H. Lombardo. 1991. Who's pro-choice and why. *Ethology and Sociobiology* 13: 49–71.

Betzig, L., and S. Weber. 1993. Polygyny in American politics. *Politics and the Life Sciences* 12: 45–52.

Bideau, A. 1980. A demographic and social analysis of widowhood and remarriage: The example of the Castellany of Thoissey-en-Dombes, 1670–1840. *Journal of Family History* 5(1): 28–43.

Bideau, A., and A. Perrenoud. 1981. Remariage et fécondité. Contribution à l'étude des mécanismes de récuperation despopulationsanciennes. In J. Dupâquier, E. Hélin, P. Laslett, M. Livi-Bacci, and S. Sogner (eds.), *Marriage and Remarriage in Populations of the Past*, 547–559. New York: Academic Press.

Binmore, K. 1992. *Fun and Games: A Text on Game Theory*. Lexington, Mass.: D. C. Heath.

———. 1994. *Game Theory and the Social Contract*. Vol. 1, *Playing Fair*. Cambridge, Mass.: MIT Press.

Birdsall, N. 1980. Population growth and poverty in the developing world. *Population Bulletin* 35(5): 346.

Birkhead, T. R., and A. P. Møller. 1992. *Sperm Competition in Birds: Evolutionary Causes and Consequences*. London: Academic Press.

Black, J. 2005. *War Since 1945*. Chicago: University of Chicago Press.

Blanchard, R. 2001. Fraternal birth order and the maternal immune hypothesis of male homosexuality. *Hormones and Behavior* 40: 105–114.

Blanchard, R., and P. Klassen. 1997. H-Y antigen and homosexuality in men. *Journal of Theoretical Biology* 185 (3): 373–378.

Blanchard, R., M. Kurban, T. Blak, P. Klassen, R. Dickey, and J. Cantor. 2012. Sexual attraction to others: A comparison of two models of alloerotic responding in men. *Archives of Sexual Behavior* 41: 13–29.

Bliege Bird, R. L., and D. W. Bird. 1997. Delayed reciprocity and tolerated theft: The behavioral ecology of food-sharing strategies. *Current Anthropology* 38: 49–78.

Blomberg, S. B., G. Hess, and D. Tan. 2011. Terrorism and the economics of trust. *Journal of Peace Research* 48(3): 383–398.

Blood, R. O. 1960. Resolving family conflicts. *Conflict Resolution* 4(2): 209–219.

Blum, D. 1997. *Sex on the Brain: The Biological Differences between Men and Women*. New York: Viking.

Blurton Jones, N. B. 1984. A selfish origin for human food sharing: Tolerated theft. *Ethology and Sociobiology* 5: 1–3.

———. 1986. Bushman birth spacing: A test for optimal interbirth intervals. *Ethology and Sociobiology* 7: 91–105.

———. 1987a. Bushman birth spacing: Direct tests of some simple predictions. *Ethology and Sociobiology* 8: 183–203.

———. 1987b. Tolerated theft, suggestions about the ecology and evolution of sharing, hoarding, and scrounging. *Social Science Information* 26: 31–54.

———. 1989. The costs of children and the adaptive scheduling of births: Towards a sociobiological perspective on demography. In A. Rasa, C. Vogel, and E. Voland (eds.), *Sexual and Reproductive Strategies*, 265–282. Kent, U.K.: Croom Helm.

———. 1997. Too good to be true? Is there really a tradeoff between number and care of offspring in human reproduction? In L. L. Betzig (ed.), *Human Nature: A Critical Reader*, 83–86. Oxford: Oxford University Press.

Blurton Jones, N., and M. J. Konner. 1973. Sex differences in the behavior of Bushman and London two- to five-year-olds. In J. Crook and R. Michael (eds.), *Comparative Ecology and Behavior of Primates*, 689–749. New York: Academic Press.

Blurton Jones, N., and R. M. Sibley. 1978. Testing adaptiveness of culturally determined behavior: Do Bushman women maximize their reproductive success by spacing births widely and foraging seldom? In N. Blurton Jones and V. Reynolds (eds.), *Human Behavior and Adaptation*, 135–157. London: Taylor and Francis.

Bocklandt, S., S. Horvath, E. Vilain, and D. Hamer. 2006. Extreme skewing of X chromosome inactivation in mothers of homosexual men. *Human Genetics* 118(6): 691–694.

Bodley, J. H. 1990. *Victims of Progress*. 3rd ed. Mountain View, Calif.: Mayfield Publishing.

Boesch, C. 1996. Social groupings in Taï chimpanzees. In W. C. McGrew, L. F. Marchant, and T. Nishida (eds.), *Great Ape Societies*, 101–113. Cambridge: Cambridge University Press.

Boesch, C., and U. Reichard. 2003. *The Evolution of Monogamy*. Cambridge: Cambridge University Press.

Bongaarts, J. 1978. A framework for analyzing the proximate determinants of fertility. *Population and Development Review* 4(1): 105–132.

———. 1982. The fertility inhibiting effects of the intermediate fertility variables. *Studies in Family Planning* 13(6–7): 179–189.

Bonner, J. T. 1980. *The Evolution of Culture in Animals*. Princeton, N.J.: Princeton University Press.

Boomsma, J., M. Beekman, C. Cormwallis, A. Griffin, L. Holman, W. Hughes, L. Keller, B. Oldroyd, and F. Ratnieks. 2011. Only full-sibling families evolved eusociality. *Nature* 471: E4–E5.

Boone, J. L., III. 1986. Parental investment and elite family structure in preindustrial states: A case study of late medieval–early modern Portuguese genealogies. *American Anthropologist* 88: 859.

———. 1988. Parental investment, social subordination and population processes among the 15th and 16th century Portuguese nobility. In L. Betzig, M. Borgerhoff Mulder, and P. Turke (eds.), *Human Reproductive Behaviour: A Darwinian Perspective*, 201–219. Cambridge: Cambridge University Press.

Borgerhoff Mulder, M. 1987. On cultural and biological success: Kipsigis evidence. *American Anthropologist* 89: 617–634.

Borgerhoff Mulder, M. 1988a. Kipsigis bridewealth payments. In L. Betzig, M. Borgerhoff Mulder, and P. Turke (eds.), *Human Reproductive Behaviour: A Darwinian Perspective*, 65–82. Cambridge: Cambridge University Press.

———. 1988b. Reproductive success in three Kipsigis cohorts. In T. H. Clutton-Brock (ed.), *Reproductive Success*, 419–435. Chicago: University of Chicago Press.

———. 1990. Kipsigis women's preferences for wealthy men: Evidence for female choice in mammals? *Behavioral Ecology and Sociobiology* 27: 255–264.

———. 1991. Human behavioral ecology. In J. R. Krebs and N. B. Davies (eds.), *Behavioural Ecology*, 69–98. 3rd ed. London: Blackwell Scientific.

———. 1992. Reproductive decisions. In E. A. Smith and B. Winterhalder (eds.), *Evolutionary Ecology and Human Behavior*, 339–374. New York: Aldine de Gruyter.

———. 1995. Bridewealth and its correlates. *Current Anthropology* 36(3): 573–603, including commentary.

———. 1997. Marrying a married man: A postscript. In L. Betzig (ed.), *Human Nature: A Critical Reader*, 115–117. Oxford: Oxford University Press.

———. 1998a. Brothers and sisters: How sibling interactions affect optimal parental allocations. *Human Nature* 9: 119–162.

———. 1998b. The demographic transition: Are we any closer to an evolutionary explanation? *Trends in Ecology and Evolution* 13: 266–270.

———. 2009. Serial monogamy as polygyny or polyandry? Marriage in the Tanzanian Pimbwe. *Human Nature* 20(2): 130–150.

———. 2010. The unusual women of Mpimbwe: Why sex differences in humans are not universal. In P. M. Kaeppeler and J. B. Silk (eds.). *Mind the Gap*, 85–108. Heidelberg: Springer-Verlag.

Borgerhoff Mulder, M., and K. Rauch. 2009. Sexual conflict in humans: Variations and solutions. *Evolutionary Anthropology* 18: 201–214.

Borgerhoff Mulder, M., S. Bowles, T. Hertz, A. Bell, J. Beise, G. Clark, I. Fazzio, M. Gurven, K. Hill, P. L. Hooper, W. Irons, H. Kaplan, D. Leonetti, B. Low, F. Marlowe, S. Naidu, D. Nolin, P. Piraino, R. Quinlan, R. Sear, M. Shenk, E. A. Smith, and P. Weissner. 2009. The intergenerational transmission of wealth and the dynamics of wealth in pre-modern societies. *Science* 326(5953): 682–688.

Borgerhoff Mulder, M., I. Fazzio, W. Irons, R. L. McElreath, S. Bowles, A. Bell, T. Hertz, and L. Hazzah. 2010. Pastoralism and wealth inequality: Revisiting an old question. *Current Anthropology* 51(1): 35–48.

Borgia, G. 1979. Sexual selection and the evolution of mating systems. In M. F. Blum and N. Blum (eds.), *Sexual Selection and Reproductive Competition in Insects*, 19–80. New York: Academic Press.

———. 1985. Bower quality, number of decorations and mating success of male satin bowerbirds (*Ptilinorynchus violaceous*): An experimental analysis. *Animal Behaviour* 33: 266–271.

Boswell, J. 1990. *The Kindness of Strangers: Abandonment of Children in Western Europe from Late Antiquity to the Renaissance*. New York: Vintage Books. First published by Pantheon, 1988.

Bowles, S., E. A. Smith, and M. Borgerhoff Mulder. 2010. The emergence and persistence of inequality in premodern societies. *Current Anthropology* 51(1): 7–17.

Boyd, R., and P. J. Richerson. 1985. *Culture and the Evolutionary Process*. Chicago: University of Chicago Press.

————. 1988. The evolution of reciprocity in sizeable groups. *Journal of Theoretical Biology* 132: 337–356.

————. 1990. Culture and cooperation. In J. J. Mansbridge (ed.), *Beyond Self-Interest*, 111–132. Chicago: University of Chicago Press.

————. 1992. Punishment allows the evolution of cooperation (or anything else) in sizeable groups. *Ethology and Sociobiology* 13: 171–195.

————. 1996. Why culture is common, but cultural evolution is rare. *Proceedings of the British Academy* 88: 77–93.

————. 2005. Solving the puzzle of human cooperation. In S. Levinson and P. Jaisso (eds.), *Evolution and Culture*, 105–132. Cambridge, Mass.: MIT Press.

————. 2009. Culture and the evolution of human cooperation. *Philosophical Transactions of the Royal Society B* 364: 3281–3288.

Boyd, R., and J. B. Silk. 1997. *How Humans Evolved*. New York: Norton.

Boyd, R., P. J. Richerson, and J. Henrich. 2011a. The cultural niche: Why social learning is essential for human adaptation. *Proceedings of the National Academy of Sciences USA* 108: 10918–10925.

————. 2011b. Rapid cultural adaptation can facilitate the evolution of large-scale cooperation. *Behavioral Ecology and Sociobiology* 65: 431–444.

Brändström, A. 1984. De Kärlekslösa mödrarna: Spädbarnsdödligheten I Sverige under 1800-talet med särskild hänsyn till Nedertorneå. *Acta Universitatis Umensis* 62: 1–271.

Brewer, M., A. Ratcliffe, and S. Smith. 2009. Does welfare reform affect fertility? Evidence from the UK. Working paper WP08/09, Institute for Fiscal Studies, London.

Brogan, P. 1990. *The Fighting Never Stopped: A Comprehensive Guide to World Conflict Since 1945*. New York: Vintage Books.

Bro-Jørgensen. 2011. Queuing in space and time reduces the lek paradox on an antelope lek. *Evolutionary Ecology* 25: 1385–1395.

Brosnan, S. F. 2013. Justice- and fairness-related behaviors in non-human primates. *Proceedings of the National Academy of Sciences USA* 110: 10416–10423.

Brosnan, S. F., and F.B.M. de Waal. 2003. Monkeys reject unequal pay. *Nature* 425(6955): 297–299.

————. 2004a. A concept of value during experimental exchange in brown capuchin monkeys (*Cebus apella*). *Folia Primatologica* 75: 317–330.

————. 2004b. Reply to: Animal behavior: Fair refusal by capuchin monkeys. *Nature* 428: 140.

————. 2004c. Socially learned preferences for differentially rewarded tokens in the brown capuchin monkey (*Cebus apella*). *Journal of Comparative Psychology* 118: 133–139.

Brosnan, S. F., O. D. Jones, M. Gardner, S. P. Lambeth, and S. J. Schapiro. 2012. Evolution and the expression of biases: Situational value changes the endowment effect in chimpanzees. *Evolution and Human Behavior* 33: 378–386.

Brosnan, S. F., H. C. Schiff, and F.B.M. de Waal. 2005. Tolerance for inequity may increase with social closeness in chimpanzees. *Proceedings of the Royal Society B* 1560: 253–258.

Brown, G. and J. Silk. 2002. Reconsidering the null hypothesis: Is maternal rank associated with birth sex ratios in primates? *Proceedings of the National Academy of Sciences USA* 99(17): 11252–11255.

Brown, G., K. Laland, and M. Borgerhoff Mulder. 2009. Bateman's principles and human sex roles. *Trends in Ecology and Evolution* 246: 297–304.

Brown, J. L. 1983. Cooperation—a biologist's dilemma. In J. S. Rosenblatt (ed.), *Advances in Behavior*, 1–37. New York: Academic Press.

———. 1997. A theory of mate choice based on heterozygosity. *Behavioral Ecology* 8: 60–65.

Brunner, H. G. 1996. MAOA deficiency and abnormal behaviour: Perspectives on an association. *Ciba Foundation Symposium* 194: 155–164 (discussion 164–167).

Brunner, H. G., M. Nelen, X. O. Breakfield, H. H. Rogers, and B. A. Oost. 1993. Abnormal behavior associated with a point mutation in the structural gene for monoamine oxidase A. *Science* 262: 578–580.

Budiansky, S. 1995. *Nature's Keepers: The New Science of Nature Management.* New York: Free Press.

Bueno de Mesquita, B. 1981. *The War Trap.* New Haven, Conn.: Yale University Press.

Bugos, P. E., and L. M. McCarthy. 1984. Ayoreo infanticide: A case study. In G. Hausfater and S. B. Hrdy (eds.), *Infanticide: Comparative and Evolutionary Perspectives*, 503–520. New York: Aldine de Gruyter.

Bull, J., and R. Vogt. 1979. Temperature-dependent sex determination in turtles. *Science* 206(4423): 1186–1188.

Burke, S., and K. Collins. 2001. Gender differences in leadership styles and management skills. *Women in Management Review* 16(5): 244–257.

Burley, N. 1979. The evolution of concealed ovulation. *American Naturalist* 114: 835–858.

Burt, A., and R. Trivers. 2006. *Genes in Conflict: The Biology of Selfish Genetic Elements.* Cambridge, Mass: Belknap Press.

Buss, D. M. 1985. Human mate selection. *American Scientist* 73: 47–51.

———. 1989. Sex differences in human mate preferences: Evolutionary hypotheses tested in 37 cultures. *Behavioral and Brain Sciences* 12: 1–49.

———. 1994. *The Evolution of Desire.* New York: Basic Books.

———. 1999. *Evolutionary Psychology: The New Science of the Mind.* Boston: Allyn and Bacon.

Buss, D. M., and L. A. Dedden, LA. 1990. Derogation of competitors. *Journal of Social and Personal Relationships* 7: 395–422.

Buss, D. M., and D. P. Schmitt. 1993. Sexual strategies theory: An evolutionary perspective on human mating. *Psychological Review* 100: 204–232.

Bygott, J. D. 1974. Agonistic behavior and dominance in wild chimpanzees. Ph.D. diss. University of Cambridge, Cambridge, U.K.

Bygott, J. D., B.C.R. Bertram, and J. P. Hanby. 1979. Male lions in large coalitions gain reproductive advantages. *Nature* 282: 839–841.

Byrne, R. 1993. Do larger brains mean greater intelligence? *Behavioral and Brain Sciences* 16: 696–697.

———. 1995. *The Thinking Ape: Evolutionary Origins of Intelligence.* Oxford: Oxford University Press.

Byrne, R., and A. Whiten. 1988. *Machiavellian Intelligence: Social Expertise and the Evolution of Intellect in Monkeys, Apes, and Humans.* Oxford: Clarendon Press.

———. 1992. Cognitive evolution in primates: Evidence from tactical deception. *Man* 27: 609–627.

Cabello y Robles, D. 1961. A description of the Comanche Indians in 1786 by the governor of Texas. *West Texas Historical Association Yearbook* 37: 177–182.

Cabourdin, G. 1981. Le remariage en France sous l'ancien régime (seizième–dix-huitième siècles). In J. Dupâquier, E. Hélin, P. Laslett, M. Livi-Bacci, and S. Sogner (eds.), *Marriage and Remarriage in Populations of the Past*, 273–285. New York: Academic Press.

Cadoret, R. J., L. D. Leve, and E. Devor. 1997. Genetics of aggressive and violent behavior. *Psychiatric Clinics of North America* 20: 301–322.

Cain, M. 1985. On the relationship between landholding and fertility. *Population Studies* 39: 5–15.

Cairns, R. B. 1996. Aggression from a developmental perspective: Genes, environments and interactions. *Ciba Foundation Symposium* 194: 45–56 (discussion 57–60).

Caldwell, J. C. 1976. Toward a restatement of demographic transition theory. *Population and Development Review* 2: 321–366.

———. 1982. *Theory of Fertility Decline*. New York: Academic Press.

———. 1983. Direct economic costs and benefits of children. In R. A. Bulatao and R. D. Lee (eds.), *Determinants of Fertility in Developing Countries*. Vol. 1, *Supply and Demand for Children*, 458–493. New York: Academic Press.

Campbell, A. 1995. A few good men: Evolutionary psychology and female adolescent aggression. *Ethology and Sociobiology* 16: 99–123.

Campbell, D. T. 1975. Conflicts between biological and social evolution and between psychology and moral tradition. *American Psychologist* 30: 21–37.

———. 1979. Comments on the sociobiology of ethics and moralizing. *Behavioral Science* 24: 37–45.

———. 1983. Legal and primary group controls. In M. Gruter and P. Bohanon (eds.), *Law, Biology, and Culture: The Evolution of Law*, 159–171. Santa Barbara, Calif.: Ross-Ericson.

Campbell, K. L., and J. W. Wood. 1988. Fertility in traditional societies. In P. Diggory, M. Potts, and S. Teper (eds.), *Natural Human Fertility: Social and Biological Determinants*, 39–69. London: Macmillan.

Cancian, F. A. 1973. *What Are Norms? A Study of Beliefs and Action in a Maya Community*. Cambridge: Cambridge University Press.

Caprara, G., G. Alessandri, and N. Eisenberg. 2012. Prosociality: The contributions of traits, values, and self-efficacy beliefs. *Journal of Personality and Social Psychology* 102(6): 1289–1303.

Carcopino, J. 1941. *Daily Life in Ancient Rome*. London: Routledge and Sons.

Carniero, R. L. 1970. A theory of the origin of the state. *Science* 169: 733–738.

Case, T. J. 1978. Endothermy and parental care in the terrestrial vertebrates. *American Naturalist* 112: 861–874.

Cashdan, E. 1985. Coping with risk: Reciprocity among the Basarwa of northern Botswana. *Man* 20: 454–474.

———. 2001. Ethnocentrism and xenophobia: A cross-cultural study. *Current Anthropology* 42(5): 760–765.

———. 2012. In-group loyalty or out-group avoidance? Isolating the links between pathogens and in-group assortative sociality. *Behavioral and Brain Sciences* 35: 82.

Cashdan, E., and M. Steele. 2013. Pathogen prevalence, group bias, and collectivism in the Standard Cross-Cultural Sample. *Human Nature* 24: 59–75.

Cashdan, E., F. W. Marlowe, A. Crittenden, C. Porter, and B. Wood. 2012. Sex differences in spatial cognition among Hadza foragers. *Evolution and Human Behavior* 33: 274–284.

Cassels, R. 1984. Funal extinctions and prehistoric man in New Zealand and the Pacific islands. In P. S. Martin and R. G. Klein (eds.), *Quaternary Extinctions: A Prehistoric Revolution*, 553–573. Tucson: University of Arizona Press.

Cavalli-Sforza, L. L., and M. W. Feldman. 1981. *Cultural Transmission and Evolution.* Princeton, N.J.: Princeton University Press.

Cavalli-Sforza, L. L., M. W. Feldman, K. H. Chen, and S. Dornbusch. 1982. Theory and observation in cultural transmission. *Science* 218: 19–27.

Cesarini, D., C. Dawes, M. Johannesson, P. Lichtenstein, and P. Wallace. 2009. Genetic variation in preferences for giving and risk taking. *Quarterly Journal of Economics* 124: 809–842.

Chagnon, N. 1979. Is reproductive success equal in egalitarian societies? In N. A. Chagnon and W. Irons (eds.), *Evolutionary Biology and Human Social Behavior: An Anthropological Perspective*, 374–401. North Scituate, Mass.: Duxbury Press.

———. 1982. Sociodemographic attributes of nepotism in tribal populations: Man the rule-breaker. In Kings' College Sociobiology Group (eds.), *Current Problems in Sociobiology*, 291–318. Cambridge: Cambridge University Press.

———. 1988. Life histories, blood revenge, and warfare in a tribal population. *Science* 239: 985–992.

———. 1997. *Yanomamö.* 5th ed. Fort Worth, Tex.: Harcourt Brace.

Chagnon, N. A., and R. Hames. 1979. Protein deficiency and tribal warfare in Amazonia: New data. *Science* 203: 910–913.

Chamberlin, E. R. 1965. *Everyday Life in Renaissance Times.* London: B. T. Batsford.

Chapais, B. 1983. Dominance, relatedness, and the structure of female relationships in rhesus monkeys. In R. A. Hinde (ed.), *Primate Social Relationships: An Integrated Approach*, 209–219. Oxford: Blackwell Scientific.

Chapman, C. A., J. Rothman, and J. Lambert. 2012. Food as a selective force in primates. In J. Mitani, J. Call, P. Kappeler, R. A. Palombit, and J. B. Silk (eds.), *The Evolution of Primate Societies*, 149–168. Chicago: University of Chicago Press.

Chapman, E., S. Baron-Cohen, B. Auyeung, R. Knickmeyer, K. Taylor, and G. Hackett. 2006. Fetal testosterone and empathy: Evidence from the Empathy Quotient (EQ) and the "reading the mind in the eyes" test. *Social Neuroscience* 1: 135–148.

Charlesworth, B. 1978. The population genetics of anisogamy. *Journal of Theoretical Biology* 73: 347–357.

———. 1980. *Evolution in Age-structured Populations.* Cambridge: Cambridge University Press.

Charlton, S.E.M. 1984. *Women in Third World Development.* Boulder, Colo.: Westview Press.

Charnov, E. L. 1982. *The Theory of Sex Allocation.* Princeton, N.J.: Princeton University Press.

———. 1991. Evolution of life history variation among female mammals. *Proceedings of the National Academy of Sciences USA* 88: 1134–1137.

———. 1993. *Life History Invariants: Some Explorations of Symmetry in Evolutionary Ecology.* Oxford: Oxford University Press.

Cheney, D. 1987. Interactions and relationships between groups. In B. B. Smuts, D. L. Cheney, R. M. Seyfarth, R. W. Wrangham, and T. T. Struhsaker (eds.), *Primate Societies*, 267–281. Chicago: University of Chicago Press.

Cheney, D., and R. Seyfarth. 1987. The influence of inter-group competition on the survival and reproduction of female vervet monkeys. *Behavioral Ecology and Sociobiology* 21: 375–386.

———. 1992. *How Monkeys See the World*. Chicago: University of Chicago Press.

Cheney, D., and R. W. Wrangham. 1987. Predation. In B. B. Smuts, D. L. Cheney, R. M. Seyfarth, R. W. Wrangham, and T. T. Struhsaker (eds.), *Primate Societies*, 227–239. Chicago: University of Chicago Press.

Cheng, P. W., K. J. Holyoak, R. E. Nisbett, and L. M. Oliver. 1986. Pragmatic versus syntactic approaches to training deductive reasoning. *Cognitive Psychology* 18: 293–328.

Childs, G. M. 1949. *Umbundu Kinship and Character*. Oxford: Oxford University Press.

Chisholm, J., and V. Burbank. 1991. Monogamy and polygyny in southeast Arnhem Land: Male coercion and female choice. *Ethology and Sociobiology* 12: 291–313.

Chitty, D. 1996. *Do Lemmings Commit Suicide? Beautiful Hypotheses and Ugly Facts*. Oxford: Oxford University Press.

Chudek, M., and J. Henrich. 2011. Culture-gene coevolution, norm-psychology, and the emergence of human prosociality. *Trends in Cognitive Science* 15(5): 218–226.

Chung, W., and A. Auger. 2013. Gender differences in neurodevelopment and epigenetics. *European Journal of Physiology* 465: 573–584.

Cigno, A. 1998. Fertility decisions when infant survival is endogenous. *Journal of Population Economics* 11: 21–28.

Clark, D. C., S. J. DeBano, and A. J. Moore. 1997. The influence of environmental quality on sexual selection in *Nauphoeta cineria* (Dictyoptera: Blaberidae). *Behavioral Ecology* 8: 46–53.

Clarke, A. L., and B. Low. 1992. Ecological correlates of human dispersal in 19th century Sweden. *Animal Behaviour* 44: 677–693.

Clastres, P. 1972a. *Chronique des Indiens Guayaki. Ce que Sevant les Aché, Chasseurs. Nomades du Paraguay*. Paris: Pion.

———. 1972b. The Guayaki. In M. Bicchieri (ed.), *Hunters and Gatherers Today*, 138–174. New York: Holt, Reinhart, and Winston.

Clay, J. W. 1988. *Indigenous Peoples and Tropical Forests*. Cambridge, Mass.: Cultural Survival.

Clutton-Brock, T. H. 1983. Selection in relation to sex. In D. S. Bendall (ed.), *From Molecules to Men*, 457–481. Cambridge: Cambridge University Press.

——— (ed.). 1988. *Reproductive Success: Studies of Individual Variation in Contrasting Breeding Systems*. Chicago: University of Chicago Press.

———. 1991. *The Evolution of Parental Care*. Princeton, N.J.: Princeton University Press.

———. 2009. Sexual selection in females. *Animal Behaviour* 77: 3–11.

Clutton-Brock, T. H., and P. Harvey. 1976. Evolutionary rules and primate societies. In P.P.G. Bateson and R. A. Hinde (eds.), *Growing Points in Ethology*, 195–237. Cambridge: Cambridge University Press.

———. 1978. Mammals, resources, and reproductive strategies. *Nature* 273: 191–195.

Clutton-Brock, T. H., and G. A. Parker. 1995a. Sexual coercion in animal societies. *Animal Behaviour* 49: 1345–1365.

———. 1995b. Punishment in animal societies. *Nature* 373: 209–216.

Clutton-Brock, T. H., S. Albon, and F. Guinness. 1986. Great expectations: Dominance, breeding success and offspring sex ratio in red deer. *Animal Behaviour* 34: 460–471.

Clutton-Brock, T. H., F. E. Guinness, and S. D. Albon. 1982. *Red Deer: The Ecology of Two Sexes.* Wildlife Behavior and Ecology Series. Chicago: University of Chicago Press.

Coale, A. J., and T. J. Trussell. 1974. Model fertility schedules: Variations in the age structure of childbearing in human populations. *Population Index* 40: 185–258. (See also Erratum, *Population Index* 41: 572.)

———. 1978. Technical note: Finding the two parameters that specify a model schedule of fertility. *Population Index* 44: 203–213.

Coale, A. J., and S. C. Watkins. 1986. *The Decline of Fertility in Europe.* Princeton, N.J.: Princeton University Press.

Cobey, K., C. Klipping, and A. Buunk. 2013. Hormonal contraceptive use lowers female intrasexual competition in pair-bonded women. *Evolution and Human Behavior* 34: 294–298.

Cohen, J. E. 1995. *How Many People Can the Earth Support?* New York: Norton.

———. 1998. Can a more equal world support more or fewer people than a less equal one?: Theoretical perspectives. Harvard Institute for International Development, Development Discussion Paper 628, Harvard University, Cambridge, Mass.

Cohen, M., R. Riolo, and R. Axelrod. 2001. The role of social structure in the maintenance of cooperative regimes. *Rationality and Society* 13(1): 5–32.

Cohen, R. 1967. *The Kanuri.* New York: Rinehart and Winston.

Coleman D., and R. Schofield (eds.). 1986. *The State of Population Theory: Forward from Malthus.* London: Blackwell Scientific.

Collier, J. F., and M. Rosaldo. 1981. Politics and gender in simple societies. In S. B. Ortner and H. Whitehead (eds.), *Sexual Meanings: The Cultural Construction of Gender and Sexuality.* Cambridge: Cambridge University Press.

Colwell, R. K. 1974. Predictability, constancy, and contingency of periodic phenomena. *Ecology* 55(5): 1148–1153.

Comfort, A. 1956. *Ageing: The Biology of Senescence.* New York: Holt, Rinehart, and Winston.

Confer, J., J. Easton, D. Fleischman, C. Goetz, D. Lewis, C. Perilloux, and D. Buss. 2010. Evolutionary psychology: Controversies, questions, prospects, and limitations. *American Psychologist* 65(2): 110–126.

Confer, J., C. Perilloux, and D. Buss. 2010. More than just a pretty face: Men's priority shifts toward bodily attractiveness in short-term versus long-term mating contexts. *Evolution and Human Behavior* 31: 348–353.

Connor, R. 1986. Pseudo-reciprocity: Investment in mutualism. *Animal Behaviour* 34: 1562–1566.

———. 1995a. Altruism among non-relatives: Alternatives to the "prisoner's dilemma." *Trends in Ecology and Evolution* 10: 84–86.

———. 1995b. The benefits of mutualism: A conceptual framework. *Biology Review* 70: 427–457.

———. 2010. Cooperation beyond the dyad: on simple models and a complex society. *Philosophical Transactions of the Royal Society B* 365(1553): 2687–2697.

Connor, R., R. A. Smolker, and A. F. Richards. 1992. Two levels of alliance formation among male bottlenose dolphins (*tursiops* sp.). *Proceedings of the National Academy of Sciences USA* 89: 987–990.

Cook, K., R. Hardin, and M. Levi. 2007. *Cooperation Without Trust?* Russell Sage Foundation Series on Trust, Book 9. New York: Russell Sage Foundation.

Cooney, T. M., and P. Uhlenburg. 1992. Support from parents over the life course: The adult child's perspective. *Social Forces* 71: 63–84.

Cords, M. 1987. Forest guenons and patas monkeys: Male-male competition in one-male groups. In B. B. Smuts, D. L. Cheney, R. M. Seyfarth, R. W. Wrangham, and T. T. Struhsaker (eds.), *Primate Societies*, 98–11. Chicago: University of Chicago Press.

Corsini, C. A. 1981. Why is remarriage a male affair? Some evidence from Tuscan villages during the eighteenth century. In J. Dupâquier, E. Hélin, P. Laslett, M. Livi-Bacci, and S. Sogner (eds.), *Marriage and Remarriage in Populations of the Past*, 385–395. New York: Academic Press.

Cosgrove, K., C. Mazure, and J. Staley. 2007. Evolving knowledge of sex differences in brain structure, function, and chemistry. *Biological Psychiatry* 62(8): 847–855.

Cosmides, L., and J. Tooby. 1992. Cognitive adaptations for social exchange. In J. Barkow, L. Cosmides, and J. Tooby (eds.), *The Adapted Mind*, 163–228. New York: Oxford University Press.

———. 1996. Are humans good intuitive statisticians after all? Rethinking some conclusions from the literature on judgement under uncertainty. *Cognition* 58: 1–73.

Costanza, R. 1987. Social traps and environmental policy. *Bioscience* 37: 407–12.

Costanza, R., F. Andrade, P. Antunes, M. van denBelt, D. Boesch, D. Boersma, D. Boesch, F. Catarino, S. Hanna, K. Limburg, B. Low, M. Molitor, J. Pereira, S. Rayner, R. Santos, J. Wilson, M. Young. 1998. Principles for sustainable governance of the oceans. *Science* 281(5374): 198–199.

Costanza, R., I. Kubiszewski, E Giovannini, H. Lovins, J. McGlade, K. Pickett, K. Ragnarsdóttir, D. Roberts, R. De Vogli, and R. Wilkinson. 2014. Time to leave GDP behind. *Nature* 505: 283–287 (including supplemental information).

Costanza, R., B. Low, E. Ostrom, and J. Wilson (eds.) 2001a. *Institutions, Ecosystems, and Sustainability*. New York: Lewis Publishers.

———. 2001b. Future directions. In R. Costanza, B. Low, E. Ostrom, and J. Wilson (eds.), *Institutions, Ecosystems, and Sustainability*, 233–245. New York: Lewis Publishers.

Cowlishaw, G., and R. I. Dunbar. 1991. Dominance rank and mating success in male primates. *Animal Behaviour* 41: 1045–1056.

Cowlishaw, G., and R. Mace. 1996. Cross-cultural patterns of marriage and inheritance: A phylogenetic approach. *Ethology and Sociobiology* 17: 87–97.

Crimmins, E. M., and R. A. Easterlin. 1984. The estimation of natural fertility: A micro approach. *Social Biology* 31: 160–70.

Cronin, H. 1991. *The Ant and the Peacock*. Cambridge: Cambridge University Press.

Cronk, L. 1989. Low socioeconomic status and female-biased parental investment: The Mukogodo example. *American Anthropologist* 91: 414–429.

———. 1991a. Human behavioral ecology. *Annual Review of Anthropology* 20: 25–53.

———. 1991b. Intention versus behavior in parental sex preferences among the Mukogodo of Kenya. *Journal of Biosocial Sciences* 23: 229–240.

———. 1991c. Low socioeconomic status and female-biased parental investment: The Mukogodo example. *American Anthropologist* 91: 414–429.

———. 1991d. Preferential parental investment in daughters over sons. *Human Nature* 2: 387–417.

———. 1991e. Wealth, status, and reproductive success among the Mukogodo of Kenya. *American Anthropologist* 93(2): 345–360.

———. 1993. Parental favoritism toward daughters. *American Scientist* 81: 272–279.

Cronk, L. 1994. Evolutionary theories of morality and the manipulative use of signals. *Zygon* 29: 81–101.

———. 1995. Is there a role for culture in human behavioral ecology? *Ethology and Sociobiology* 16: 181–205.

———. 1999. Female-biased parental investment and growth performance among Mukogodo children. In L. Cronk, N. Chagnon, and W. Irons (eds.), *Adaptation and Human Behavior: An Anthropological Perspective*, 203–221. Hawthorne, N.Y.: Aldine de Gruyter.

Cronk, L., and B. Leech. 2013. *Meeting at Grand Central: Understanding the Social and Evolutionary Roots of Cooperation*. Princeton: Princeton University Press.

Crook, J. H., and S. J. Crook. 1988. Tibetan polyandry: Problems of adaptation and Fitness. In L. Betzig, M. Borgerhoff Mulder, and P. Turke (eds.), *Human Reproductive Behavior: A Darwinian Approach*, 97–114. Cambridge: Cambridge University Press.

Crosby, A. 1986. *Ecological Imperialism: The Biological Expansion of Europe*. Cambridge: Cambridge University Press.

Cross, J. 1983. *A Theory of Adaptive Economic Behavior*. Cambridge: Cambridge University Press.

Cross, J., and M. Guyer. 1980. *Social Traps*. Ann Arbor: University of Michigan Press.

Crow, J. F. 1958. Some possibilities for measuring selection intensities in man. *Human Biology* 30: 1–13.

Crowfoot, M., and R. Wrangham. 2010. Intergroup aggression in primates and humans: The case for a unified theory. In P. M. Kaeppeler and J. B. Silk (eds.), *Mind the Gap*, 171–196. Heidelberg: Springer-Verlag.

Croze, H., A. K. Hillman, and E. M. Lang. 1981. Elephants and their habits: How do they tolerate each other? In C. W. Fowler and T. Smith (eds.), *Dynamics of Large Mammal Populations*, 297–316. New York: Wiley.

Curtis, H. J. 1963. Biological mechanisms underlying the aging process. *Science* 141: 686–694.

Cyrus, K., S. Schwarz, and M. Hassebrauck. 2011. Systematic cognitive biases in courtship context: Women's commitment-skepticism as a life-history strategy? *Evolution and Human Behavior* 32: 13–20.

Daly, M., and M. Wilson. 1983. *Sex, Evolution, and Behavior*. 2nd ed. Boston: Willard Grant.

———. 1984. A sociobiological analysis of human infanticide. In G. Hausfater and S. B. Hrdy (eds.), *Infanticide: Comparative and Evolutionary Perspectives*, 487–502. New York: Aldine de Gruyter.

———. 1985. Child abuse and other risks of not living with both parents. *Ethology and Sociobiology* 6: 197–210.

———. 1987. Children as homicide victims. In R. J. Gelles and Jane B. Lancaster (eds.), *Child Abuse and Neglect: Biosocial Dimensions*, 201–214. New York: Aldine de Gruyter.

———. 1988. *Homicide*. Hawthorne, N.Y.: Aldine de Gruyter.

Darwin, C. (1859) 1987. *On the Origin of Species by Means of Natural Selection*. Facsimile of the first edition, with an introduction by Ernst Mayr. Cambridge, Mass.: Harvard University Press.

———. 1871. *The Descent of Man and Selection in Relation to Sex*. 2 vols. London: John Murray.

Das Gupta, M., and P. N. Mari Bhat. 1997. Fertility decline and increased manifestation of sex bias in India. *Population Studies* 51: 307–315.

Davidson, M. and R. J. Burke. 2011. *Women in Management Worldwide*. 2nd ed. Farnham, Surrey, U.K.: Gower Publishing.

Davies, N. B. 1991. Mating systems. In J. R. Krebs and N. B. Davies (eds.), *Behavioral Ecology: An Evolutionary Approach*, 263–294. 3rd ed. London: Blackwell Scientific.

———. 1992. *Dunnock Behaviour and Social Evolution*. Oxford: Oxford University Press.

Davies, N. B., J. R. Krebs, and S. A. West. 2012. *An Introduction to Behavioural Ecology*. 4th ed. Chichester, U.K.: Wiley-Blackwell.

Dawes, R. 1988. *Rational Choice in an Uncertain World*. Fort Worth, Tex.: Harcourt Brace Jovanovich.

Dawkins, R. 1979. Twelve misunderstandings of kin selection. *Zeitschrift für Tierpsychologie* 47: 61–76.

———. 1982. *The Extended Phenotype: The Gene as the Unit of Selection*. Oxford: Oxford University Press.

———. 1986. *The Blind Watchmaker*. New York: Norton.

———. 2006. *The Selfish Gene*. 30th anniversary ed. Oxford: Oxford University Press.

Dawkins, R., and J. R. Krebs. 1978. Animal signals: Information or deception? In J. R. Krebs and N. B. Davies (eds.), *Behavioural Ecology: An Evolutionary Approach*, 282–309. Oxford: Blackwell Scientific.

Deaner, R., and B. Smith, B. 2013. Sex differences in sports across 50 societies. *Cross-Cultural Research* 47: 268–309.

Deaner, R., D. Geary, D. Puts, S. Ham, J. Kruger, E. Fles, B. Winegard, and T. Grandis. 2012. A sex difference in the predisposition for physical competition: Males play sports much more than females even in the contemporary US. *PLoS ONE* 7(11): e49168. doi: 10.1371/journal.pone.0049168.

De Bellis, M., M. Keshavan, S. Beers, J. Hall, K. Frustaci, A. Masalehdan, J. Noll, and A. Boring. 2001. Sex differences in brain maturation during childhood and adolescence. *Cerebral Cortex* 11(6) 552–557.

DeBruine, L., B. C. Jones, J. Crawford, L. Welling, and A. Little. 2010. The health of a nation predicts their mate preferences cross-culturally. Cultural variation in women's preferences for masculinized male faces. *Proceedings of the Royal Society B* 277: 2405–2410.

DeFries, J. C., R. P. Corley, R. C. Johnson, S. G. Vandenberg, and J. R. Wilson. 1982. Sex-by-generation and ethnic group-by-generation interactions in the Hawaii Family Study of Cognition. *Behavior Genetics* 12: 223–230.

DeNault, L. K., and D. A. McFarland. 1995. Reciprocal altruism between male vampire bats, *Desmodus rotundus*. *Animal Behaviour* 49: 855–856.

Denevan, W. M. 1992. The pristine myth: The landscape of the Americas in 1492. *Annals of the Association of American Geographers* 82: 369–385.

Denig, E. T. 1961. *Five Indian Tribes of the Upper Missouri: Sioux, Arickaras, Assiniboines, Cree, Crows*. Edited and with an introduction by John C. Ewers. Norman: University of Oklahoma Press.

Dennett, D. C. 1991. *Consciousness Explained*. Boston: Little, Brown.

———. 1995. *Darwin's Dangerous Idea*. London: Penguin Press.

deRuiter, J., J. van Hoof, and W. Scheffrahn. 1994. Social and genetic aspects of paternity in wild long-tailed macaques (*Macaca fascicularis*). *Behaviour* 129: 203–224.

Deutscher, I. 1973. *What We Say/What We Do: Sentiments and Acts*. Glenview: Scott, Foresman.

de Waal, F. 1982. *Chimpanzee Politics*. New York: Harper and Row.

———. 1984. Sex differences in the formation of coalitions among chimpanzees. *Ethology and Sociobiology* 5: 239–255.

———. 1986. The integration of dominance and social bonding in primates. *Quarterly Review of Biology* 61: 459–479.

———. 1989. *Peacemaking among Primates*. Cambridge, Mass.: Harvard University Press.

———. 1996a. Conflict as negotiation. In W. C. McGrew, L. Marchant, and T. Nishida (eds.), *Great Ape Societies*, 159–172. Cambridge: Cambridge University Press.

———. 1996b. *Good Natured: The Origins of Right and Wrong in Humans and Other Animals*. Cambridge, Mass.: Harvard University Press.

———. 1997a. The chimpanzee's service economy: Food for grooming. *Evolution and Human Behavior* 18: 375–386.

———. 1997b. *The Forgotten Ape*. Berkeley: University of California Press.

Dewsbury, D. A. 1982. Dominance rank, copulatory behavior, and differential reproduction. *Quarterly Review of Biology*. 57: 135–159.

De Young, R. 1993. Changing behavior and making it stick: The conceptualization and management of conservation behavior. *Environment and Behavior* 25: 485–505.

Diamond, J. 1992. *The Third Chimpanzee: The Evolution and Future of the Human Animal*. New York: HarperCollins.

———. 1997. *Why Sex Is Fun*. New York: Basic Books.

———. 2004. *Collapse: How Societies Choose to Fail or Succeed*. New York: Viking Adult.

Dickemann, M. 1979. The reproductive structure of stratified societies: A preliminary model. In N. A. Chagnon and W. Irons (eds.), *Evolutionary Biology and Human Social Organization: An Anthropological Perspective*, 331–367. North Scituate, Mass.: Duxbury Press.

———. 1981. Paternal confidence and dowry competition: A biocultural analysis of purdah. In R. D. Alexander and D. W. Tinkle (eds.), *Natural Selection and Social Behavior*, 417–438. New York: Chiron Press.

Dienske, H. 1986. A comparative approach to the question of why human infants develop so slowly. In J. G. Else and P. C. Lee (eds.), *Primate Ontogeny, Cognition, and Social Behavior*, 147–154. London: Cambridge University Press.

Divale, W. 1973. *Warfare in Primitive Societies: A Bibliography*. Santa Barbara, Calif.: American Bibliographic Center Clio.

Divale, W., and M. Harris. 1976. Population, warfare, and the male supremacist complex. *American Anthropologist* 80(1): 21–41.

Dixon, N. F. 1976. *On the Psychology of Military Incompetence*. London: Cape.

Dixson, B., and R. Brooks. 2013. The role of facial hair in women's perceptions of men's attractiveness, health, masculinity, and parenting abilities. *Evolution and Human Behavior* 34: 236–241.

Dobzhansky, T. 1961. Discussion. In J. S. Kennedy (ed.), *Insect Polymorphism*, 111. London: Royal Entomological Society.

Dominey, W. J. 1980. Female mimicry in male bluegill sunfish—a genetic polymorphism? *Nature* 284: 546–548.

Dorjahn, V. R. 1958. Fertility, polygyny, and their interrelations in Temne society. *American Anthropologist* 60: 838–860.

Dosmukhambetova, D., and A. Manstead. 2011. Strategic reactions to unfaithfulness: female self-presentation in the context of mate attraction is linked to uncertainty of paternity. *Revolution and Human Behavior* 32: 106–117.

Downhower, J. F., and E. L. Charnov. 1998. A resource range invariance rule for optimal offspring size predicts patterns of variability in parental phenotypes. *Proceedings of the National Academy of Sciences USA* 95: 6208–6211.

Drake, M. 1969. *Population and Society in Norway: 1735–1865.* Cambridge: Cambridge University Press.

Draper, P. 1976. Social and economic constraints on child life among the !Kung. In R. B. Lee and I. DeVore (eds.), *Kalihari Hunter-Gatherers,* 199–217. Cambridge, Mass.: Harvard University Press.

———. 1980. The interaction of behavior variables in the development of dominance relations. In D. R. Omark, F. F. Strayer, and D. G. Freedman (eds.), *Dominance Relations: An Ethological View of Human Conflicts and Social Interaction,* 427–439. New York: Garland STPM Press.

Draper, P., and E. Cashdan. 1988. Technological change and child behavior among the !Kung. *Ethnology* 27(4): 339–365.

Drury, W. H. 1973. Population changes in New England seabirds. *Bird Banding* 44: 267–313.

Duchworth, A., and M. Seligman. 2006. Self-discipline gives girls the edge: Gender in self-discipline, grades, and achievement scores. *Journal of Educational Psychology* 98: 198–208.

Dugatkin, L. A. 1996. Interface between culturally based preferences and genetical preferences: Female mate choice in *Poecilia reticulata. Proceedings of the National Academy of Sciences USA* 93(7): 2770–2773.

———. 1997. *Cooperation among Animals: An Evolutionary Approach.* Oxford: Oxford University Press.

———. 1998. Game theory and cooperation. In L. A. Dugatkin and H. K. Reeve (eds.), *Game Theory and Animal Behavior,* 38–63. Oxford: Oxford University Press.

Dugatkin, L. A., and H. K. Reeve. 1994. Behavioral ecology and the levels-of-selection debate: Dissolving the group selection controversy. *Advances in the Study of Behavior* 23: 101–133.

——— (eds.). 1998. *Game Theory and Animal Behavior.* Oxford: Oxford University Press.

Dugatkin, L. A., L. Farrand, R. Wilkins, and D. S. Wilson. 1994. Altruism, tit for tat, and "outlaw" genes. *Evolutionary Ecology* 8: 431–437.

Dunbar, R.I.M. 1983. Structure of gelada baboon reproductive units, 4. Integration at group level. *Zeitschrift für Tierpsychologie* 63: 265–283.

———. 1987. Demography and reproduction. In B. B. Smuts, D. L. Cheney, R. M. Seyfarth, R. W. Wrangham, and T. T. Struhsaker (eds.), *Primate Societies,* 240–249. Chicago: University of Chicago Press.

———. 1988. *Primate Social Systems.* London: Croom Helm.

———. 1992. Neocortex size as a constraint on group size in primates. *Journal of Human Evolution* 20: 469–493.

Dunbar, R.I.M., and E. P. Dunbar. 1977. Dominance and reproductive success among female gelada baboons. *Nature* 266: 351–352.

Dunbar, R.I.M., A. Clark, and N. L. Hurst. 1995. Conflict and cooperation among the Vikings: Contingent behavioral decisions. *Ethology and Sociobiology* 16: 233–246.

Duncan, G., W. J. Yeung, J. Brooks-Gunn, and J. R. Smith. 1998. How much does childhood poverty affect the life chances of children? *American Sociological Review* 63: 406–423.

Duncan, N. 2004. It's important to be nice, but it's nicer to be important: girls, popularity, and sexual competition. *Sex Education: Sexuality, Society, and Learning* 4(2): 137–152.

Dunnigan, J. F., and A. Bay. 1986. *A Quick and Dirty Guide to War: Briefings on Present and Potential Wars*. Rev. ed. New York: Quill/William Morrow.

Dupâquier, J. 1972. De l'animal à l'homme: Le mechanisme autorégulateur des populations traditionelles. *Revue de l'Institut Sociologique de l'Université Libre de Bruxilles* 2: 177–211.

Durham, W. H. 1976. Resource competition and human aggression, part 1: A review of primitive war. *Quarterly Review of Biology* 51: 385–415.

———. 1991. *Coevolution: Genes, Culture, and Human Diversity*. Stanford, Calif.: Stanford University Press.

Dweck, C. S. 1975. The role of expectations and attributions in the alleviation of learned helplessness. *Journal of Personality and Social Psychology* 31: 674–685.

Dweck, C. S., and T. E. Goetz. 1978. Attributions and learned helplessness. *New Directions in Attribution Research* 2: 157–179. Hillsdale, N.J.: Erlbaum.

Dweck, C. S., and C. B. Wortman. 1982. Learned helplessness, anxiety, and achievement motivation. In W. H. Krohne and L. Lauz (eds.), *Achievement, Stress, and Anxiety*, 93–125. New York: Hemisphere.

Dweck, C. S., W. Davidson, S. Nelson, and B. Enna. 1978. Sex differences in learned helplessness: II. The contingencies of evaluative feedback in the classroom. *Developmental Psychology* 14: 268–276.

Dyson-Hudson, R., and E. A. Smith. 1978. Human territoriality. *American Anthropologist* 80: 21–42.

Eals, M., and I. Silverman. 1994. The hunter-gatherer theory of spatial sex differences: Proximate factors mediating the female advantage in recall of object arrays. *Ethology and Sociobiology* 15: 95–105.

Early, J. D., and J. F. Peters. 1990. *The Population Dynamics of the Mucajai Yanomamö*. New York: Academic Press.

Easterlin, R. 1978. The economics and sociology of fertility: A synthesis. In C. Tilly (ed.), *Historical Studies of Changing Fertility*, 57–134. Princeton, N.J.: Princeton University Press.

Easterlin, R., and E. Crimmins. 1985. *The Fertility Revolution: A Supply-Demand Analysis*. Chicago: University of Chicago Press.

Ebert, D., and W. D. Hamilton. 1996. Sex against virulence: The coevolution of parasitic diseases. *Trends in Ecology and Evolution* 11: 79–82.

Eckel, C. C., and P. J. Grossman. 1988. Are women less selfish than men? Evidence from dictator experiments. *Economic Journal* 108: 726–735.

———. 2008. Differences in the economic decisions of men and women: Experimental evidence. In *Handbook of Experimental Economics Results*, Vol. 1, 509–519. New York: Elsevier.

Eckholm, E. 2014. Access to abortion falling as states pass restrictions. *New York Times*, April 1.

Economos, A. C. 1980. Brain-life span conjecture: A re-evaluation of the evidence. *Gerontology* 26: 82–89.

Ecuyer-Dub, I., and M. Robert. 2004. Spatial ability and home-range size: Examining the relationship in Western men and women (*Homo sapiens*). *Journal of Comparative Psychology* 118(2): 217–231.

Edgerton, R. B. 1992. *Sick Societies: Challenging the Myth of Primitive Harmony*. New York: Free Press.

Edwards, A., and B. McKibben. 2010. *Thinking Beyond Sustainability: Pathways to a Resilient Society*. Gabriola Island, B.C., Canada: New Society Publishers.

Elena, S., and R. E. Lenski. 2003. Evolution experiments with microorganisms: The dynamics and genetic bases of adaptation. *Nature Reviews Genetics* 4(6): 457–469.

Ehrlich, P. R. 1997. *A World of Wounds: Ecologists and the Human Dilemma*. Oldendorf/Luhe, Germany: Ecology Institute.

Ehrlich, P. R., and A. H. Ehrlich. 1981. *Extinction: The Causes and Consequences of the Disappearance of Species*. New York: Random House.

Ehrlich, P., and J. Holdren. 1971. Impact of population growth. *Science* 171:1212–1217.

Eibl-Eibesfeldt, I. 1979. *The Biology of Peace and War*. London: Viking Press.

Elgar, M. A., and N. E. Pierce. 1988. Mating success and fecundity in an ant-tended Lycaenid butterfly. In T. H. Clutton-Brock (ed.), *Reproductive Success: Studies of Individual Variation in Contrasting Breeding Systems*, 59–75. Chicago: University of Chicago Press.

Ellis, L. 1993. Dominance and reproductive success among non-human animals: A cross-species comparison. *Ethology and Sociobiology* 16: 257–333.

Ellison, P. 1991. Reproductive ecology and human fertility. In C.G.N. Mascie-Taylor and G. W. Lasker (eds.), *Applications of Biological Anthropology to Human Affairs*, 14–54. Cambridge: Cambridge University Press.

Elton, G. R. 1975. Taxation for war and peace in early Tudor England. In J. M. Winter (ed.), *War and Economics in Development*, 33–48. Cambridge: Cambridge University Press.

Ember, C. R. 1978. Myths about hunter-gatherers. *Ethnology* 17: 439–448.

———. 1981. A cross-cultural perspective on sex differences. In R. H. Monroe, R. L. Monroe, and B. B. Whiting (eds.), *Handbook of Cross-Cultural Human Development*, 531–580. New York: Garland Press.

Ember, C. R., and M. Ember. 1992. Resource unpredictability, mistrust, and war: A cross-cultural study. *Journal of Conflict Resolution* 36(2): 242–262.

Emlen, S. T. 1982a. The evolution of helping: I. An ecological constraints model. *American Naturalist* 119: 29–39.

———. 1982b. The evolution of helping: II. The role of behavioral conflict. *American Naturalist* 119: 40–53.

———. 1995. An evolutionary theory of the family. *Proceedings of the National Academy of Science USA* 92: 8092–8099.

———. 1997. Predicting family dynamics in social vertebrates. In J. R. Krebs and N. B. Davies (eds.), *Behavioural Ecology*, 228–253. 4th ed. Oxford: Blackwell Scientific.

Emlen, S. T., and L. W. Oring. 1977. Ecology, sexual selection, and the evolution of mating systems. *Science* 197: 215–223.

Enders, W., T. Sandler, and K. Gaibulloev. 2011. Domestic versus transnational terrorism: Data, decomposition, and dynamics. *Journal of Peace Research* 48(s): 319–337.

Endler, J. A. 1986. *Natural Selection in the Wild*. Princeton, N.J.: Princeton University Press.

Epstein, J., and R. Axtel. 1996. *Growing Artificial Societies from the Bottom Up (Complex Adaptive Systems)*. Cambridge, Mass: Brookings Institute MIT Press.

Essock-Vitale, S. 1984. The reproductive success of wealthy Americans. *Ethology and Sociobiology* 5: 45–49.

Essock-Vitale, S., and M. T. McGuire. 1988. What 70 million years hath wrought: Sexual histories and reproductive success of a random sample of American women. In in L. L. Betzig, M. Borgerhoff Mulder, and P. Turke (eds.), *Human Reproductive Behavior: A Darwinian Perspective*, 221–236. Cambridge: Cambridge University Press.

Ewers, J. C. 1958. *The Blackfeet*. Norman: University of Oklahoma Press.

Fadiman, J. A. 1982. *An Oral History of Tribal Warfare: The Meru of Mt. Kenya*. Athens: Ohio University Press.

Fahlgren, K. 1956. *Skellefteå & Sockens Historia*. Uppsala, Sweden: Almqvist and Wiksells Boktryckeri AB.

Fairbanks, L. A., and M. T. McGuire. 1986. Determinants of fecundity and reproductive success in captive vervet monkeys. *American Journal of Primatology* 7: 27–38.

Falconer, D. S. 1981. *Introduction to Quantitative Genetics*. 2nd ed. London: Longman.

Farran, C. 1963. *Matrimonial Laws of the Sudan*. London: Buttersworth.

Fathauer, G. 1961. Trobrianders. In D. Schneider and K. Gough (eds.), *Matrilineal Kinship*, 234–269. Berkeley: University of California Press.

Fedigan, L. M. 1983. Dominance and reproductive success in primates. *Yearbook of Physical Anthropology* 26: 91–129.

Fehr, E., and U. Fishbacher. 2004. Third-party punishment and social norms. *Evolution and Human Behavior* 25: 63–87.

Fehr, E., and S. Gächter. 1998. Reciprocity and economics: The economic implications of *Homo reciprocans*. *European Economic Review* 42: 845–859.

Fehr, E., S. Gächter, and G. Kirchsteiger. 1997. Reciprocity as a contract enforcement device: Experimental evidence. *Econometrica* 65: 833–860.

Feldman, M. W., L. L. Cavalli-Sforza, and L. A. Zhivotovsky. 1994. On the complexity of cultural transmission and evolution. In G. Cowan, D. Pines, and D. Melzer (eds.), *Complexity: Metaphors, Models, and Reality*, 47–62. SFI Studies in the Science of Complexity, Proceedings, vol. 19. Reading, Mass.: Addison Wesley.

Fernandez-Duque, E., A. Di Fiore, and M. Huck. 2012. The behavior, ecology, and social evolution of New World monkeys. In J. Mitani, J. Call, P. Kappeler, R. A. Palombit and J. B. Silk (eds.), *The Evolution of Primate Societies*, 43–54. Chicago: University of Chicago Press.

Ferriere, R., and R. Michod. 2011. Inclusive fitness in evolution. *Nature* 471: E6–E8.

Ferrill, A. 1985. *The Origins of War*. London: Thames and Hudson.

Festing, M.F.W., and D. K. Blackmore. 1971. Life span of specified pathogen-free (MRC Category 4) mice and rats. *Laboratory Animal Bulletin* 5: 179–192.

Fichtel, C. 2012. Predation. *In* J. Mitani, J. Call, P. Kappeler, R. A. Palombit, and J. B. Silk (eds.), *The Evolution of Primate Societies*, 169–194. Chicago: University of Chicago Press.

Fincher, C., and R. Thornhill. 2012. Parasite stress promotes in-group assortative sociality: The cases of strong family ties and heightened religiosity. *Behavioral and Brain Science* 35: 61–119 (including commentary).

Fincher, C., R. Thornhill, D. Murray, and M. Schaller. 2008.Pathogen prevalence predicts human cross-cultural variability in individualism/collectivism. *Proceedings of the Royal Society B* 275: 1279–1285,

Fisher, A. 1984. *Africa Adorned.* New York: Abrams.

Fisher, M. L., A. Cox, and F. Gordon, 2009. Self-promotion versus competitor derogation: The influence of sex and romantic relationship status on intrasexual competition strategy selection. *Journal of Evolutionary Psychology* 7(4): 287–808.

Fisher, R. A. 1958. *The Genetical Theory of Natural Selection.* New York: Dover Books.

Fiske, P., P. T. Rintamäki, and E. Karvonen. 1998. Mating success in lekking males: A meta-analysis. *Behavioral Ecology* 9: 328–338.

Flandrin, J. L. 1979. *Families in Former Times: Kinship, Household, and Sexuality.* Translated by Richard Southern. Cambridge: Cambridge University Press.

Flemming, T., O. Jones, L. Mayo, T. Stoinski, and S. Brosnan. 2012. The endowment effect in orangutans. *International Journal of Comparative Psychology* 25: 285–298.

Flinn, M. V. 1981. Uterine versus agnatic kinship variability and associated cousin marriage preferences. In R. D. Alexander and D. W. Tinkle (eds.), *Natural Selection and Social Behavior,* 439–475. New York: Chiron Press.

———. 1997. Culture and the evolution of social learning. *Evolution and Human Behavior* 18: 23–67.

Flinn, M. V., and B. S. Low. 1986. Resource distribution, social competition, and mating patterns in human societies. In D. Rubenstein and R. Wrangham (eds.), *Ecological Aspects of Social Evolution,* 217–243. Princeton, N.J.: Princeton University Press.

Fossey, D. 1983. *Gorillas in the Mist.* Boston: Houghton Mifflin.

Fowler, C., and T. Smith (eds.), *Dynamics of Large Mammal Populations,* 99–120. New York: Wiley.

Fox, M., R. Sear, J. Beise, G. Ragsdale, E. Voland, and L. Knapp. 2010. Grandma plays favorites: X-chromosome relatedness and sex-specific childhood mortality. *Proceedings of the Royal Society B* 277: 567–573.

Frank, R. 1985. *Choosing the Right Pond.* Oxford: Oxford University Press.

———. 1988. *Passions within Reason.* New York: Norton.

———. 2004. *What Price the Moral High Ground?: Ethical Dilemmas in Competitive Environments.* Princeton, NJ: Princeton University Press.

Frank, S. A. 1986. Hierarchical selection theory and sex ratios. I. General solutions for structured populations. *Theoretical Population Biology* 29: 312–342.

———. 1990. When to copy or avoid an opponent's strategy. *Journal of Theoretical Biology* 170: 41–46.

———. 1995. Mutual policing and repression of competition in the evolution of cooperative groups. *Nature* 377: 520–522.

———. 1996. The design of natural and artificial adaptive systems. In Michael R. Rose and George V. Lauder (eds.), *Adaptation,* 451–505. San Diego: Academic Press.

———. 1998. *Foundations of Social Evolution.* Princeton, N.J.: Princeton University Press.

Frater, J. 2007. Top 11 methods of interrogation. http://listverse.com/2007/08/23/top-11-methods-of-interrogation/ (accessed February 18, 2014).

Freedman, D. G. 1974. *Human Infancy: An Evolutionary Perspective.* Hillsdale, N.J.: Erlbaum.

Freedman, D. G. 1980. Sexual dimorphism and the status hierarchy. In D. R. Omark, F. F. Strayer, and D. G. Freedman (eds.), *Dominance Relations*, 261–271. New York: Garland Press.

Freedman, D. S., and A. Thornton. 1982. Income and fertility: The elusive relationship. *Demography* 19(1): 65–78.

Friedlander, L. 1907. *Roman Life and Manners under the Early Empire*, vol. 1. 2nd ed. Reprinted 1979. New York: Arno Press.

Fuchs, R. 1984. *Abandoned Children: Foundlings and Child Welfare in Nineteenth-Century France*. Albany: State University of New York Press.

Gagneaux, P., D. Woodruff, and C. Boesch. 1997. Furtive mating in female chimpanzees. *Nature* 387: 358–359.

Gaibulloev, K., and T. Sandler. 2011. The adverse effects of transnational and domestic terrorism on growth in Africa. *Journal of Peace Research* 48(3): 355–371.

Galloway, P. R. 1986. Differentials in demographic responses to annual price variations in prerevolutionary France: A comparison of rich and poor areas in Rouen, 1681–1787. *European J. Pop.* 2(1986): 269–305.

Gallup, G. G. 1982. Permanent breast enlargement in human females: A socio-biological analysis. *Journal of Human Evolution* 11: 597–601.

———. 1986. Unique features of human sexuality in the context of human evolution. In D. Byrne and K. Kelley (eds.), *Alternative Approaches to the Study of Sexual Behavior*, 13–42. Hillsdale, N.J.: Erlbaum.

Gangestad, S., and R. Thornhill. 2009. Human oestrus. *Proceedings of the Royal Society B* 275: 991–1000.

Gangestad, S., R. Thornhill, and C. E. Garver-Apgar. 2010. Men's facial masculinity changes in their female partners' sexual interests across the ovulatory cycle, whereas men's intelligence does not. *Evolution and Human Behavior* 31: 412–424.

Gangestad, S. W., R. Thornhill, and R. A. Yeo. 1994. Facial attractiveness, developmental stability, and fluctuating asymmetry. *Ethology and Sociobiology* 15: 73–85.

Gardiner, A., S. A. West, and G. Wild. 2011. The genetical theory of kin selection. *Journal of Evolutionary Biology* 24: 1020–1043.

Gardner, J. F. 1986. *Women in Roman Law and Society*. Bloomington: Indiana University Press.

Garenne, M., and E. van de Walle. 1989. Polygyny and fertility among the Sereer of Senegal. *Population Studies* 43: 267–283.

Gat, A. 2010. Why war? Motivations for fighting in the human state of nature. In P. M. Kaeppeler and J. B. Silk (eds.), *Mind the Gap*, 197–222. Heidelberg: Springer-Verlag.

Gaulin, S.J.C. 1980. Sexual dimorphism in the post-reproductive lifespan: Possible causes. *Human Evolution* 9: 227–232.

Gaulin, S.J.C., and J. S. Boster. 1990. Dowry as female competition. *American Anthropologist* 92: 994–1005.

———. 1997. When are husbands worth fighting for? In L. Betzig (ed.), *Human Nature: A Critical Reader*, 372–374. Oxford: Oxford University Press.

Gaulin, S.J.C., and R. W. Fitzgerald. 1986. Sex differences in spatial ability: An evolutionary hypothesis and test. *American Naturalist* 127: 74–88.

Gaulin, S.J.C., and H. A. Hoffman. 1988. Evolution and development of sex differences in spatial ability. In L. Betzig, M. Borgerhoff Mulder, and P. Turke (eds.), *Human Reproductive Behaviour: A Darwinian Perspective*, 129–152. Cambridge: Cambridge University Press.

Gaulin, S.J.C., and C. J. Robbins. 1991. Trivers-Willard effect in contemporary North American society. *American Journal of Physical Anthropology* 85: 61–68.

Gaulin, S.J.C., D. H. McBurney, and S. L. Brakeman-Wartell. 1997. Matrilateral biases in the investment of aunts and uncles: A consequence and measure of paternity uncertainty. *Human Nature* 8: 139–151.

Gaunt, D. 1983. The property and kin relations of retired farmers. In R. Wall, J. Robin, and P. Laslett (eds.), *Family Forms in Historic Europe*, 249–279. Cambridge: Cambridge University Press.

———. 1987. Rural household organization and inheritance in Northern Europe. *Journal of Family History* 12: 121–141.

Gauthier, A. H., and J. Hatzius. 1997. Family benefits and fertility: An econometric analysis. *Population Studies* 51: 295–306.

Geary, D. C. 2010. *Male, Female: The Evolution of Human Sex Differences*. Washington, D.C.: American Psychological Association.

Georgellis, Y., and H. J. Wall. 1992. The fertility effect of dependent tax exemptions: Estimates for the United States. *Applied Economics* 24: 1139–1145.

Gerger, T., and G. Hoppe. 1980. Education and society: The geographer´s view. *Acta Universitatis Stockholmensis* 1: 1–124.

Geronimus, A. T. 1996a. Black/white differences in the relationship of maternal age to birthweight: A population-based test of the Weathering Hypothesis. *Social Science and Medicine* 42(4): 589–597.

———. 1996b. What teen mothers know. *Human Nature* 7(4): 323–352.

Geronimus, A. T., J. Bound, and T. A. Waidmann. 1997. Health inequality, family caretaking systems, and population variation in fertility timing. Paper presented at 1997 annual meeting of the Population Association of America, Washington, D. C.

Geronimus, A. T., J. Bound, T. A. Waidmann, M. M. Hillemeier, and P. B. Burns. 1996. Excess mortality among blacks and whites in the United States. *New England Journal of Medicine* 335: 1552–1558.

Gewertz, D. 1981. A historical reconstruction of female dominance among the Chambri of Papua New Guinea. *American Ethnologist* 8: 94–106.

———. 1983. *Sepik River Societies: A Historical Ethnography of the Chambri and Their Neighbors*. New Haven, Conn.: Yale University Press.

Gewertz, D., and F. K. Errington. 1991. *Twisted Histories, Altered Contexts: Representing the Chambri in a World System*. Cambridge: Cambridge University Press.

Ghiselin, M. T. 1969. The evolution of hermaphroditism among animals. *Quart. Rev. Biol.* 44: 189–208.

Gibbard, A. 1992. *Apt Feelings and Wise Choices*. Cambridge, Mass.: Harvard University Press.

Gibson, K. 2009. Differential parental investment in families with both adopted and genetic children. *Evolution and Human Behavior* 30: 184–189.

Gies, F. 1984. *The Knight in History*. New York: Harper and Row.

Gigerenzer, G., P. Todd, and the ABC Research Group. 2000. *Simple Heuristics That Make Us Smart*. Oxford: Oxford University Press.

Gilby, I., L. Brent, E. Wroblewski, R. Rudicell, B. Hahn, J. Goodall, and A. Pusey. 2013. Fitness benefits of coalitionary aggression in male chimpanzees. *Behavioral Ecology and Sociobiology* 67:373–381. doi: 10.1007/s00265-012-1457-6.

Gintis, H. 2003. Solving the puzzle of prosociality. *Rationality and Society* 15: 155–187.

Gintis, H., E. Smith, and S. Bowles. 2001. Costly signaling and cooperation. *Journal of Theoretical Biology* 213(1): 103–119.

Gleditsch, K., N. Metternich, and A. Ruggeri. 2014. Data and progress in peace and conflict research. *Journal of Peace Research* 51(2): 301–314.

Glick, P. C., and S.-L. Lin. 1986. Recent changes in divorce and remarriage. *Journal Marriage and Family* 48: 737–747.

Gluckman, P., and M. Hanson. 2006. *Mismatch: The Lifestyle Diseases Timebomb*. Oxford: Oxford University Press.

Godin, J.-G., and L. A. Dugatkin. 1996. Female mating preference for bold males in the guppy, *Poecilia reticulata*. *Proceedings of the National Academy of Sciences USA* 93(19): 10262–10267.

Goetz, A. T., and T. K. Shackelford. 2009. Sexual conflict in humans: Evolutionary consequences of asymmetric parental investment and paternity uncertainty. *Animal Biology* 59: 449–456.

Goldizen, A. W. 1987. Tamarins and marmosets: Communal care of offspring. In B. B. Smuts, D. L. Cheney, R. M. Seyfarth, R. W. Wrangham, and T. T. Strusaker (eds.), *Primate Societies*, 34–43. Chicago: University of Chicago Press.

Goldstein, A. P. 1989. Aggression reduction: Some vital steps. In J. Groebel and R. A. Hinde (eds.), *Aggression and War: Their Biological and Social Bases*, 112–131. Cambridge: Cambridge University Press.

Goldstein, M. 1976. Fraternal polyandry and fertility in a high Himalayan valley in Northwest Nepal. *Human Ecology* 4: 223–233.

Goodale, J. C. 1971. *Tiwi Wives*. Seattle: University of Washington Press.

Goodall, J. 1986. *The Chimpanzees of Gombe: Patterns of Behavior*. Cambridge, Mass.: Harvard University Press.

Goodall, J., A. Bandora, E. Bergmann, C. Busse, H. Matama, E. Mpongo, A. Pierce, and D. Riss. 1979. Intercommunity interactions in the chimpanzee population of the Gombe National Park. In D. A. Hamburg and E. R. McCown (eds.), *The Great Apes*, 13–53. Menlo Park, Calif.: Benjamin Cummings.

Goose, S. D. 1988. Low-intensity warfare: The warriors and their weapons. In M. T. Klare and Peter Kornbluh (eds.), *Low-Intensity Warfare: Counterinsurgency, Proinsurgency, and Anti-Terrorism in the Eighties*, 80–111. New York: Pantheon.

Gorer, G. 1967. *Himalayan Village*. 2nd ed. New York: Basic Books.

Gould, S. J., and R. C. Lewontin. 1979. The spandrels of San Marcos and the Panglossian paradigm: A critique of the adaptationist programme. *Proceedings of the Royal Society B* 205: 581–598.

Gouzoules, H., S. Gouzoules, and L. Fedigan. 1982. Behavioral dominance and reproductive success in female Japanese monkeys (*Macaca fuscata*). *Animal Behaviour* 30: 1138–1150.

Gowaty, P. A. (ed.). 1997. *Feminism and Evolutionary Biology*. New York: Chapman and Hall.

Grafen, A. 1984. Natural selection, kin selection and group selection. In J. R. Krebs and N. B. Davies (eds.), *Behavioural Ecology: An Evolutionary Approach*, 62–84. 2nd ed. Oxford: Blackwell Scientific.

——. 1991. Modelling in behavioural ecology. In J. R. Krebs and N. B. Davies (eds.), *Behavioural Ecology: An Evolutionary Approach*, 5–31. Oxford: Blackwell Scientific.

Grammer, K., B. Fink, and N. Neave. 2005. Human pheromones and sexual attraction. *European Journal of Obstetrics and Gynecology and Reproductive Biology* 118(2): 135–142.

Gray, P. B., and K. G. Anderson. 2010. *Fatherhood: Evolution and Paternal Behavior.* Cambridge, Mass.: Harvard University Press.

Gray, R. 1983. The impact of health and nutrition on natural fertility. In R. A. Bulatao and R. D. Lee (eds., with P. E. Hollerbach and John Bongaarts), *Determinants of Fertility in Developing Countries,* 139–162. New York: Academic Press.

Greenfield, M. D., R. Danka, J. Gleason, B. Harris, and Y. Zhou. 2012. Genotype × environment interaction, environmental heterogeneity, and the lek paradox. *Journal of Evolutionary Biology* 25(4): 601–613.

Greenwood, P. J. 1980. Mating systems, philopatry, and dispersal in birds and mammals. *Animal Behavior* 28: 1140–1162.

Gregory, A., J. Light-Hausermann, F. Rijsdijk, and T. Eley. 2009. Behavioral genetic analyses of prosocial behavior in adolescents. *Developmental Science* 12: 165–174.

Griffith, J. D. 1980. Economy, family and remarriage: Theory of remarriage and application to preindustrial England. *Journal of Family Issues* 1(4): 479–496.

Griskevicius, V., A. Cantú, and M. van Vugt. 2012. The evolutionary bases for sustainable behavior: Implications for marketing, policy, and social entrepreneurship. *Journal of Policy and Marketing* 31(1): 115–128.

Groebel, J., and R. A. Hinde. 1989. A multi-level approach to the problems of aggression and war. In Jo Groebel and Robert A. Hinde (eds.), *Aggression and War: Their Biological and Social Bases,* 223–229. Cambridge: Cambridge University Press.

Gurven, M. 2004. Does market penetration affect economic game behavior? In Henrich, J., R. Boyd, S. Bowles, C. Camerer, E. Fehr, H. Gintis (eds.), *Foundations of Human Sociality: Economic Experiments and Ethnographic Evidence from Fifteen Small-Scale Societies,* 194–231. Oxford: Oxford University Press.

———. 2012. Human survival and life history in evolutionary perspective. In J. Mitani, J. Call, P. Kappeler, R. A. Palombit, and J. B. Silk (eds.), *The Evolution of Primate Societies,* 293–314. Chicago: University of Chicago Press.

Gurven, M., and H. Kaplan. 2007. Longevity among hunter-gatherers: A cross-cultural examination. *Population and Development Review* 33(2): 321–365.

Gurven, M., M. Borgerhoff Mulder, P. L. Hooper, H. Kaplan, R. Quinlan, R. Sear, E. Schniter, C. von Rueden, S. Bowles, T. Hertz, and A. Bell. 2010. Domestication alone does not lead to inequality: Intergenerational wealth transmission among horticulturalists. *Current Anthropology* 51(1): 49–64.

Gurven, M., H. Kaplan, and K. Hill. 2004. Why do foragers share and sharers forage? Explorations of social dimensions of foraging. *Research in Economic Anthropology* 23: 19–43.

Guttentag, M., and P. Secord. 1983. *Too Many Women? The Sex Ratio Question.* New York: Sage.

Gutteridge, W. (ed.). 1986. *Contemporary Terrorism.* New York: Facts on File.

Guttmacher Institute. 2014. State policies in brief: An overview of abortion laws as of 1 January 2014. http://www.guttmacher.org/media/presskits/abortion-US/statsandfacts.html.

Hackett, J. 1983. *The Profession of Arms.* New York: Macmillan.

Hackworth, D. H. 1989. *About Face: The Odyssey of an American Warrior.* With Julie Sherman. New York: Simon and Schuster.

Haddix, K. 2001. Leaving your wife and your brothers: When polyandrous marriages fall apart. *Evolution and Human Behavior* 22: 47–60.

Haig, D. 1992. Genomic imprinting and the theory of parent-offspring conflict. *Seminars in Developmental Biology* 3: 153–160.

———. 1993. Genetic conflicts in human pregnancy. *Quarterly Review of Biology* 68(4): 495–532.

———. 2002. *Kinship and Genomic Imprinting.* New Brunswick, N.J.: Rutgers University Press.

Haig, D. 2004. Genomic imprinting and kinship: How good is the evidence? *Annual Review of Genetics* 38: 553–585.

Hajnal, J. 1965. European marriage patterns in perspective. In D. V. Glass and D.E.C. Eversley (eds.), *Population in History: Essays in Historical Demography,* 101–143. London: Edward Arnold.

Haldane, J.B.S. 1932. *The Causes of Evolution.* New York: Harper and Brothers.

———. 1955. Population genetics. *New Biology* 18: 34–51.

Hale, J. R. 1985. *War and Society in Renaissance Europe, 1450–1620.* New York: St. Martin's Press.

Hallett, J. P. 1984. *Fathers and Daughters in Roman Society.* Princeton, N.J.: Princeton University Press.

Halperin, S. 1979. Temporary association patterns in free-ranging chimpanzees. In D. Hamburg and E. McCown (eds.), *The Great Apes,* 491–499. Menlo Park, Calif.: Benjamin Cummings.

Halpern, D. 2006. Assessing gender gaps in learning and academic achievement. In P. Alexander and P. Winne (eds.), *Handbook of Educational Psychology,* 635–653. Mahwah, N.J.: Erlbaum.

Halpern, D., and M. LaMay. 2000. The smarter sex: A critical review of sex differences in science and mathematics. *Educational Psychology Review* 12: 229–246.

Hamer, D., and Peter Copeland. 1998. *Living with Our Genes.* New York: Doubleday.

Hamer, D., S. Hu, V. Magnuson, N. Hu, and A. Pattatucci. 1993. A linkage between DNA markers on the X chromosome and male sexual orientation. *Science* 261(5119): 321–327.

Hames, R. B. 1979. A comparison of the shotgun and the bow in Neotropical forest hunting. *Human Ecology* 7: 219–252.

———. 1988a. The allocation of parental care among the Ye'kwana. In L. Betzig, M. Borgerhoff Mulde, and P. Turke (eds.), *Human Reproductive Behaviour: A Darwinian Perspective,* 237–252. Cambridge: Cambridge University Press.

———. 1988b. Game conservation or efficient hunting? In B. J. McCay and J. M. Acheson (eds.), *The Question of the Commons: The Culture and Ecology of Communal Resources,* 92–107. Tucson: University of Arizona Press.

———. 1989. Time, efficiency, and fitness in the Amazonian protein quest. *Research in Economic Anthropology* 11: 43–85.

———. 1991. Wildlife conservation in tribal societies. In M. L. Oldfield and J. B. Alcorn (eds.), *Biodiversity: Culture, Conservation, and Ecodevelopment,* 172–199. Boulder, Colo.: Westview Press.

———. 1996. Costs and benefits of monogamy and polygyny for Yanomamö women. *Ethology and Sociobiology* 17: 181–199.

———. 2007. The ecological savage debate. *Annual Review of Anthropology* 36: 177–1990.

Hames, R., and R. Blanchard. 2012. Anthropological data regarding the adaptiveness of hebephilia. *Archives of Sexual Behavior* 41: 745–747.

Hamilton, W. D. 1964. The genetical evolution of social behaviour I, II. *Journal of Theoretical Biology* 7: 1–52.

———. 1966. The moulding of senescence by natural selection. *Journal of Theoretical Biology* 12: 12–45.

———. 1970. Selfish and spiteful behaviour in an evolutionary model. *Nature* 228: 1218–1220.

———. 1971. Geometry for the selfish herd. *Journal of Theoretical Biology* 31: 295–311.

———. 1975. Innate social aptitudes of man: An approach from evolutionary genetics. In R. Fox (ed.), *ASA Studies 4: Biosocial Anthropology*, 133–153. London: Malaby Press.

———. 1979. Wingless and fighting males in fig wasps and other insects. In M. S. and N. A. Blum (eds.), *Sexual Selection and Reproductive Competition in Insects*, 167. New York: Academic Press.

———. 1980. Sex versus non-sex versus parasite. *Oikos* 35: 282–290.

———. 1996. *Narrow Roads of Gene Land: The Collected Papers of W. D. Hamilton*. Vol. 1, *Evolution of Social Behaviour*. Oxford: W. H. Freeman.

Hamilton, W. D., and M. Zuk. 1982. Heritable true fitness and bright birds: A role for parasites? *Science* 218: 384–387.

Hamilton, W. D., P. A. Henderson, and N. Moran. 1981. Fluctuation of environment and coevolved antagonist polymorphism as factors in the maintenance of sex. in R. D. Alexander and D. W. Tinkle (eds.), *Natural Selection and Social Behavior: Recent Research and Theory*, 262–381. New York: Chiron Press.

Hammel, E. A., S. Johansson, and C. Gunsberg. 1983. The value of children during industrialization: Sex ratios in childhood in nineteenth-century America. *Journal of Family History* 8(4): 400–417.

Hammerstein, P. 1996. Darwinian adaptation, population genetics and the streetcar theory of evolution. *Journal of Mathematical Biology* 34: 511–532.

Hampson, E., and D. Kimura. 1988. Reciprocal effects of hormonal fluctuations on human motor and perceptual-spatial skills. *Behavioral Neuroscience* 102: 456–459.

Hannesson, R., K. Salvanes, and D. Squires. 2008. *Technological Change and the Tragedy of the Commons: The Lofoten Fishery over Hundred and Thirty Years*. Discussion paper SAM 5 2008. Trondheim, Norway: Institutt for Samfunnsøkonomi.

Hanson, V. D. 1989. *The Western Way of War: Infantry Battle in Ancient Greece*. Oxford: Oxford University Press.

Harcourt, A. H. 1978. Strategies of emigration and transfer by primates, with particular reference to gorillas. *Zeitschrift für Tierpsychologie* 48: 401–420.

Hardin, G. 1968. The tragedy of the commons. *Science* 162: 1243–1248.

Harpending, H. 1998. Commentary on D. S. Wilson's "Hunting, sharing and multilevel selection." *Current Anthropology* 39: 88–89.

Harpending, H., P. Draper, and A. Rogers. 1987. Human sociobiology. *Yearbook of Physical Anthropology* 30: 127–150.

———. 1990. Fitness in stratified societies. *Ethology and Sociobiology* 11: 497–509.

Harris, J. A., P. A. Vernon, and D. I. Boomsma. 1998. The heritability of testosterone: A study of Dutch adolescent twins and their parents. *Behavioral Genetics* 28: 165–171.

Harris, M. 1979. *Cultural Materialism*. New York: Random House.

Hart, C. W., and A. R. Pilling. 1960. *The Tiwi of North Australia*. New York: Holt.

Hart, S. L. 1997. Beyond greening: Strategies for a sustainable world. *Harvard Business Review* 75: 66–76.

Hartung, J. 1982. Polygyny and the inheritance of wealth. *Current Anthropology* 23: 1–12.

———. 1983. In defense of Murdock: A reply to Dickemann. *Current Anthropology* 24(1): 125–126.

———. 1992. Getting real about rape. *Behavioral and Brain Sciences* 15(2): 390–392.

———. 1997. If I had it to do over. In L. Betzig (ed.), *Human Nature: A Critical Reader*, 344–348. Oxford: Oxford University Press.

Harvey, P., and A. H. Harcourt. 1984. Sperm competition, testis size and breeding systems in primates. In R. L. Smith (ed.), *Sperm Competition and the Evolution of Animal Mating Systems*, 589–600. New York: Academic Press.

Harvey, P., and R. M. Zammuto. 1985. Patterns of mortality and age at first reproduction in natural populations of mammals. *Nature* 315: 319–320.

Harvey, P., R. D. Martin, and T. H. Clutton-Brock. 1987. Life histories in comparative perspective. In B. B. Smuts, D. L. Cheney, R. M. Seyfarth, R. W. Wrangham, and T. Struhsaker (eds.), *Primate Societies*, 181–196. Chicago: University of Chicago Press.

Hasegawa, T., and M. Hiraiwa-Hasegawa. 1983. Opportunistic and restrictive matings among wild chimpanzees in the Mahale Mountains. *Journal of Ethology* 1: 75–85.

Hausfater, G., and S. B. Hrdy. 1984. *Infanticide: Comparative and Evolutionary Perspectives*. New York: Aldine de Gruyter.

Havanon, N., J. Knodel, and W. Sittitrai. 1992. The impact of family on wealth accumulation in rural Thailand. *Population Studies* 46: 37–51.

Hawkes, K. 1990. Why do men hunt? Some benefits for risky strategies. In E. Cashdan (ed.), *Risk and Uncertainty in Tribal and Peasant Economies*, 145–166. Boulder, Colo.: Westview Press.

———. 1991. Showing off: Tests of an hypothesis about men's foraging goals. *Ethology and Sociobiology* 12: 29–54.

———. 1992. Sharing and collective action. In E. A. Smith and B. Winterhalder (eds.), *Evolutionary Ecology and Human Behavior*, 269–300. New York: Aldine de Gruyter.

———. 1993. Why hunter-gatherers work: An ancient version of the problem of public goods. *Current Anthropology* 34: 341–361.

Hawkes, K., and E. L. Charnov. 1988. Human fertility: Individual or group benefit? *Current Anthropology* 20: 469–471.

Hawkes, K., R. L. Bliege Bird, and D. W. Bird. 1998. Commentary on D. S. Wilson's "Hunting, sharing and multilevel selection." *Current Anthropology* 39: 89–90.

Hawkes, K., J. F. O'Connell, and N. G. Blurton Jones. 1997. Hadza women's time allocation, offspring provisioning, and the evolution of long postmenopausal life spans. *Current Anthropology* 18(4): 551–577.

Hawkes, K., J. F. O'Connell, N. G. Blurton Jones, H. Alvarez, and E. L. Charnov. 1998. Grandmothering, menopause, and the evolution of human life histories. *Proceedings of the National Academy of Sciences USA* 95: 1336–1339.

———. 1999. The grandmother hypothesis and human evolution. In L. Cronk, N. Chagnon, and W. Irons (eds.), *Adaptation and Human Behavior: An Anthropological Perspective*, 237–258. Hawthorne, N.Y.: Aldine de Gruyter.

Hawkes, K., J. F. O'Connell, and L. Rogers. 1997. The behavioral ecology of modern hunter-gatherers, and human evolution. *Trends in Ecology and Evolution* 12: 29–32.

Hawkes, K., A. R. Rogers, and E. L. Charnov. 1995. The male's dilemma: Increased off-spring production is more paternity to steal. *Evolutionary Ecology* 9: 662–677.

Hayami, A. 1980. Class differences in marriage and fertility among Tokugawa villagers in Mino Province. *Keio Economic Studies.* 17(1): 1–16.

Heinemann, U. R. 1990. *Eunuchs for the Kingdom of Heaven.* New York: A. Deutsch, Ltd.

Heinen, J., and B. Low. 1992. The behavioral ecology of environmental conservation. *Environmental Conservation* 19(2): 105–116.

Heinrich, B. 1989. *Ravens in Winter.* New York: Summit.

Heinsohn, R., and C. Packer. 1995. Complex cooperative strategies in groups-territorial African lions. *Science* 269: 1260–1262.

Heithaus, M., A. Frid, A. Wirsing, and B. Worm. 2008. Predicting ecological consequences of marine top predator declines. *Trends in Ecology and Evolution* 24: 202–210.

Hemelrijk, C. K. 1994. Support for being groomed in long-tailed macaques, *Macaca fascicularis. Animal Behaviour* 48: 479–481.

Henn, W., K. D. Zang, and D. Skuse. 1997. Mosiacism in Turner's syndrome. *Nature* 390: 569.

Henrich, J., R. Boyd, S. Bowles, C. Camerer, E. Fehr, and H. Gintis (eds.). 2004. *Foundations of Human Sociality: Economic Experiments and Ethnographic Evidence from Fifteen Small-Scale Societies.* Oxford: Oxford University Press.

Henrich, J., R. Boyd, and P. J. Richerson. 2008. Five misunderstandings about cultural evolution. *Human Nature* 19: 119–137.

———. 2012. The puzzle of monogamous marriage. *Philosophical Transactions of the Royal Society B* 367(1589): 657–669.

Henrich, J., S. J. Heine, and A. Norenzayan. 2010. The weirdest people in the world? *Behavioral and Brain Sciences* 33(2–3): 61–83.

Henrich, N., and J. Henrich. 2007. *Why Humans Cooperate: A Cultural and Evolutionary Explanation.* New York: Oxford University Press.

Henry, L. 1961. Some data on natural fertility. *Social Biology* 8: 81–91.

Herre, E., and W. Wcislo. 2011. In defence of inclusive fitness theory. *Nature* 471: E8–E9.

Herrmann, E., and M. Tomasello. 2012. Human cultural cognition. In J. Mitani, J. Call, P. Kappeler, R. A. Palombit, and J. B. Silk (eds.), *The Evolution of Primate Societies,* 701–714. Chicago: University of Chicago Press.

Hershkovits, M. J., and F. S. Hershkovits. 1934. *Rebel Destiny.* Freeport, N.Y.: Books for Libraries Press.

Hewlett, B. S., and L. L. Cavalli-Sforza. 1986. Cultural transmission among Aka Pygmies. *American Anthropologist* 88: 922–934.

Hiatt, L. R. 1981. Polyandry in Sri Lanka: A test case for parental investment theory. *Man* 15: 583–602.

Hibbing, J. R., K. B. Smith, and J. R. Alford. 2013. *Predisposed: Liberals, Conservatives, and the Biology of Political Differences.* New York: Routledge.

Hill, C. M., and H. L. Ball. 1996. Abnormal births and other "ill omens": The adaptive case for infanticide. *Human Nature* 7: 381–401.

Hill, D. 1987. Social relationships between adult male and female rhesus macaques: 1. Sexual consortships. *Primates* 28: 439–456.

Hill, E., and B. Low. 1991. Contemporary abortion patterns: A life-history approach. *Ethology and Sociobiology* 13: 35–48.

Hill, J. 1984. Prestige and reproductive success in man. *Ethology and Sociobiology* 5: 77–95.

Hill, K., and A. M. Hurtado. 1991. The evolution of premature reproductive senescence and menopause in human females: An evaluation of the "Grandmother Hypothesis." *Human Nature* 2: 313–350.

————. 1996. *Ache Life History: The Ecology and Demography of a Foraging People*. New York: Aldine de Gruyter.

Hill, K., and A. M. Hurtado. 2009. Cooperative breeding in South American hunter-gatherers. *Proceedings of the Royal Society B* 276 (1674): 3863–3870.

Hill, K., and H. Kaplan. 1988a. Tradeoffs in male and female reproductive strategies among the Ache: Part 1. In L. Betzig, M. Borgerhoff Mulder, and P. Turke (eds.), *Human Reproductive Behaviour: A Darwinian Perspective*, 277–289. Cambridge: Cambridge University Press.

————. 1988b. Tradeoffs in male and female reproductive strategies among the Ache: Part 2. In L. Betzig, M. Borgerhoff Mulder, and P. Turke (eds.), *Human Reproductive Behaviour: A Darwinian Perspective*, 291–305. Cambridge: Cambridge University Press.

Hill, K., M. Barton, and M. Hurtado. 2009. The emergence of human uniqueness: Characters underlying behavioral modernity. *Evolutionary Anthropology* 18: 187–200.

Hines, M. 2009. Gendered behavior across the lifespan. In R. Lerner, M. Lamb, and A. Freund (eds.), *The Handbook of Lifespan Development*, Vol. 2, 341–378. Hoboken, N.J.: Wiley.

————. 2010. Sex-related variation in human behavior and the brain. *Trends in Cognitive Science* 14: 448–456.

————. 2011. Gender development and the human brain. *Annual Review of Neuroscience* 34: 69–88.

Hines, M., C. Brook, and G. Conway. 2004. Androgen and psychosexual development: Core gender identity, sexual orientation and recalled childhood gender role behavior in women and men with congenital adrenal hyperplasia (CAH). *Journal of Sex Research* 41: 75–81.

Hirschberger, S. 2001. Biological factors in the development of sexual orientation. In A. R. D'Augelli and C. J. Patterson (eds.), *Lesbian, Gay, and Bisexual Identities and Youth: Psychological Perspectives*, 27–51. New York: Oxford University Press.

Hoekstra, R. F. 1987. The evolution of sexes. In S. Stearns (ed.), *The Evolution of Sex and Its Consequences*, 59–92. Basel: Birkhäuser.

Hofbauer, J., and K. Sigmund. 1998. *Evolutionary Games and Population Dynamics*. Cambridge: Cambridge University Press.

Hofman, M. A. 1983. Energy metabolism, brain size, and longevity in mammals. *Quarterly Review of Biology* 58: 495–512.

Höglund, J., and R. V. Alatalo. 1995. *Leks*. Princeton, N.J.: University Press.

Holekamp, K. E., and P. W. Sherman. 1989. Why male ground squirrels disperse. *American Scientist* 77: 232–239.

Holland, J. 1995. *Hidden Order: How Adaptation Builds Complexity*. Reading Mass.: Helix Books.

Holm, P., B. Hersoug, and S. Rånes. 2000. Revisiting Lofoten: Co-managing fish stocks or fishing space? *Human Organization* 59: 353–364.

Holmes, R. 1985. *Acts of War: The Behavior of Men in Battle*. New York: Free Press.

Hooks, B. L., and P. A. Green. 1993. Cultivating male allies: A focus on primate females, including *Homo sapiens*. *Human Nature* 4: 81–107.

Howard, R. D. 1979. Estimating reproductive success in natural populations. *American Naturalist* 114: 221–231.

Howell, N. 1979. *Demography of the Dobe !Kung*. New York: Academic Press.

Hrdy, S. B. 1974. Male-male competition and infanticide among the lemurs (*Presbytis entellus*) of Abu Rajasthan. *Folia Primatologia* 22: 19–58.

———. 1976. The care and exploitation of nonhuman primate infants by conspecifics other than the mother. *Advances in the Study of Behavior* 6: 101–158.

———. 1978. Allomaternal care and the abuse of infants among Hanuman langurs. In D. J. Chivers and P. Herbert (eds.), *Recent Advances in Primatology*, vol. 1, 169–172. New York: Academic Press.

———. 1979. Infanticide among animals: A review classification and implications for the reproductive strategies of females. *Ethology and Sociobiology* 1: 13–40.

———. 1981. *The Woman That Never Evolved*. Cambridge, Mass.: Harvard University Press.

———. 1992. Fitness tradeoffs in the history and evolution of delegated mothering with special reference to wet-nursing, abandonment, and infanticide. *Ethology and Sociobiology* 13: 409–442.

———. 1997a. Mainstreaming Medea. In L. L. Betzig (ed.), *Human Nature: A Critical Reader*, 423–426. Oxford: Oxford University Press.

———. 1997b. Raising Darwin's consciousness: Female sexuality and the prehominid origins of patriarchy. *Human Nature* 8: 1–49.

———. 1999. *Mother Nature: A History of Mothers, Infants, and Natural Selection*. Cambridge: Cambridge University Press.

———. 2000. The optimal number of fathers: Evolution, demography, and history in the shaping of female mate preferences. In D. LeCroy and P. Moller (eds.), *Evolutionary Perspectives on Human Reproductive Behavior*, 75–96. Annals of the New York Academy of Sciences 907. New York: New York Academy of Sciences.

———. 2009. *Mothers and Others: The Evolutionary Origins of Mutual Understanding*. Cambridge, Mass.: Harvard University Press.

Hrdy, S. B., and D. S. Judge. 1993. Darwin and the puzzle of primogeniture. *Human Nature* 4: 1–45.

Hu, S., A. Pattitucci, C. Patterson, et al.1995. Linkage between sexual orientation and chromosome Xq28 in males but not females. *Nature Genetics* 11(3): 248–256.

Hughes, A. 1986. Reproductive success and occupational class in eighteenth-century Lancashire, England. *Social Biology* 33: 109–115.

———. 1988. *Evolution and Human Kinship*. Oxford: Oxford University Press.

Hull, T. H. 1990. Recent trends in sex ratios at birth in China. *Population and Development Review* 16(1): 63–83.

Hultquist, P. 2013. Power parity and peace? The role of relative power in civil war settlement. *Journal of Peace Research* 50(5): 623–634.

Humphrey, N. K. 1976. The social function of intellect. In P.P.G. Bateson and R. A. Hinde (eds.), *Growing Points in Ethology*, 303–317. London: Cambridge University Press.

———. 1983. *Consciousness Regained: Chapters in the Development of Mind*. Oxford: Oxford University Press.

Hunn, E. 1982. Mobility as a factor limiting resource use in the Columbian Plateau of North America. In N. Williams and E. Hunn (eds.), *Natural Resource Managers: North America and Australian Hunter-Gatherers*, 17–43. Boulder: Westview Press.

Huntingford, F., and A. Turner. 1987. *Animal Conflict*. London: Chapman and Hall.

Hurst, L. D. 1991. Sex, slime, and selfish genes. *Nature* 354: 23–2 4.

———. 1992. Intragenomic conflict as an evolutionary force. *Proceedings of the Royal Society B* 244: 91–99.

Hurst, L. D. 1995. Selfish genetic elements and their role in evolution: The evolution of sex and some of what that entails. *Philosophical Transactions of the Royal Society B* 349: 321–332.

———. 1996. Why are there only two sexes? *Proceedings of the Royal Society B* 263: 415–422.

Hurst, L. D., and W. D. Hamilton. 1992. Cytoplasmic fusion and the nature of sexes. *Proceedings of the Royal Society B* 247: 189–207.

Hurst, L. D., and G. McVean. 1997. Growth effects of uniparental disomies and the conflict theory of genomic imprinting. *Trends in Genetics* 13: 436–443.

Hurtado, M., and K. Hill. 1990. Seasonality in a foraging society: Variation in diet, work effort, fertility, and the sexual division of labor among the Hiwi of Venezuela. *Journal of Anthropological Research* 46: 293–346.

Hurtado, M., K. Hawkes, and K. Hill. 1985. Female subsistence strategies among Ache hunter-gatherers of eastern Paraguay. *Human Ecology* 13: 1–28.

Hurtado, M., K. Hill, H. Kaplan, and I. Hurtado. 1992. Tradeoffs between food acquisition and child care among Hiwi and Ache women. *Human Nature* 3: 185–216.

Huth, P. K. 1988. *Extended Deterrence and the Prevention of War*. New Haven, Conn.: Yale University Press.

Hutt, C. 1972. *Males and Females*. London: Penguin.

Hyde, J. S. 1996. Where are the gender differences? Where are the gender similarities? In D. Buss and N. Malamuth (eds.), *Sex, Power, Conflict: Evolutionary and Feminist Perspectives*, 107–118. New York: Oxford University Press.

Imhof, A. E. 1981. Remarriage in rural populations and in urban middle and upper strata in Germany from the sixteenth to the twentieth century. In J. Dupâquier, J. E. Hélin, P. Laslett, M. Livi-Bacci, and S. Sogner (eds.), *Marriage and Remarriage in Populations of the Past*, 335–346. New York: Academic Press.

Inger, G. 1980. *Svensk Rattshistoria*. Lund, Sweden: Liber Laromedel.

Inglehart, R., and P. Norris. 2003. *Rising Tide: Gender Equality and Cultural Change Around the World*. Cambridge: Cambridge University Press.

Inoue, M., F. Mitsunaga, H. Ohsawa, A. Takenaka, Y. Sugiyama, S. Gaspard, and O. Takenaka. 1991. Male mating behaviour and paternity discrimination by DNA fingerprinting in a Japanese macaque group. *Folia Primatologia* 56: 202–210.

Irons, W. 1975. *The Yomut Turkmen: A Study of Social Organization among a Central Asian Turkic-Speaking Population*. Museum of Anthropology, Anthropological Papers 58. Ann Arbor: University of Michigan.

———. 1979a. Emic and reproductive success. In N. A. Chagnon and W. Irons (eds.), *Evolutionary Biology and Human Social Behavior: An Anthropological Perspective*, 257–272. North Scituate, Mass.: Duxbury Press.

———. 1979b. Natural selection, adaptation, and human social behavior. In N. A. Chagnon and W. Irons (eds.), *Evolutionary Biology and Human Social Behavior: An Anthropological Perspective*, 4–38. North Scituate, Mass.: Duxbury Press.

———. 1980. Is Yomut social behavior adaptive? In G. Barlow and J. Silverman (eds.), *Sociobiology: Beyond Nature/Nurture?* 417–473. Boulder, Colo.: Westview Press.

———. 1983a. The cultural and reproductive success hypothesis: A synthesis of empirical tests. Paper presented at annual meeting of American Anthropological Association, 1993.

———. 1983b. Human female reproductive strategies. In Samuel K. Wasser (ed.), *Social Behavior of Female Vertebrates*, 169–213. New York: Academic Press.

———. 1991. How did morality evolve? *Zygon* 26: 49–85.

———. 1996a. In our own self image: The evolution of morality, deception, and religion. *Skeptic* 4: 50–61.

———. 1996b. Morality, religion, and human nature. In W. M. Richardson and W. Wildman (eds.), *Religion and Science: History, Method, Dialogue*, 375–399. New York: Routledge.

———. 1998. Adaptively relevant environments versus the environment of evolutionary adaptedness. *Evolutionary Anthropology* 6: 194–204.

Isler, K., and C. van Schaik. 2014. How humans evolved large brains: Comparative evidence. *Evolutionary Anthropology* 23: 65–75.

Iwasa, Y. 1998. The conflict theory of genomic imprinting: How much can be explained? *Current Topics in Developmental Biology* 40: 255–293.

Jacobson, H., and W. Zimmerman. 1993. *Behavior, Culture, and Conflict in World Politics*. Ann Arbor: University of Michigan Press.

Jaeggi, A., J. Burkhart, and C. van Schaik. 2010. On the psychology of cooperation in humans and other primates: Combining the natural history and experimental evidence of prosociality. *Philosophical Transactions of the Royal Society B* 365: 2723–2735.

James, T. W., and D. Kimura. 1997. Sex differences in remembering the locations of objects in an array: Location-shifts versus location-exchanges. *Evolution and Human Behavior* 18: 155–163.

Johansson, S., and Ola Nygren. 1991. The missing girls of China: A new demographic account. *Population and Development Review* 17(1): 35–51.

Johnson, G. 1986. Kin selection, socialization, and patriotism: An integrating theory. *Politics and the Life Sciences* 4:128–139.

Johnson, G., S. H. Ratwik, and T. J. Sawyer. 1987. The evocative significance of kin terms in patriotic speech. In V. Reynolds, V. Falger, and I. Vine (eds.), *The Sociobiology of Ethnocentrism: Evolutionary Dimensions of Xenophobia, Discrimination, Racism, and Nationalism*, 137–174. London: Croom Helm.

Johnson, N. E., and S. Lean. 1985. Relative income, race and fertility. *Population Studies* 39: 99–112.

Johnson, R. L., and E. Kapsalis. 1998. Menopause in free-ranging rhesus macaques: Estimated incidence, relation to body condition, and adaptive significance. *International Journal of Primatology* 19: 751–765.

Johnson, S. B., and R. C. Johnson. 1991. Support and conflict of kinsmen in Norse earldoms, Icelandic families, and the English royalty. *Ethology and Sociobiology* 12: 211–220.

Johnstone, R., and M. Cant. 2010. The evolution of menopause in cetaceans and humans: The role of demography. *Proceedings of the Royal Society B* 277: 3765–3771.

Jolly, A. 1966. Lemur social behavior and primate intelligence. *Science* 153: 501–506.

———. 1985. *The Evolution of Primate Behavior*. 2nd ed. New York: Macmillan.

Jones, B.., L. DeBruine, A. Little, C. Conway, L. Welling, and F. Smith. 2007. Sensation seeking and men's face preferences. *Evolution and Human Behavior* 28: 439–446.

Jones, B., L. DeBruine, D. Perrett, A. Little, D. Feinberg, and M. J. I. Smith. 2008. Effects of menstrual cycle phase on face preferences. *Archives of Sexual Behavior* 37: 78–84.

Jones, D. 1995. Sexual selection, physical attractiveness, and facial neoteny. *Current Anthropology* 36: 723–748 (including commentary).

Jones, D. 1996. An evolutionary perspective on physical attractiveness. *Evolutionary Anthropology* 5: 97–109.

Jones, D., and K. Hill. 1993. Criteria of facial attractiveness in five populations. *Human Nature* 4: 271–296.

Jones, E. C. 1975. The post-reproductive phase in mammals. In P. van Keep and C. Lauritzen (eds.), *Frontiers of Hormone Research*, vol. 3, 1–20. Basel: Karger.

Jones, J. H. 2009. The force of selection on the human life cycle. *Evolution and Human Behavior* 30(5): 305–314.

———. 2011. Primates and the evolution of slow life histories. *Current Biology* 21(18) R708–R17. doi: 10.1016/j.cub.2011.08.025.

Jones, J. H., and B. Ferguson. 2009. Demographic and social predictors of intimate partner violence in Colombia. *Human Nature* 20: 185–203.

Jörberg, L. 1972. *A History of Prices in Sweden: 1732–1914*. 2 vols. Lund, Sweden: CWK Gleerup.

———. 1975. Structural change and economic growth in nineteenth-century Sweden. In Steven Koblik (ed.), *Sweden's Development from Poverty to Affluence, 1750–1970*, 92–135. Minneapolis: University of Minnesota Press.

Jorgensen, W. 1980. *Western Indians*. San Francisco: Freeman.

Josephson, S. C. 1993. Status, reproductive success, and marrying polygynously. *Ethology and Sociobiology* 14: 391–396.

Judge, D. S. 1995. American legacies and the variable life histories of women and men. *Human Nature* 6: 291–323.

Judge, D. S., and S. B. Hrdy. 1992. Allocation of accumulated resources among close kin: Inheritance in Sacramento, California, 1890–1984. *Ethology and Sociobiology* 13: 495–522.

Junger, S. 1997. *The Perfect Storm*. New York, Norton,

Just, W. 1970. *Military Men*. New York: Knopf.

Kachel, A. F., L. S. Premo, and J-J. Hublin. 2011. Grandmothering and natural selection. *Proceedings of the Royal Society B* 278: 384–391.

Kadlec, J. A., and W. Drury. 1968. Structure of the New England herring gull population. *Ecology* 49: 644–676.

Kagan, J. 1981. Universals in human development. In R. H. Monroe, R. L. Monroe, and B. B. Whiting (eds.), *Handbook of Cross-cultural Human Development*, 53–62. New York: Garland Press.

Kahneman, D. 2011. *Thinking, Fast and Slow*. New York: Farrar, Straus and Giroux.

Kahneman, D., and A. Tversky. 1982. On the study of statistical intuitions. In D. Kahneman, P. Slovic, and A. Tversky (eds.), *Judgements under Uncertainty: Heuristics and Biases*, 493–508. Cambridge: Cambridge University Press.

Kano, T. 1996. Male rank order and copulation rate in a unit-group of bonobos at Wamba, Zaïre. In W. C. McGrew, L. Marchant, and T. Nishida (eds.), *Great Ape Societies*, 135–145. Cambridge: Cambridge University Press.

Kaplan, H. 1994. Evolutionary and wealth flows theories of fertility: Empirical tests and new models. *Population and Development Review* 20: 753–791.

———. 1997. The evolution of the human life course. In K. Wachter and C. Finch (eds.), *Between Zeus and Salmon: The Biodemography of Longevity*. Washington, D.C.: National Academy Press.

Kaplan, H., and K. Hill. 1985. Hunting ability and reproductive success among male Ache foragers: Preliminary results. *Current Anthropology* 26: 131–133.

Kaplan, H., and J. B. Lancaster. 1999. The life histories of men in Albuquerque: An evolutionary-economic analysis of parental investment and fertility in modern society. In L. Cronk, N. Chagnon, and W. Irons (eds.), *Adaptation and Human Behavior: An Anthropological Perspective*. Hawthorne, N.Y.: Aldine de Gruyter.

Kaplan, H., M. Gurven, J. Winkling, P. Hooper, and J. Stieglitz. 2010. Learning, menopause, and the human adaptive complex. *Annals of the New York Academy of Sciences* 1204: 30–42.

Kaplan, H., J. B. Lancaster, J. A. Bock, and S. E. Johnson. 1995. Fertility and fitness among Albuquerque men: A competitive labour market theory. In R.I.M. Dunbar (ed.), *Human Reproductive Decisions*, 96–136. London: St. Martin's Press.

Kappeler, P. 2012. Mate choice. In J. Mitani, J. Call, P. Kappeler, R. A. Palombit, and J. B. Silk (eds.), *The Evolution of Primate Societies*, 367–386. Chicago: University of Chicago Press

Karlsen, K. F. 1987. *The Devil in the Shape of a Woman: Witchcraft in Colonial New England*. New York: Norton.

Karsten, R. 1923. *Blood Revenge, War, and Victory Feasts among the Jibaro Indians of Eastern Ecuador*. Smithsonian Institution, Bureau of American Ethnology Bulletin 79. Washington, D.C.: U.S. Government Printing Office.

Kasarda, J. D., J.O.G. Billy, and K. West. 1986. *Status Enhancement and Fertility: Reproductive Responses to Social Mobility and Educational Opportunity*. New York: Academic Press.

Keegan, J. 1987. *The Mask of Command*. London: Jonathan Cape.

Keegan, J., and A. Wheatcroft. 1986. *Zones of Conflict: An Atlas of Future Wars*. New York: Simon and Schuster.

Keeley, L. H. 1996. *War before Civilization: The Myth of the Peaceful Savage*. Oxford: Oxford University Press.

Kelly, N., and S. Alonzo. 2009. Will male advertisement be a reliable indicator of parental care, if offspring survival depends on male parental care? *Proceedings of the Royal Society B* 276: 3175–3183.

Kenrick, D., M. Trost, and V. Sheets. 1996. Power, harassment, and trophy mates: The feminist advantages of an evolutionary perspective. In D. Buss and N. Malamuth (eds.), *Sex, Power, Conflict: Evolutionary and Feminist Perspectives*, 29–53. New York: Oxford University Press.

Keohane, R. O. 1984. *After Hegemony: Cooperation and Discord in the World Political Economy*. Princeton, N.J.: Princeton University Press.

Keohane, R. O., and E. Ostrom (eds.). 1995. *Local Commons and Global Interdependence: Heterogeneity and Cooperation in Two Domains*. London: Sage.

Keverne, E. B., R. Fundele, M. Narashima, S. C. Barton, and M. A. Surani. 1996. Genomic imprinting and the differential roles of parental genomes in brain development. *Developmental Brain Research* 92: 91–100.

Keyfitz, N. 1985. *Applied Mathematical Demography*. 2nd ed. New York: Springer-Verlag.

Kirkpatrick, M. 1996. Genes and adaptation: A pocket guide to the theory. In M. Rose and G. Lauder (eds.), *Adaptation*, 125–146. New York: Academic Press.

Klare, M. T., and P. Kornbluh (eds.). 1988. *Low-Intensity Warfare: Counterinsurgency, Proinsurgency, and Anti-terrorism in the Eighties*. New York: Pantheon.

Knafo, A., and R. Plomin. 2006. Prosocial behavior from early to middle childhood: Genetic and environmental influences on stability and change. *Developmental Psychology* 42: 771–786.

Knodel, J. 1981. Remarriage and marital fertility in Germany during the eighteenth and nineteenth centuries: An exploratory analysis based on German village genealogies. In J. Dupaquier, E. Helin, P. Laslett, and M. Livi-Bacci (eds.), *Marriage and Remarriage in Populations of the Past*, 581–590. New York: Academic Press.

———. 1988. *Demographic Behavior in the Past: German Village Populations in the 18th and 19th Centuries*. Cambridge: Cambridge University Press.

Knodel J., and K. A. Lynch. 1985. The decline of remarriage: Evidence from German village populations in the eighteenth and nineteenth centuries. *Journal of Family History* 10(1): 34–59.

Knodel, J., and M. Wongsith. 1991. Family size and children's education in Thailand: Evidence from a national sample. *Demography* 28: 119–131.

Knodel, J., N. Havanon, and W. Sittitrai. 1990. Family size and the education of children in the context of rapid fertility decline. *Population and Development Review* 16(1): 31–62.

Knodel, J., C. Saengtienchai, B. S. Low, and R. Lucas. 1997. An evolutionary perspective on Thai sexual attitudes. *Journal of Sex Research* 34(3): 292–303.

Kohlberg, L. L. 1981. *Essays on Moral Development*. Vol. 1, *The Philosophy of Moral Development*. San Francisco: Harper and Row.

———. 1984. *Essays on Moral Development*. Vol. 2, *The Psychology of Moral Development*. San Francisco: Harper and Row.

Kokko, K., and L. Pulkkinen. 2005. Stability of aggressive behavior from childhood to middle age in women and men. *Aggressive Behavior* 31: 485–497

Konner, M. J. 1981. Evolution of human behavior development. In R. H. Monroe, R. L. Monroe, and B. B. Whiting (eds.), *Handbook of Cross-Cultural Development*, 3–51. New York: Garland Press.

Kottak, C. 1990. Culture and "economic development." *American Anthropologist* 92: 723–731.

Kottak, C., and A. C. G. Costa. 1993. Ecological awareness, environmentalist action, and international conservation strategy. *Human Organization* 52: 335–343.

Krakoff, S. 2012. Planetary identity formation and the relocalization of environmental law. *Florida Law Review* 64(1): 87–140.

Kramer, K. 2005. Children's help and the pace of reproduction: Cooperative breeding in humans. *Evolutionary Anthropology* 14(6): 224–237.

———. 2008. Early sexual maturity among Pumé foragers of Venezuela: Fitness implications of teen motherhood. *American Journal of Physical Anthropology*. 136: 338–350.

Kramer, K., and P. Ellison. 2010. Pooled energy budgets: Resituating human energy allocation trade-offs. *Evolutionary Anthropology* 19: 136–137.

Krebs, J. R., and N. B. Davies (eds.). 1991. *Behavioral Ecology: An Evolutionary Approach*. 3rd ed. London: Blackwell Scientific.

———. 1993. *An Introduction to Behavioral Ecology.* 3rd ed. Sunderland, Mass.: Sinauer Associates.

——— (eds.). 1997. *Behavioral Ecology: An Evolutionary Approach.* 4th ed. London: Blackwell Scientific.

Krech, S., III. 1999. *Myth and History: The Ecological Indian.* New York: W. W. Norton.

Krushnick, G. 2013. Access to resources shaped maternal decision making: Evidence from a factorial vignette experiment. *PLoS ONE* 8(9): e75539. doi: 10.1371/journal. pone.0075539.

Kruuk, H. 1972. *The Spotted Hyena: A Study of Predation and Social Behavior.* Chicago: University of Chicago Press.

Kumm, J., and M. W. Feldman. 1997. Gene-culture coevolution and sex ratios II. Sex-chromosomal distorters and cultural preferences for offspring sex. *Theoretical Population Biology* 52: 1–15.

Kurland, J. 1979. Matrilines: The primate sisterhood and the human avunculate. In N. Chagnon and W. Irons (eds.), *Evolutionary Biology and Human Social Behavior*, 145–180. North Scituate, Mass.: Duxbury.

Kurland, J. A., and S.J.C. Gaulin. 1984. The evolution of male parental investment: Effects of genetic relatedness and feeding ecology on the allocation of reproductive effort. In D. M. Taub (ed.), *Primate Paternalism*, 259–306. New York: Van Nostrand Reinhold.

Kurlansky, M. 1998. *Cod: A Biography of the Fish that Changed the World.* Penguin Books.

Kurzban, R. 2011. *Why Everyone (Else) Is a Hypocrite.* Princeton, N.J.: Princeton University Press.

Kurzban, R., and H. C. Barrett. 2012. Origins of cumulative culture. *Science* 335: 1056–1057.

Lack, D. 1947. The significance of clutch size. *Ibis* 89: 302–352.

———. 1966. *Population Studies of Birds.* Oxford: Oxford University Press.

Laland, K. N., J. Kumm, and M. W. Feldman. 1994. Gene-culture coevolutionary theory: A test case. *Current Anthropology* 36(1): 131–156 (including commentary and response).

Lancaster, J. B. 1986. Human adolescence and reproduction: An evolutionary perspective. In J. B. Lancaster and B. A. Hamburg (eds.), *School-Age Pregnancy and Parenthood*, 17–38. New York: Aldine de Gruyter.

———. 1991. A feminist and evolutionary biologist looks at women. *Yearbook of Physical Anthropology* 34: 1–11.

Lancaster, J. B., and B. J. King. 1992. An evolutionary perspective on menopause. In V. Kerns and J. K. Brown (eds.), *In Her Prime: New Views of Middle-Aged Women*, 7–15. 2nd ed. Urbana: University of Illinois Press.

Langergraber, K. E. 2012. Cooperation among kin. In J. Mitani, J. Call, P. Kappeler, R. A. Palombit, and J. B. Silk (eds.), *The Evolution of Primate Societies*, 491–513. Chicago: University of Chicago Press.

Langston, N., S. Rohwer, and D. Gori. 1997. Experimental analyses of intra- and intersexual competition in red-winged blackbirds. *Behavioral Ecology* 8: 524–533.

Långström, N., Q Rahman, E. Carlström, and P. Lichtenstein. 2010. Genetic and environmental effects on same-sex sexual behavior: A population study of twins in Sweden. *Archives of Sexual Behavior* 39(1): 75–80.

Larrick, R. P., R. E. Nisbett, and J. N. Morgan. 1993. Who uses the cost-benefit rules of choice? Implications for the normative status of microeconomic theory. *Organizational Behavior and Human Decision Processes* 56: 331–347.

Lass-Hennemann, J., C. Deuter, L. Kuehl, A. Schulz, T. Blumenthal, and H. Schachinger. 2010. Effects of stress on human mating preferences: Stressed individuals prefer dissimilar mates. *Proceedings of the Royal Society B* 277: 2175–2183.

Lavely, W., and R. B. Wong. 1998. Revising the Malthusian narrative: The comparative study of population dynamics in late Imperial China. *Journal of Asian Studies* 57: 714–748.

Lawson, D., and R. Mace. 2009. Trade-offs in modern parenting: A longitudinal study of sibling competition for parental care. *Evolution and Human Behavior* 30: 170–183.

———. 2010. Optimizing modern family size: Trade-offs between fertility and the economic costs of reproduction. *Human Nature* 21: 39–61.

———. 2011. Parental investment and the optimization of human family size. *Proceedings of the Royal Society B* 366: 333–343.

Lawson, D., A. Alvergne, and M. Gibson. 2012. The life-history trade-off between fertility and child survival. *Proceedings of the Royal Society B* 279:4755–4764.

Leacock, E. 1955. Matrilocality in a simple hunting economy. *Southwestern Journal of Anthropology* 11: 31–47.

Leacock, E., and R. Lee. 1982. *Politics and History in Band Societies*. Cambridge: Cambridge University Press.

Le Boeuf, B., and J. Reiter. 1988. Lifetime reproductive success in Northern elephant seals. In T. H. Clutton-Brock (ed.), *Reproductive Success: Studies of Individual Variation in Contrasting Breeding Systems*, 344–383. Chicago: University of Chicago Press.

Ledyard, J. O. 1995. Public goods: A survey of experimental research. In J. H. Kagel and A. E. Roth (eds.), *The Handbook of Experimental Economics*, 111–194. Princeton, N.J.: Princeton University Press.

Lee, J., and C. Campbell. 1997. *Fate and Fortune in Rural China: Social Organization and Population Behavior in Liaoning, 1774–1873*. Cambridge: Cambridge University Press.

Lee, J., and S. Guo (eds.). 1994. *Qingdai huanhzu renkou xingwei yu shehui huan-jing* (Population behavior and social setting of the Qing imperial lineage). Beijing: Beijing University Press.

Lee, J., C. Campbell, and F. Wang. 1993. The last emperors: An introduction to the demography of the Qing imperial lineage. In D. Rehler and R. Schofield (eds.), *New and Old Methods in Historical Demography*, 361–382. Oxford: Oxford University Press.

Lee, Richard. 1979a. *The !Kung San*. London: Cambridge University Press.

———. 1979b. Politics, sexual and non-sexual, in an egalitarian society. In E. Leacock and R. Lee (eds.), *Politics and History in Band Societies*, 37–59. Cambridge: Cambridge University Press.

Lee, Ronald D. 1990. Comment: The second tragedy of the commons. *Population and Development Review* 16 (suppl.): 315–322.

Leigh, D. 2004. UK forces taught torture methods. *The Guardian*, May 8, 2004.

Leigh, E. 1971. *Adaptation and Diversity: Natural History and the Mathematics of Evolution*. San Francisco: Freeman, Cooper & Co.

Leimar, O. 1996. Life-history analysis of the Trivers and Willard sex-ratio problem. *Behavioral Ecology* 7: 316–325.

Lenski, R. E., and M. Travisano. 1994. Dynamics of adaptation and diversification: A 10,000-generation experiment with bacterial populations. *Proceedings of the National Academy of Sciences USA* 91: 6808–6814.

Lessells, C. M. 1991. The evolution of life history strategies. In J. R. Krebs and N. B. Davies (eds.), *Behavioural Ecology*, chapter 2. 3rd ed. Oxford: Blackwell Scientific.

Lessells, C. M., and J. McNamara. 2012. Sexual conflict over parental investment in repeated bouts: Negotiation reduces overall care. *Proceedings of the Royal Society B* 279: 1506–1514.

Lesthaeghe, R., and C. Wilson. 1986. Modes of production, secularization and the pace of the fertility decline in western Europe, 1870–1930. In A. J. Coale and S. C. Watkins (eds.), *The Decline of Fertility in Europe*, 261–292. Princeton, N.J.: Princeton University Press.

Letendre, K., C. Fincher, and R. Thornhill. 2012. Infectious disease, collectivism, and warfare. In T. Schackelford and V. Weekes-Shackelford (eds.), *The Oxford Handbook of Evolutionary Perspectives on Violence, Homicide, and War*, 351–371.

Leutenegger, W. 1979. Evolution of litter size in primates. *American Naturalist* 114: 881–903.

Lever, J. 1976. Sex differences in the games children play. *Social Problems* 23: 478–487.

———. 1978. Sex differences in the complexity of children's play and games. *American Sociological Review* 43: 471–483.

Levine, N., and J. Silk. 1997. Why polyandry fails: Sources of instability in polyandrous marriages. *Current Anthropology* 38: 375–398.

Levy, M. K. 2012. "Boys fight, girls fight": Adolescent girls speak about girls' aggression. *Girlhood Studies*, 5(2), 45–64.

Lewellen, T. C. 1983. *Political Anthropology*. South Hadley, Mass.: Bergen and Garvey.

Lewis, K. 1998. Pathogen resistance as the origin of kin altruism. *Journal of Theoretical Biology* 193: 359–363.

Lewis, O. 1959. *Five Families: Mexican Case Studies in the Culture of Poverty*. New York: Basic Books.

———. 1970. *Anthropological Essays*. New York: Random House.

Lewontin, R. C. 1970. The units of selection. *Annual Review of Ecology and Systematics* 1: 1–18.

Lext, G. 1968. *Mantalsskrivningen i Sverige fore 1860*. Meddelanden från Ekonomisk-Historiska Institutionen vid Göteborgs Universitet 13. Göteborg: Göteborgs Universitet.

Li, L., and M. K. Choe. 1997. A mixture model for duration data: Analysis of second births in China. *Demography* 34: 189–197.

Lie, H., L. Simmons, and G. Rhodes. 2010. Genetic dissimilarity, genetic diversity, and mate preferences in humans. *Evolution and Human Behavior* 31: 48–58.

Lin, Y-H. 2011. Constructing alternative femininity: The gender identity of "Bad Girls" in Taiwan. In L. E. Bass and D. A. Kinney (eds.), *The Well-Being, Peer Cultures and Rights of Children*, 159–179. Bingley, U.K.: Emerald Group Publishing.

Lindenfors, P. 2002. Sexually antagonistic selection on primate size. *Journal of Evolutionary Biology* 15: 595–607.

Lindert, P. H. 1978. *Fertility and Scarcity in America*. Princeton, N.J.: Princeton University Press.

Linton, R. 1939. Marquesan culture and analysis of Marquesan culture. In A. Kardiner (ed.), *The Individual and His Society*, 137–250. New York: Columbia University Press.

Lips, J. E. 1947. Naskapi law. *Transactions of the American Philosophical Society* 37: 379–491.

Logan, M., and H. N. Qirko. 1996. An evolutionary perspective on maladaptive traits and cultural conformity. *American Journal of Human Biology* 8: 615–629.

Lo-Johannsson, F.A.I. 1981. *Sveriges Rikes Lag: Gillard och Antagen Parisdagen År, 1734, faksimilutgava*. Malmö, Sweden: Giulunds.

Lonsdorf, E. V., and S. Ross. 2012. Socialization and development of behavior. In J. Mitani, J. Call, P. Kappeler, R. A. Palombit, and J. B. Silk (eds.), *The Evolution of Primate Societies*, 245–268. Chicago: University of Chicago Press.

Lorenz, K. 1966. *On Aggression*. New York: Harcourt Brace Jovanovich.

Lovelock, J. 1988. *The Ages of Gaia*. New York: Norton.

Lovelock, J., and L. Margulis. 1974. Biological modulation of the earth's atmosphere. *Icarus* 21: 471–489.

Low, B. 1978. Environmental uncertainty and the parental strategies of marsupials and placentals. *American Naturalist* 112: 197–213.

———. 1979. Sexual selection and human ornamentation. In N. Chagnon and W. Irons (eds.), *Evolutionary Biology and Human Social Behavior*, 462–486. North Scituate, Mass.: Duxbury Press.

———. 1988a. Measures of polygyny in humans. *Current Anthropology* 29(1): 189–194.

———. 1988b. Pathogen stress and polygyny in humans. In L. Betzig, M. Borgerhoff Mulder, and P. Turke (eds.), *Human Reproductive Behaviour: A Darwinian Perspective*, 115–128. Cambridge: Cambridge University Press.

———. 1989a. Cross-cultural patterns in the training of children: An evolutionary perspective. *Journal of Comparative Psychology* 103: 311–319.

———. 1989b. Occupational status and reproductive behavior in 19th century Sweden: Locknevi parish. *Social Biology* 36: 82–101.

———. 1990a. Fat and deception. *Ethology and Sociobiology* 11: 67–74.

———. 1990b. Human responses to environmental extremeness and uncertainty: A cross-cultural perspective. In E. Cashdan (ed.), *Risk and Uncertainty in Tribal and Peasant Economies*, 229–255. Boulder, Colo.: Westview Press.

———. 1990c. Land ownership, occupational status, and reproductive behavior in 19th-century Sweden: Tuna parish. *American Anthropologist* 92(2): 457–468.

———. 1990d. Marriage systems and pathogen stress in human societies. *American Zoologist* 30: 325–339.

———. 1990e. Sex, power, and resources: Ecological and social correlates of sex differences. *Journal of Contemporary Sociology* 27: 45–71.

———. 1991. Reproductive life in nineteenth-century Sweden: An evolutionary perspective on demographic phenomena. *Ethology and Sociobiology* 12: 411–448.

———. 1992. Sex, coalitions, and politics in preindustrial societies. *Politics and the Life Sciences* 11(1): 63–80.

———. 1993a. Ecological demography: A synthetic focus in evolutionary anthropology. *Evolutionary Anthropology* 1993: 106–112.

———. 1993b. An evolutionary perspective on war. In H. Jacobson and W. Zimmerman (eds.), *Behavior, Culture, and Conflict in World Politics*, 13–56. Ann Arbor: University of Michigan Press.

———. 1994a. Human sex differences in behavioral ecological perspective. *Analyse & Kritik* 16: 38–67.

———. 1994b. Men in the demographic transition. *Human Nature* 5(3): 223–253.

———. 1996a. The behavioral ecology of conservation in traditional societies. *Human Nature* 7: 353–379.

———. 1996b. Men, women and sustainability. *Population and Environment* 18(2): 111–141.

———. 2001. Sex, wealth, and fertility: Old rules, new environments. In L. Cronk, N. Chagnon, and W. Irons (eds.), *Adaptation and Human Behavior: An Anthropological Perspective*, 323–344. New York: Aldine.

———. 2003. Ecological and social complexities of human monogamy. In C. Boesch and U. Reichart (eds.), *The Evolution of Monogamy*, 161–176. Cambridge: Cambridge University Press.

———. 2005. Families: An evolutionary ecological approach. In J. L. Roopnarine, and U. Gielen (eds.), *Families in Global Perspective*, 14–32. Boston: Allyn & Bacon.

———. 2007. Ecological and Sociocultural influences on mating and marriage systems. In R. Dunbar and L. Barrett (eds.), *The Oxford Handbook of Evolutionary Psychology*, 449–462. Oxford University Press, Oxford.

———. 2009. Marriage and family: The evolutionary ecological context. In H. Elizabeth Peters and Claire Kamp Dush (eds.), *Marriage and Family: Complexities and Perspectives*, 33–56. New York: Columbia University Press.

———. 2012. Gender equity issues in evolutionary perspective. In C. Roberts (ed.), *Applied Evolutionary Ecology*, 131–148. Oxford: Oxford University Press.

———. 2013a. Fertility: Life history and ecological aspects. In M. L. Fisher, J. R. Garcia, and R. Sokol Chang (eds.), *Evolution's Empress: Darwinian Perspectives on the Nature of Women*, 222–242. New York: Oxford University Press.

———. 2013b. Noble savages or consumate consumers: The behavioral ecology of building a green economy. In R. Richardson (ed.), *Building a Green Economy*, 45–68. East Lansing: Michigan State University Press.

Low, B., and A. L. Clarke. 1991. Occupational status, land ownership, migration, and family patterns in 19th-century Sweden. *Journal of Family History* 16(2): 117–138.

———. 1992. Resources and the life course: Patterns in the demographic transition. *Ethology and Sociobiology* 13: 463–494.

———. 1993. Historical perspectives on population and environment: Data from 19th-century Sweden. In G. Ness, W. Drake, and S. Brechen (eds.), *Population-Environment Dynamics: Ideas and Observations*, 13–55. Ann Arbor: University of Michigan Press.

Low, B., and J. Heinen. 1993. Population, resources, and environment. *Population and Environment* 15(1): 7–41.

Low, B., R. D. Alexander, and K. M. Noonan. 1987. Human hips, breasts and buttocks: Is fat deceptive? *Ethology and Sociobiology* 8: 249–257.

———. 1988. Response to Judith Anderson's comments on Low, Alexander, and Noonan (1987). *Ethology and Sociobiology* 9: 325–328.

Low, B., A. L. Clarke, and K. Lockridge. 1992. Toward an ecological demography. *Population and Development Review* 18(1): 1–31.

Low, B., A. Hazel, N. Parker, and K. B. Welch. 2008. Influences on women's reproductive lives: Unexpected ecological underpinnings. *Cross-Cultural Research* 42: 201–219.

Low, B., A. Hazel, N. Parker, and K. B. Welch. 2013. Life expectancy, fertility, and women's lives: A life-history perspective. *Cross-Cultural Research* 47(2): 198–225.

Low, B., C. P. Simon, and K. G. Anderson. 2002. An evolutionary perspective on demographic transitions: Modeling multiple currencies. *American Journal of Human Biology* 14(1): 149–167.

———. 2003. The biodemography of modern women: Tradeoffs when resources become limiting. In J. Rogers and H-P. Kohler (eds.), *The Biodemography of Human Fertility and Reproduction*, 105–134. Boston: Kluwer Press.

Low, B., C. P. Simon, and D. Finkbeiner. 2005. Favored places in the selfish herd: Positional costs and benefits in fish schools. In L. Bookner, S. Forrest, M. Mitchell, and R. Riolo (eds.), *Perspectives on Adaptation in Natural and Artificial Systems*, 213–238. Oxford: Oxford University Press.

Lucas, M., and E. Koff. 2013. How conception risk affects competition and cooperation with attractive women and men. *Evolution and Human Behavior* 34: 16–22.

Lukaszewski, A., and J. Roney. 2010. Kind toward whom? Mate preferences for personality traits are target specific. *Evolution and Human Behavior* 31: 29–38.

Luker, K. 1996. *Dubious Conceptions: The Politics of Teenage Pregnancy.* Cambridge, Mass.: Harvard University Press.

Lumsden, C. J., and E. O. Wilson. 1981. *Genes, Mind, and Culture: The Coevolutionary Process.* Cambridge, Mass.: Harvard University Press.

Lutz, W. 1994. *The Future Population of the Earth.* London: Earthscan Press.

MacArthur, R. H., and E. O. Wilson. 1967. *The Theory of Island Biogeography.* Princeton, N.J.: Princeton University Press.

Maccoby, E. E. 2002. Gender and group process: A developmental perspective. *Current Directions in Psychological Science* 11: 54–58.

Maccoby, E. E., and C. N. Jacklin. 1974. *The Psychology of Sex Differences.* Palo Alto, Calif.: Stanford University Press.

Mace, G. M. 1990. Birth, sex ratio, and infant mortality rates in captive western lowland gorillas. *Folia Primatologia* 55(3–4): 156–165.

Mace, R. 1996. Biased parental investment and reproductive success in Gabbra pastoralists. *Behavioral Ecology and Sociobiology* 38: 75–81.

———. 1998. The coevolution of human fertility and wealth inheritance strategies. *Philosophical Transactions of the Royal Society B* 353: 389–397.

———. 2000. Evolutionary ecology of human life history. *Animal Behaviour* 59(1): 1–10.

Mace, R. 2013a. Cooperation and conflict between women in the family. *Evolutionary Anthropology* 22: 251–258.

———. 2013b. The cost of children. *Nature* 499: 32–33.

Mace, R., and A. Alvergne. 2012. Female reproductive competition within families in rural Gambia. *Proceedings of the Royal Society B* 279: 2219–2227.

Mace, R., and H. Colleran. 2009. Kin influence on the decision to start using modern contraception: A longitudinal study from rural Gambia. *American Journal of Human Biology* 21: 472–477.

MacKellar, F. L. 1996. On human carrying capacity. *Population and Development Review* 22: 145–156.

———. 1997. Population and fairness. *Population and Development Review* 23: 359–376.

Macunovich, D. J. 1998. Fertility and the Easterlin hypothesis: An assessment of the literature. *Journal of Population Economics* 11: 53–111.

Maechling, C., Jr. 1988. Counterinsurgency: The first ordeal by fire. In M. T. Klare, and P. Kornbluh (eds.), *Low-Intensity Warfare: Counterinsurgency, Proinsurgency, and Anti-terrorism in the Eighties*, 21–48. New York: Pantheon.

Mailath, G., and L. Samuelson. 2006. *Repeated Games and Reputation: Long-Run Relationships*. New York: Oxford University Press.

Malamuth, N. 1996. The confluence model of sexual aggression: Feminist and evolutionary perspectives. In D. Buss and N. Malamuth (eds.), *Sex, Power, Conflict: Evolutionary and Feminist Perspectives*, 269–295. New York: Oxford University Press.

Mallett, M. 1987. *The Borgias*. Chicago: Academy Press.

Malmström, A. 1981. *Successionsratt II*. Uppsala, Sweden: Iustrus Forlag.

Mancebo, F. 2013. The pitfalls of sustainability policies: Insights into plural sustainabilities. *Challenges in Sustainability* 1(1): 29–40.

Manners, R. A. 1967. The Kipsigis of Kenya: Culture change in a "model" East African tribe. In R. Needham (ed.), *Contemporary Change in Traditional Societies*, 207–359. Urbana: University of Illinois Press.

Manson, J., and R. Wrangham. 1991. Intergroup aggression in chimpanzees and humans. *Current Anthropology* 32: 369–390.

Marlowe, F. 2004. Dictators and ultimatums in an egalitarian society of hunter-gatherers. In Henrich, J., R. Boyd, S. Bowles, C. Camerer, E. Fehr, and H. Gintis (eds.), *Foundations of Human Sociality: Economic Experiments and Ethnographic Evidence from Fifteen Small-Scale Societies*, 194–231. Oxford: Oxford University Press.

———. 2012. The socioecology of human reproduction. In J. Mitani, J. Call, P. Kappeler, R. A. Palombit, and J. B. Silk (eds.), *The Evolution of Primate Societies*, 469–486. Chicago: University of Chicago Press.

Marsh, H., and T. Kasuya. 1984. Changes in the ovaries of the short-finned pilot whale, *Globocephala macrorhynchus*, with age and reproductive activity. *Report of the International Whaling Commission* 1984: 259–310.

———. 1986. Evidence for reproductive senescence in female cetaceans. *Report of the International Whaling Commission* Special Issue 8: 83–95.

Martin, P. S., and R. G. Klein. 1984. *Quaternary Extinctions: A Prehistoric Revolution*. Tucson: University of Arizona Press.

Martin, R. D. 1983. Human brain evolution in an ecological context. 52nd James Arthur Lecture on the Evolution of the Human Brain. New York: American Museum of Natural History.

Mason, K. O. 1997. Explaining fertility transitions. *Demography* 34: 443–454.

Mathew, S., and R. Boyd. 2011. Punishment sustains large-scale cooperation in prestate warfare. *Proceedings of the National Academy of Sciences* 108(28): 11375–11380.

Mathews, G. B., A. Fane, B. Conway, C. Brook, and M. Hines. 2009. Personality and congenital adrenal hyperplasia: Possible effects of prenatal androgen exposure. *Hormones and Behavior* 55: 285–291.

Mathews, P., and R. Sear. 2013. Family and fertility: Kin influence on the progression to a second birth in the British Household Panel Study. *PLoS ONE* 8(3): e56941. doi: 10.1371/journal.pone.0056941.

Maynard Smith, J. 1969. The status of neo-Darwinism. In C. H. Waddington (ed.), *Towards a Theoretical Biology*. Vol. 2, *Sketches*, 82–89. Edinburgh: Edinburgh University Press.

———. 1974. The theory of games and the evolution of animal conflict. *Journal of Theoretical Biology* 47: 209–221.

Maynard Smith, J. 1978. *The Evolution of Sex*. Cambridge: Cambridge University Press.

———. 1979. Game theory and the evolution of behaviour. *Proceedings of the Royal Society B* 205: 475–488.

———. 1982. *Evolution and the Theory of Games*. Cambridge: Cambridge University Press.

Maynard Smith, J., and G. A. Parker. 1976. The logic of asymmetrical contests. *Animal Behavior* 24: 159–175.

Maynard Smith, J., and G. R. Price. 1973. The logic of animal conflicts. *Nature* 246: 15–18.

Mazur, L. A. (ed.). 1994. *Beyond the Numbers: A Reader on Population, Consumption, and the Environment*. Washington, D.C.: Island Press.

McBurney, D. H., S.J.C. Gaulin, T. Devineni, and C. Adams. 1997. Superior spatial memory of women: Stronger evidence for the gathering hypothesis. *Evolution and Human Behavior* 18: 165–174.

McCabe, K., S. Rassenti, and V. L. Smith. 1998. Reciprocity, trust, and payoff privacy in extensive form bargaining. *Games and Economic Behavior* 24: 10–24.

McCandless, B., A. Roberts, and T. Starnes. 1972. Teachers' marks, achievement test scores, and aptitude relations with respect to social class, race, and sex. *Journal of Educational Psychology* 63: 153–159.

McCulloch, M. 1952. *The Ovimbundu of Angola*. London: International African Institute.

McCullough, J. M., and E. Y. Barton. 1991. Relatedness and mortality risk during a crisis year: Plymouth colony 1620–1621. *Ethology and Sociobiology* 12: 195–209.

McCullough, M. E., E. Carter, C. N. DeWall, and C. Corrales. 2012. Religious cognition down-regulates sexually-selected, characteristically male behaviors in men, but not in women. *Evolution and Human Behavior* 33: 562–568.

McElreath, R., and R. Boyd. 2007. *Mathematical Models of Social Evolution*. Chicago: University of Chicago Press.

McGrew, W. C., L. F. Marchant, and T. Nishida (eds.). 1996. *Great Ape Societies*. Cambridge: Cambridge University Press.

McInnis, R. M. 1977. Childbearing and land availability: Some evidence from individual household data. In R. Lee (ed.), *Population Patterns in the Past*, 201–227. New York: Academic Press.

McIntyre, M., and C. P. Edwards. 2009. The early development of gender differences. *Annual Review of Anthropology* 38: 83–97.

McNeill, William H. 1963. *The Rise of the West: A History of the Human Community*. Chicago: University of Chicago Press.

———. 1982. *The Pursuit of Power: Technology, Armed Force, and Society Since A.D. 1000*. Chicago: University of Chicago Press.

Mead, M. 1935. *Sex and Temperament in Three Primitive Societies*. New York: William Murrow.

Mealey, L. 1985. The relationship between social status and biological success: A case study of the Mormon religious hierarchy. *Ethology and Sociobiology* 6: 249–257.

Mealey, L., and W. Mackey. 1990. Variation in offspring sex ratio in women of differing social status. *Ethology and Sociobiology* 11: 83–95.

Mech, D. 1977. Productivity, mortality, and population trends of wolves in northeastern Minnesota. *Journal of Mammalogy* 58: 559–574.

Meehan, C., R. Quinlan, and C. Malcolm. 2013. Cooperative breeding and maternal energy expenditure among Aka foragers. *American Journal of Human Biology* 25: 42–57.

Mendels, F. F. 1981. *Industrialization and Population Pressure in Eighteenth-Century Flanders*. New York: Arno Press.

Merrigan, P., and Y. St.-Pierre. 1998. An econometric and neoclassical analysis of the timing and spacing of births in Canada from 1950 to 1990. *Journal of Population Economics* 11(1): 29–51.

Mesquida, C. G., and N. I. Wiener. 1996. Human collective aggression: A behavioral ecology perspective. *Ethology and Sociobiology* 17: 247–262.

Mesterton-Gibbons, M., and L. Dugatkin. 1992. Cooperation among unrelated individuals: Evolutionary factors. *Quarterly Review of Biology* 67: 267–281.

Meyer-Bahlburg, H., C. Dolezal, S. Baker, and M. New. 2008. Sexual orientation in women with classical or non-classical congenital adrenal hyperplasia as a function of degree of prenatal androgen excess. *Archives of Sexual Behavior* 37: 85–99.

Michaels, D., and C. Monforton. 2005. Manufacturing uncertainty: Contested science and the protection of the public's health and environment. *American Journal of Public Health* 95(51), Suppl. 1: S39–S48.

Michod, R. E., and B. R. Levin (eds.). 1988. *The Evolution of Sex*. Sunderland, Mass.: Sinauer.

Mildvan, A., and B. Strehler. 1960. A critique of theories of mortality. In B. L. Strehler (ed.), *The Biology of Aging*, 45–78. New York: American Institute of Biological Sciences.

Miles, D. R., and G. Carey. 1997. Genetic and environmental architecture of human aggression. *Journal of Personality and Social Psychology* 72: 207–217.

Millard-Ball, A. 2012. Do city climate plans reduce emissions? *Journal of Urban Economics* 71: 289–311.

Miller, E. 1975. War, taxation, and the English economy in the late 13th and early 14th centuries. In J. M. Winter (ed.), *War and Economics in Development*, 11–32. Cambridge: Cambridge University Press.

Miller, G. 2000. *The Mating Mind: How Sexual Choice Shaped the Evolution of Human Nature*. New York: Doubleday.

Milner, R., M. Jennions, and P. Backwell. 2010. Safe sex: Male-female coalitions and precopulatory mate-guarding in a fiddler crab. *Biology Letters* 6(2): 180–182.

Mitani, J. 1984. The behavioral regulation of monogamy in gibbons (*Hylobates muelleri*). *Behavioral Ecology and Sociobiology* 15: 225–229.

Mitani, J., J. Call, P. Kappeler, R. A. Palombit, and J. B. Silk (eds.) 2012. *The Evolution of Primate Societies*. Chicago: University of Chicago Press

Mitchell, N. J., and F. Janzen. 2010. Temperature-dependent sex determination and contemporary climate change. *Sexual Development* 4: 129–140.

Mitterauer, M., and R. Sieder. 1982. *The European Family*. Oxford: Blackwell.

Moir, A., and D. Jessel. 1991. *Brain Sex: The Real Difference between Men and Women*. New York: Lyle Stewart.

Møller, A. P. 1987. Variation in badge size in male house sparrows *Passer domesticus*—evidence for status signalling. *Animal Behaviour* 35: 1637–1644.

———. 1989. Viability costs of male tail ornaments in a swallow. *Nature* 339: 132–135.

———. 1994. *Sexual Selection and the Barn Swallow*. Oxford: Oxford University Press.

Monroe, R. L., and R. H. Monroe. 1975. *Cross-Cultural Human Development*. Monterey, Calif.: Brooks/Cole.

Moore, D., S. Wigby, S. English, S. Wong, T. Székely, and F. Harrison. 2013. Selflessness is sexy: Reported helping behavior increases desirability of men and women as long-term sexual partners. *BMC Evolutionary Biology* 13: 182. http://www.biomedcentral .com/1471-2148/13/182 (accessed December 2013).

Moore, J. H. 1990. The reproductive success of Cheyenne war chiefs: A counter-example to Chagnon. *Current Anthropology* 31: 169-173.

Moore, T., and D. Haig. 1991. Genomic imprinting in mammalian development: A parental tug-of-war. *Trends in Genetics* 7: 45-49.

Morell, V. 1998. A new look at monogamy. *Science* 281: 1982-1983.

Morin, P., J. Wallis, J. Moore, and D. Woodruff. 1994. Paternity exclusion in a community of chimpanzees using hypervariable simple sequence repeats. *Molecular Ecology* 3: 469-478.

Mosk, C. 1983. *Patriarchy and Fertility: Japan and Sweden 1880-1960*. New York: Academic Press.

Moss, M. 2013. *Salt Sugar Fat: How the Food Giants Hooked Us*. New York: Random House.

Mueller, U. 1993. Social status and sex. *Nature* 363: 490.

Mueller, U., and A. Mazur. 1996. Facial dominance in West Point cadets predicts military rank 20+ years later. *Social Forces* 74: 823-850.

———. 1998. Reproductive constraints on dominance competition in male *Homo sapiens*. *Evolution and Human Behavior* 19: 387-396.

Muller, M. N., and M. E. Thompson. 2012. Mating, parenting, and male reproductive strategies. In J. Mitani, J. Call, P. Kappeler, R. A. Palombit, and J. B. Silk (eds.), *The Evolution of Primate Societies*, 387-411. Chicago: University of Chicago Press.

Murdock, G. P. 1967. *Ethnographic Atlas*. Pittsburgh: University of Pittsburgh Press.

———. *Atlas of World Cultures*. Pittsburgh: University of Pittsburgh Press.

Murdock, G. P., and C. Provost. 1973. Factors in the division of labor by sex. *Ethnology* 12: 203-225.

Murdock, G. P., and D. White. 1969. Standard cross-cultural sample. *Ethnology* 8: 329-369.

Murphy, L., L. Sievert, K. Begum, T. Sharmeen, E. Puleo, O. Chowdhury, S. Muttukrishna, and G. Bentley. 2013. Life course effects on age at menopause among Bangladeshi Sedentees and migrants to the UK. *American Journal of Human Biology* 25: 83-93.

Myers, R. A., and B. Worm. 2003. Rapid worldwide depletion of predatory fish communities. *Nature* 423: 280-283.

———. 2005. Extinction, survival or recovery of large predatory fishes. *Philosophical Transactions of the Royal Society B* 360(1453): 13-20.

Nakahashi, W., and M. Feldman. 2014. Evolution of division of labor: Emergence of different activities among group members. *Journal of Theoretical Biology* 348: 65-79.

Nesse, R. M. 1990. Evolutionary explanations of emotions. *Human Nature* 1: 261-289.

Nesse, R. M., and K. C. Berridge, 1997. Psychoactive drug use in evolutionary perspective. *Science* 278: 63-66.

Nettle, D., A. Colléony, and M. Cockerill. 2011. Variation in cooperative behavior within a single city. *PLoS ONE* 6(10): e26922. doi: 10.1371/journal.pone.0026922.

Newman, M. 2010. *Networks: An Introduction*. Oxford: Oxford University Press.

Newton-Fisher, N. 2006. Female coalitions against male aggression in wild chimpanzees of the Budongo Forest; *International Journal of Primatology* 27: 1589–1599.

Nice, M. M. 1962. Development of behavior in precocial birds. *Transactions of the Linnaean Society of New York* 8: 1–211.

Niefert, M. R., J. M. Seacat, and W. E. Jobe. 1985. Lactation failure due to insufficient glandular development of the breast. *Pediatrics* 76: 823–828.

Nisbett, R. E. 1980. *Human Inference: Strategies and Shortcomings of Social Judgement.* Englewood Cliffs, N.J.: Prentice Hall.

Nisbett, R. E. and D. Cohen. 1996. *Culture of Honor: The Psychology of Violence in the South.* Boulder, Colo.: Westview Press.

Nisbett, R. E., G. T. Fong, D. R. Lehman, and P. W. Cheng. 1987. Teaching reasoning. *Science* 238: 625–631.

Nishida, T. 1979. The social structure of chimpanzees of the Mahale Mountains. In D. A. Hamburg and E. R. McCown (eds.), *The Great Apes*, 73–122. Menlo Park, Calif.: Benjamin Cummings.

———. 1983. Alpha status and agonistic alliance in wild chimpanzees (*Pan troglodytes schweinfurthii*). *Primates* 24: 318–336.

Nishida, T., and M. Hiraiwa-Hasegawa. 1987. Chimpanzees and bonobos: Cooperative relationships among males. In B. B. Smuts, D. L. Cheney, R. M. Seyfarth, R. W. Wrangham, and T. T. Struhsaker (eds.), *Primate Societies*, 165–178. Chicago: University of Chicago Press.

Nishida, T., and K. Hosaka. 1996. Coalition strategies among adult chimpanzees of the Mahale Mountains, Tanzania. In W. C. McGrew, L. Marchant, and T. Nishida (eds.), *Great Ape Societies*, 114–134. Cambridge: Cambridge University Press.

Nishida, T., M. Hiraiwa-Hasegawa, and Y. Takahata. 1985. Group extinction and female transfer in wild chimpanzees in the Mahale Mountains. *Zeitschrift für Tierpsychologie* 67: 284–301.

Nishida, T., H. Takasaki, and Y. Takasaki. 1990. Demography and reproductive profiles. In T. Nishida (ed.), *The Chimpanzees of Mahale Mountains*, 63–97. Tokyo: University of Tokyo Press.

Noë, R., F.B.M. de Waal, and J.A.R.A.M. van Hoof. 1980. Types of dominance in a chimpanzee colony. *Folia Primatologica* 34: 90–110.

Norberg, A., and M. Rolén. 1979. Migration and marriage: Some empirical results from Tuna parish, 1865–1894. In J. Sundin and E. Söderlund (eds.), *Time, Space, and Man: Essays on Microdemography*. Stockholm: Almqvist and Wiksell.

Nordenskiold, E. 1949. The Cuna. In J. Steward (ed.), *Handbook of South American Indians*, vol. 4. Washington, D.C.: U.S. Government Printing Office.

Nordhaus, W. 2008. *A Question of Balance: Weighing the Options on Global Warming Policies.* New Haven: Yale University Press.

Norenzayan, A., and A. Shariff. 2008. The origin and evolution of religious prosociality. *Science* 322(5898): 58–62.

Nowak, M., and K. Sigmund. 1994. The alternating Prisoner's Dilemma. *Journal of Theoretical Biology* 168: 219–226.

Nowak, M., R. May, and K. Sigmund. 1995. The arithmetics of mutual help. *American Scientist*, June 1995, 76–81.

Nowak, M., C. Tarnita, and E. O. Wilson. 2010. The evolution of eusociality. *Nature* 466: 1057–1062.

O'Brien, D. 1977. Female husbands in southern Bantu society. In Alice Schlegel (ed.), *Sexual Stratification: A Cross-Cultural View*, 109–126. New York: Columbia University Press.

Oelschlanger, M. 1991. *The Idea of Wilderness: From Prehistory to the Age of Ecology.* New Haven, Conn.: Yale University Press.

Omark, D. R., and M. S. Edelman. 1975. A comparison status hierarchy in young children: An ethological approach. *Social Science Information* 14: 87–107.

Opie, C., Q. D. Atkinsons, R. Dunbar, and S. Schultz. 2013. Male infanticide leads to social monogamy in primates. *Proceedings of the National Academy of Sciences* 110:13328–13332.

Orians, G. H. 1969. On the evolution of mating systems in birds and mammals. *American Naturalist* 103: 589–603.

Ostergren, R. 1990. Patterns of seasonal industrial labor recruitment in a nineteenth-century Swedish parish: The case of Matfors and Tuna, 1846–1873. Demographic Data Base Report No. 5, 1–100. Umeå, Sweden.

Osterman, K, K.Bjorkqvist, K. Lagerspetz, A. Kaukianen, S. Landau, A. Fraczek, and G. Caparana. 1998. Cross-cultural evidence of female indirect aggression. *Aggressive Behavior* 24: 1–8.

Ostrom, E. 1990. *Managing the Commons.* Cambridge: Cambridge University Press.

———. 1998a. A behavioral approach to the rational choice theory of collective action. *American Political Science Review* 92: 1–22.

———. 1998b. Coping with the tragedies of the commons. Paper given at the 1998 meeting of the Society for Politics and the Life Sciences /American Political Science Association, Boston, 1998.

———. 2000. *Governing the Commons: The Evolution of Institutions for Collective Action.* Cambridge: Cambridge University Press.

Ostrom, E., and J. Walker. 2005. *Trust and Reciprocity: Interdisciplinary Lessons for Experimental Research.* Russell Sage Foundation Publications.

Ostrom, E., J. Burger, C. B. Field, R. B. Norgaard, and D. Policansky. 1999. Re-visiting the commons: Local lessons, global challenges. *Science* 284: 278–282.

Ostrom, E., R. Gardner, and J. Walker. 1994. *Rules, Games, and Common-Pool Resources.* Ann Arbor: University of Michigan Press.

Otterbein, K. 1963. Marquesan polyandry. *Marriage and Family Living* 25(2): 155–159.

———. 1970. *The Evolution of War: A Cross-Cultural Study.* Cambridge, Mass.: HRAF Press.

Outka, U., and R. Feiock. 2012. Local promises for climate mitigation: An empirical assessment. *William and Mary Environmental Law and Policy Review* 36: 635–669.

Oye, K. A. 1986. *Cooperation under Anarchy.* Princeton, N.J.: Princeton University Press.

Packer, C. 1986. The ecology of sociality in felids. In D. I. Rubenstein and R. W. Wrangham (eds.), *Ecological Aspects of Social Evolution*, 429–451. Princeton, N.J.: Princeton University Press.

Packer, C., and A. E. Pusey. 1982. Cooperation and competition within coalitions of male lions: Kin selection or game theory? *Nature* 296(5859): 740–742.

———. 1983. Adaptations of female lions to infanticide by incoming males. *American Naturalist* 121: 716–28.

———. 1984. Infanticide in carnivores. In G. Hausfater and S. B. Hrdy (eds.), *Infanticide: Comparative and Evolutionary Perspectives*. New York: Aldine de Gruyter.

Packer, C., D. A. Collins, A. Sindimwe, and J. Goodall. 1995. Reproductive constraints on aggressive competition in female baboons. *Nature* 373: 60–63.

Packer, C., M. Tatar, and A. Collins. 1998. Reproductive cessation in female mammals. *Nature* 392: 807–811.

Pagel, M. 1999. Mother and father in surprise genetic agreement. *Nature* 397: 19–20.

Palmer, C. T., B. E. Fredrickson, and C. F. Tilley. 1997. Categories and gatherings: Group selection and the mythology of cultural anthropology. *Evolution and Human Behavior* 18: 291–308.

Palumbi, S.R., D.J. Barshis, N. Traylor-Knowles, and R.A. Bay. 2014. Mechanisms of reef coral resistance to future climate change. *Science*. doi: 10.1126/science.1251336.

Pandit, S., and C. van Schaik. 2003. A model for leveling coalitions among primate males: toward a theory of egalitarianism. *Behavioral Ecology and Sociobiology* 55: 161–168.

Parish, A. R. 1996. Female relationships in bonobos (*Pan paniscus*): Evidence for bonding, cooperation, and female dominance in a male-philopatric species. *Human Nature* 7: 61–96.

Parker, G. A. 1974. Assessment strategy and the evolution of fighting behaviour. *Journal of Theoretical Biology* 47: 223–243.

———. 1984. Evolutionarily stable strategies. In J. R. Krebs and N. B. Davies (eds.), *Behavioral Ecology*, 30–61. Oxford: Blackwell Scientific.

Parker, G. A., R. R. Baker, and V.G.F. Smith. 1972. The origin and evolution of gamete dimorphism and the male-female phenomenon. *Journal of Theoretical Biology* 36: 529–553.

Parry, J. P. 1979. *Caste and Kinship in Kangara*. London: Routledge and Paul.

Partridge, L., and L. Hurst. 1998. Sex and conflict. *Science* 281: 2003–2008.

Paul, A., J. Kuester, A. Timme, and J. Arnemann. 1993. The association between rank, mating effort, and reproductive success in male Barbary macaques. *Primates* 34: 491–502.

Peacock, N. 1990. Comparative and cross-cultural approaches to the study of human female reproductive failure. In C. J. Rousseau (ed.), *Primate Life History and Evolution*, 195–220. New York: Wiley-Liss.

———. 1991. An evolutionary perspective on the patterning of maternal investment in pregnancy. *Human Nature* 2(4): 351–385.

Pebley, A., and W. Mbugua. 1989. Polygyny and fertility in sub-Saharan Africa. In R. Lesthaeghe (ed.), *Reproduction and Social Organization in Sub-Saharan Africa*, 338–364. Berkeley: University of California Press.

Pellis, S. 2002. Sex differences in play fighting revisited: Traditional and non-traditional mechanisms of sexual differentiation in rats. *Archives of Sexual Behavior* 31: 17–26.

Pellison, M. 1897. *Roman Life in Pliny's Time*. Philadelphia: George W. Jacobs & Co.

Peng, X. 1991. *Demographic Transition in China*. Oxford: Clarendon Press.

Penn, D. 2003. The evolutionary roots of our environmental problems: Toward a Darwinian ecology. *Quarterly Review of Biology* 78(3): 275–301.

Penner, A. 2003. International gender × item difficulty intersections in mathematics and science achievement tests. *Journal of Educational Psychology* 95: 650–655.

Pennington, R., and H. Harpending. 1988. Fitness and fertility among Kalahari !Kung. *American Journal of Physical Anthropology* 77: 303–319.

Pennisi, E. 1998. A genomic battle of the sexes. *Science* 281: 1984–1985.

Pérusse, D. 1993. Cultural and reproductive success in industrial societies: Testing the relationship at proximate and ultimate levels. *Behavioral in and Brain Sciences* 16: 267–322.

———. 1994. Mate choice in modern societies: Testing evolutionary hypotheses with behavioral data. *Human Nature* 5(3): 255–278.

Peter, Prince of Greece and Denmark. 1963. *A Study of Polyandry*. The Hague: Mouton.

Peters, M., L. Simmins, and G. Rhodes. 2009. Preferences across the menstrual cycle for masculinity and symmetry in photographs of male faces and bodies. *PLoS ONE* 4(1): e3148. doi: 10.137/journal.pone 0004138.

Pfister, U. 1989a. Proto-industrialization and demographic change: The Canton of Zurich revisited. *Journal of Economic History* 18: 629–62.

———. 1989b. Work roles and family structure in proto-industrial Zurich *J. Interdisciplinary Hist.* 20: 83–105.

Piaget, J. 1932. *The Moral Judgement of the Child*. New York: Free Press.

Piazza, J. 2011. Poverty, minority economic discrimination, and domestic terrorism. *Journal of Peace Research* 48(3): 339–353.

Pickrell, J., G. Coop, J. Novembre, S. Kudaravalli, J. Li, D. Absher, B.Srinivasan, G. Barsh, R. Myers, M. Feldman, and J. Pritchard. 2009. Signals of recent selection in a worldwide sample of human populations. *Genome Research* 19(5): 826–837.

Pinker, S. 2011. *The Better Angels of Our Nature: Why Violence has Declined. The Decline of Violence and its Psychological Roots*. London: Penguin.

Place, S., Peter M. Todd, Lars Penske, and J. B. Asendorpf. 2010. Humans show mate copying after observing real mate choices. *Evolution and Humans Behavior* 31: 320–325.

Plomin, R., J. C. DeFries, and G. E. McClearn. 1990. *Behavioral Genetics: A Primer*. 2nd ed. New York: Freeman.

Plous, S. 1993. *The Psychology of Judgement and Decision Making*. New York: McGraw-Hill.

Podolsky, R. H. 1985. Colony formation and attraction of the Laysan albatross and the Leach's storm petrel. Ph.D. diss. University of Michigan, Ann Arbor.

Pollan, M. 2008. *In Defense of Food: An Eater's Manifesto*. Penguin Books.

Pomiankowski, A. 1987a. The costs of choice in sexual selection. *Journal of Theoretical Biology* 128: 195–218.

———. 1987b. The 'handicap principle' does work—sometimes. *Proceedings of the Royal Society B* 127: 123–145.

Pope, T. 1990. The reproductive consequences of male cooperation in the red howler monkey: Paternity exclusion in multi-male and single-male troops using genetic markers. *Behavioral Ecology and Sociobiology* 27: 439–567.

Portney, P., and J. Weyant. 1999. *Discounting and Intergenerational Equity*. Washington D.C.: Resources for the Future.

Preston, J., R. Ritter, and J. Hernandez. 2010. Principles of religious prosociality: A review and reformulation. *Social and Personality Psychology Compass* 4(8): 574–590.

Princen, T. 2005. *The Logic of Sufficiency*. Cambridge, Mass: MIT Press.

Prokosch, M., R. Coss, J. Scheib, and S. Blozis. 2009. Intelligence and mate choice: Intelligent men are always appealing. *Evolution and Human Behavior* 30: 11–20.

Promislow, D., and P. Harvey. 1990. Living fast and dying young. A comparative analysis of life history traits among mammals. *Journal of Zoology* 220: 417–437.

Pulliam, H. R., and C. Dunford. 1980. *Programmed to Learn: An Essay on the Evolution of Culture*. New York: Columbia University Press.

Pusey, A. 2012. Magnitude and sources of variation in female reproductive performance. In J. Mitani, J. Call, P. Kappeler, R. A. Palombit, and J. B. Silk (eds.), *The Evolution of Primate Societies*, 343–366. Chicago: University of Chicago Press.

Pusey, A. E., and C. Packer. 1987. Dispersal and philopatry. In B. Smuts, D. L. Cheney, R. M. Seyfarth, R. W. Wrangham, and T. T. Struhsaker (eds.), *Primate Societies*, 250–266. Chicago: University of Chicago Press.

———. 1997. The ecology of relationships. In J. R. Krebs and N. B. Davies (eds.), *Behavioural Ecology*, chapter 11. 4th ed. Oxford: Blackwell Scientific.

Pusey, A. E., J. Williams, and J. Goodall. 1997. The influence of dominance rank on the reproductive success of female chimpanzees. *Science* 277: 828–831.

Putnam, R. D. 1993a. *Double-Edged Diplomacy: International Bargaining and Domestic Politics*. Berkeley: University of California Press.

———. 1993b. *Making Democracy Work: Civic Tradition in Modern Italy*. Princeton, N.J.: Princeton University Press.

Puts, D. 2010. Beauty and the beast: Mechanisms of sexual selection in humans. *Evolution and Human Behavior* 31: 157–175.

Puts, D., C. Jordan, and Mark Breedlove. 2006. Brother, where art thou? The fraternal birth-order effect on male sexual orientation. *Proceedings of the National Academy of Sciences USA* 103(28): 10531–10532.

Qirko, H. N. 1999. The maintenance and reinforcement of celibacy in institutionalized settings. In E. J. Sobo and S. Bell (eds.), *Anthropological Perspectives on Sexual Abstinence: Celibacy Examined*. Madison: University of Wisconsin Press.

Rabe, B. 2013. Building on sub-federal climate strategies: The challenge of regionalism. In N. Craik, I. Studer, and D. VanNijnatten, *Climate Change Policy in North America: Designing Integration in a Regional System*, 71–107. Toronto: University of Toronto Press.

Ragsdale, G., and R. A. Foley. 2012. Testing the imprinted brain: Parent-of-origin effects on empathy and systemizing. *Evolution and Human Behavior* 33: 402–410.

Rank, M. 1989. Fertility among women on welfare: Incidence and determinants. *American Sociological Review* 54: 296–304.

Ransel, D. 1988. *Mothers in Misery: Child Abandonment in Russia*. Princeton, N.J.: Princeton University Press.

Rao, V. 1993a. Dowry "inflation" in rural India: A statistical investigation. *Population Studies* 47: 283–293.

———. 1993b. The rising price of husbands: A hedonic analysis of dowry increases in rural India. *Journal of Political Economy* 101: 666–677.

———. 1997. Wife-beating in rural south India: A qualitative and econometric analysis. *Social Science and Medicine* 44: 1169–1180.

Rappaport, R. A. 1968. *Pigs for the Ancestors: Ritual in the Ecology of a New Guinea People*. New Haven, Conn.: Yale University Press.

Rattray, R. S. 1923. *Ashanti*. New York: Negro Universities Press.

Ravenholt, R. T., and J. Chao. 1984. *World Fertility Trends*. Population Reports, series J, no. 2, Department of Medical and Public Affairs. Washington, D. C.: George Washington University.

Rebar, D., and R. L. Rodriguez. 2013. Genetic variation in social influence on mate preferences. *Proceedings of the Royal Society B* 280: 20130803.

Redford, K. 1991. The ecologically noble savage. *Cultural Survival Quarterly* 15(1): 46–48.

Redpath, S., J. Young, A. Everly, W. Adams, W. Sutherland, A. Whitehouse, A. Amar, R. Lambert, J. Linnell, A. Watt, and R. Gutiérrez. 2013. Understanding and managing conservation conflicts. *Trends in Ecology and Evolution* 28: 100–109.

Reeve, H. K. 1989. The evolution of conspecific acceptance thresholds. *American Naturalist* 133: 407–435.

Reeve, H. K. 1998. Acting for the good of others: Kinship and reciprocity with some new twists. In C. Crawford and D. Krebs (eds.), *Handbook of Evolutionary Psychology: Ideas, Issues, and Applications*, 43–85. Mahwah, N.J.: Erlbaum.

Reeve, H. K., and P. Nonacs. 1992. Social contracts in wasp societies. *Nature* 359: 823–825.

Reichard, U. 2010. Extra-pair copulations in a monogamous gibbon (*Hylobates lar*). *Ethology* 100(2): 99–112.

Rhijn, J. 1973. Behavioral dimorphism in male ruffs, *Philomachus pugnax* (L.). *Behaviour* 47: 153–229.

Rice, D. 1990. *Shattered Vows: Priests Who Leave*. New York: Triumph Books.

Richards, A. I. 1951. The Bemba of north-eastern Rhodesia. In E. Colson and M. Gluckman (eds.), *Seven Tribes of Central Africa*, 164–193. Oxford: Oxford University Press.

Richardson, L. 1960. *Statistics of Deadly Quarrels*. Pittsburgh: Boxwood Press.

Richerson, P. J., and R. Boyd. 1992. Cultural inheritance and evolutionary ecology. In E. A. Smith and B. Winterhalder (eds.), *Evolutionary Ecology and Human Behavior*, 61–92. New York: Aldine de Gruyter.

———. 2005. *Not by Genes Alone: How Culture Transformed Human Evolution*. Chicago: University of Chicago Press.

Richerson, P. J., and M. Christiansen (eds.). 2013. *Cultural Evolution: Society Technology, Language and Religion*. Strüngmann Forum Reports, Vol. 12. Cambridge, Mass.: MIT Press.

Richerson, P. J., R. Boyd, and J. Henrich. 2003. The cultural evolution of human cooperation. In P. Hammerstein (ed.), *The Genetic and Cultural Evolution of Cooperation*, 357–388. Cambridge: MIT Press.

Richerson, P., R. Boyd, and B. Paciotti. 2002. An evolutionary theory of commons management. In E. Ostrom, T. Dietz, N. Dolsak, P. Stern, S. Stonich, and E. Weber (eds.), *The Drama of the Commons*, 403–442. Washington, D.C.: National Research Council.

Ricklefs, R. E. 1983. Avian postnatal development. In D. S. Farmer, J. R. King, and K. C. Parkes (eds.), *Avian Biology*, 7: 1–83. New York: Academic Press.

Ridgeway, C., E. Boyle, K. Kuipers, and D. Robinson. 1998. How do status beliefs develop? The role of resources and interactional experience. *American Sociological Review* 63: 331–350.

Ridley, M. 1993. *The Red Queen: Sex and the Evolution of Human Nature*. New York: Viking.

———. 1996. *The Origins of Virtue*. New York: Viking.

Ridley, M., and B. S. Low. 1993. Can selfishness save the environment? *Atlantic Monthly* 272(3): 76–86.

Riolo, R., M. Cohen, and R. Axelrod. 2001. Evolution of cooperation without reciprocity. *Nature* 414: 441–443.

Ritchie, M. 1990. Optimal foraging and fitness in Colombian ground squirrels. *Oecologia* 82: 56–67.

———. 1991. Inheritance of optimal foraging behavior in Colombian ground squirrels. *Evolutionary Ecology* 5:146–159.

Roberts, C., A. Little, R. Burriss, K. Cobey, K. Klapilová, J. Havlíček, B. Jones, L. DeBruine, and M. Petrie. 2014. Partner choice, relationship satisfaction, and oral contraception: The congruency hypothesis. *Psychological Science*. doi: 10.1177/095679761 452295.

Roberts, S. C., K. Klapilová, A. Little, R. Burriss, B. C. Jones, L. DeBruine, M. Petrie, and J. Havlíček. 2012. Relationship satisfaction and outcome in women who meet their partner while using oral contraception. *Proceedings of the Royal Society B* 279: 1430–1436.

Robey, B., S. O. Rutstein, and L. Morris. 1993. The fertility decline in developing countries. *Scientific American*, December 1993, 60–67.

Rodseth, L., R. W. Wrangham, A. M. Harrigan, and B. B. Smuts. 1991. The human community as a primate society. *Current Anthropology* 32: 221–241 (discussion 241–254).

Roff, D. A. 1992. *The Evolution of Life Histories: Theory and Analysis*. New York: Chapman and Hall.

———. 2001. *Life History Evolution*. Sunderland, Mass: Sinauer Associates.

Rogers, A. R. 1990. The evolutionary economics of human reproduction. *Ethology and Sociobiology* 11: 479–495.

———. 1991. Conserving resources for children. *Human Nature* 2(1): 73–82.

———. 1993. Why menopause? *Evolutionary Ecology* 7: 406–420.

———. 1995. For love or money: The evolution of reproductive and material motivations. In R. I. M. Dunbar (ed.), *Human Reproductive Decisions*, 76–95. London: St. Martin's Press.

Roscoe, J. 1923. *The Bakitara or Banyoro; The First Part of the Report to the Mackie Ethnological Expedition to Central Africa*. Cambridge: Cambridge University Press.

Rose, M. R., and G. V. Lauder (eds.). 1996. *Adaptation*. San Diego: Academic Press.

Rose, S. 1997. *Lifelines: Biology beyond Determinism*. Oxford: Oxford University Press.

Rosenblatt, P., and M. R. Cunningham. 1976. Sex differences in cross-cultural perspective. In B. B. Lloyd and J. Archer (eds.), *Explorations in Sex Differences*, 71–94. New York: Academic Press.

Røskaft, E., A. Wara, and Å. Viken. 1992. Reproductive success in relation to resource access and parental age in a small Norwegian farming parish during the period 1700–1900. *Ethology and Sociobiology* 13: 443–461.

Ross, C. 1998. Primate life histories. *Evolutionary Anthropology* 7: 54–63.

Ross, M. H. 1983. Political decision making and conflict: Additional cross-cultural codes and scales. *Ethnology* 22: 169–192.

Roth, H. L.. 1892. The natives of Borneo. *Journal of the Royal Anthropological Institute of Great Britain and Ireland* 22: 110–137.

Rowe, N. 1996. *The Pictorial Guide to the Living Primates*. Charlestown, R.I.: Pogonias Press.

Rowell, T. E. 1988. Beyond the one-male group. *Behaviour* 104: 189–210.

Royle, N., T. Pike, P. Heeb, H. Richner, and M Kölliker. 2012. Offspring social network structure predicts fitness in families. *Proceedings of the Royal Society B* 279: 4914–4922.

Rucas, S., M. Gurven, J. Winkling, and H. Kaplan. 2012. Social aggression and resource conflict across the female life-course in the Bolivian Amazon. *Aggressive Behavior* 38: 194–207.

Ruffle, B. J., and R. Sosis. 2007. Does it pay to pray? Costly ritual and cooperation. *B. E. Journal of Economic Analysis and Policy* 7: 1–35.

Ruse, M. 1982. *Darwinism Defended: A Guide to the Evolution Controversies*. Reading, Mass.: Addison-Wesley.

———. 1986. *Taking Darwin Seriously*. New York: Holt.

Ruttan, L., and M. Borgerhoff Mulder. 1999. Are East African pastoralists truly conservationists? *Current Anthropology* 40(5): 621–652.

Ryan, M. 1998. Sexual selection, receiver biases, and the evolution of sex differences. *Science* 281: 1999–2003.

Saad, G., and T. Gill. 2001. Sex differences in the ultimatum games: An evolutionary psychology perspective. *Journal of Bioeconomics* 3: 171–193.

Sacher, G. A. 1959. Relationship of lifespan to brain weight and body weight in mammals. In G.E.W. Wolstenholme and M. O'Connor (eds.), *CIBA Foundation Symposium on the Lifespan of Animals*, 115–133. Boston: Little, Brown.

Sade, D. S. 1967. Determinants of dominance in a group of free ranging rhesus monkeys. In S. A. Altmann (ed.), *Social Communication among Primates*, 99–114. Chicago: University of Chicago Press.

Salisbury, R. 1962. *From Stone to Steel: Economic Consequences of a Technological Change*. Melbourne, AU: Melbourne University Press.

Saunders, F. S. 2004. *The Devil's Broker: Seeking Gold, God, and Glory in Fourteenth-Century Italy*. London: Fourth Estate.

Schaffnit, S., and R. Sear. 2014. Wealth modifies relationships between kin and women's fertility in high-income countries. *Behavioral Ecology*. doi: 10.1093/beheco/aru059.

Schapera, I. 1930. *The Khoisan Peoples of South Africa*. London: Routledge and Kegan Paul.

Schein, E. 1957. Distinguishing characteristics of collaborators and resisters among American POWs. *J. Abnormal and Social Psychology* 55: 197–201.

Schlegel, A., and H. Barry. 1991. *Adolescence: An Anthropological Inquiry*. New York: Free Press.

Schoen, R., Y. J. Kim, C. A. Nathanson, J. Fields, and N. M. Astone. 1997. Why do Americans want children? *Population and Development Review* 23: 333–358.

Schofield, R. S., and D. Coleman. 1986. Introduction: The state of population theory. In D. Coleman and R. Schofield (eds.), *The State of Population Theory: Forward from Malthus*, 1–13. London: Blackwell Scientific.

Schülke, O., and J. Ostner. 2012. Ecological and social influences on sociality. In J. Mitani, J. Call, P. Kappeler, R. A. Palombit, and J. B. Silk (eds.), *The Evolution of Primate Societies*, 195–219. Chicago: University of Chicago Press.

Schülke, O., J. Bhagavatula, L. Vigilant, J. Ostner. 2010. Social bonds enhance reproductive success in male macaques. *Current Biology* 20(24): 2207–2210.

Schultz, T. P. 1982. Family composition and income inequality. Economic Growth Center, Paper 25. New Haven, Conn.: Yale University.

———. 1985. Changing world prices, women's wages, and the fertility transition: Sweden, 1860–1910. *J. Pol. Econ.* 93(6): 1126–1154.

Sear, R. 2008. Kin and child survival in rural Malawi. *Human Nature* 19: 277–293.

Sear, R., and D. Coall. 2011. How much does family matter? Cooperative breeding and the demographic transition. *Population and Development Review* 37 (suppl.): 81–112.

Sear, R., and R. Mace. 2008. Who keeps children alive? A review of the effects of kin on child survival. *Evolution and Human Behavior* 29(1): 1–18.

Sear, R., R. Mace, and I. A. McGregor. 2000. Maternal grandmothers improve nutritional status and survival of children in rural Gambia. *Proceedings of the Royal Society B* 267:1641–1647.

———. 2003. The effects of kin on female fertility in rural Gambia. *Evolution and Human Behavior* 24: 25–42.

Seeley, T. D. 1985. *Honeybee Ecology: A Study of Adaptation in Social Life.* Princeton, N.J.: Princeton University Press.

Segal, N. L. 1999. *Entwined Lives: Twins and What They Tell Us About Human Nature.* New York: Dutton.

Segal, N. L., and S. L. Hershberger. 1999. Cooperation and competition between twins: Findings from a Prisoner 's Dilemma game.*Evolution and Human Behavior* 20: 29–51.

Seger, J. 1981. Kinship and covariance. *Journal of Theoretical Biology* 91(1): 191–213.

Seger, J., and W. D. Hamilton. 1988. Parasites and sex. In R. Michod and B. Levin (eds.), *The Evolution of Sex*, 176–193. Sunderland, Mass: Sinauer Press.

Seger, J., and J. W. Stubblefield. 1996. Optimization and adaptation. In Michael Rose and George Lauder (eds.), *Adaptation*, 93–123. New York: Academic Press.

Seyfarth, R., and D. Cheney. 2012. Knowledge of social relations. In J. Mitani, J. Call, P. Kappeler, R. Palombit, and J. Silk (eds.), *The Evolution of Primate Societies*, 628–642. Chicago: University of Chicago Press.

Shanley, D. P., R. Sear, R. Mace, and T. B. L. Kirkwood. 2007. Testing evolutionary theories of menopause. *Proceedings of the Royal Society B* 274: 2943–2949.

Sharpe, P. 1990. The total reconstitution method: A tool for class-specific study? *Local Pop. Stud.* 44: 41–51.

Shaw, R. P., and Y. Wong. 1989. *The Genetic Seeds of Warfare: Evolution, Nationalism, and Patriotism.* Boston: Unwin Hyman.

Shenk, M. K., M. Borgerhoff Mulder, J. Beise, G. Clark, W. Irons, D. Leonetti, B. Low, S. Bowles, T. Hertz, A. Bell, and P. Piraino. 2010. Intergenerational wealth transmission among agriculturalists: Foundations of agrarian inequality. *Current Anthropology* 51(1): 65–83.

Sheppard, P., and R. Sear. 2011. Father absence predicts age at sexual maturity and reproductive timing in British men. *Biology Letters* 8: 237–240.

Sherman, P. W. 1977. Nepotism and the evolution of alarm calls. *Science* 197: 1246–1253.

———. 1981. Kinship, demography, and Belding's ground squirrel nepotism. *Behavioral Ecology and Sociobiology* 8: 251–259.

Sherman, P. W., and H. K. Reeve. 1997. Forward and backward: Alternative approaches to studying human social evolution. In L. Betzig (ed.), *Human Nature: A Critical Reader*, 147–158. Oxford: Oxford University Press.

Sherwood, J. 1988. *Poverty in Eighteenth-Century Spain: Women and Children of the Inclusa.* Toronto: University of Toronto Press.

Shively, C., and D. G. Smith. 1985. Social status and reproductive success of male *Macaca fascicularis*. *American Journal of Primatology* 9: 129–135.

Sibley, R. M., and P. Calow. 1986. *Physiological Ecology of Animals: An Evolutionary Approach*. Oxford and Boston: Blackwell Scientific.

Sibley, R. M., and R. N. Curnow. 2011. Selfishness and altruism can coexist when help is subject to diminishing returns. *Heredity* 107: 167–173.

Sichona, F. J. 1993. The polygyny-fertility hypothesis revisited: The situation in Ghana. *Journal of Biosocial Science* 25: 473–482.

Sigmund, K. 1993. *Games of Life*. Oxford: Oxford University Press.

Sikes, G. 1997. *8-Ball Chicks*. New York: Anchor.

Silk, J. 1987. Social behavior in evolutionary perspective. In B. B. Smuts, D. Cheney, R. Seyfarth, R. Wrangham, and T. Strusaker (eds.), *Primate Societies*, 318–329. Chicago: University of Chicago Press.

Silk, J., and R. Boyd. 1983. Cooperation, competition, and mate choice in matrilineal macaque groups. In S. Wasser (ed.), *Social Behavior of Female Vertebrates*, 316–348. New York: Academic Press.

Silk, J., and G. Brown. 2008. Local resource competition and local resource enhancement shape primate birth sex ratios. *Proceedings of the Royal Society B* 275: 1761–1765.

Silk, J., S. Alberts, and J. Altmann. 2004. Patterns of coalition formation by adult female baboons in Amboseli, Kenya. *Animal Behaviour* 67: 573–582.

Silk, J., J. Beehner, T. Berman, C. Crockford, A. Eng, L. Moscovic, R. Wittig, R. Seyfarth, and D. Cheney. 2010. Strong and consistent social bonds enhance the longevity of female baboons. *Current Biology* 20: 1359–1361.

Silverman, I., and M. Eals. 1992. Sex differences in spatial abilities: Evolutionary theory and data. In J. H. Barkow, L. Cosmides, and J. Tooby (eds.), *The Adapted Mind*, 523–549. New York: Oxford University Press.

Silverman, I., and K. Phillips. 1998. The evolutionary psychology of spatial sex differences. In C. Crawford and D. L. Krebs (eds.), *Handbook of Evolutionary Psychology*, 595–612. Mahwah, N.J.: Erlbaum.

Simon, J. 1974. *The Effects of Income on Fertility*. Carolina Population Center Monograph 19. Chapel Hill: University of North Carolina Press.

———. 1996. *The Ultimate Resource 2*. Princeton, N.J.: Princeton University Press.

Sing, C. F., M. Haviland, A. R. Templeton, and S. Reilly. 1995. Alternative strategies for predicting risk of atherosclerosis. In F. P. Woodford, J. Davignon, and A. Sniderman (eds.), *Atherosclerosis X*, 638–644. Amsterdam: Elsevier.

Sing, C. F., M. Haviland, A. R. Templeton, K. Zerba, and S. Reilly. 1992. Biological complexity and strategies for finding DNA variations responsible for inter-individual variation in risk of a common chronic disease, coronary artery disease. *Annals of Medicine* 24: 539–547.

Singer, J. David. 1980. Accounting for international war. *Annual Review of Sociology* (1980): 349–376.

———. 1989. The political origins of international war. In J. Groebel and R. A. Hinde (eds.), *Aggression and War: Their Biological and Social Bases*, 202–220. Cambridge: Cambridge Univ. Press.

Singh, D. 1993a. Adaptive significance of female attractiveness: Role of waist-to-hip ratio. *Journal of Personality and Social Psychology* 65: 293–307.

————. 1993b. Body shape and women's attractiveness: The critical role of waist-to-hip ratio. *Human Nature* 4: 297–321.

————. 1994. Body fat distribution and perception of desirable female body shape by young black men and women. *International Journal of Eating Disorders* 16: 289–294.

Singh, D., and S. Luis. 1995. Ethnic and gender consensus for the effect of waist-to-hip ratio on judgement of women' attractiveness. *Human Nature* 6: 51–65.

Sipe, A.W.R. 1990. *A Secret World: Sexuality and the Search for Celibacy*. New York: Brunner Mazel.

Skuse, D. H., R. S. James, D.V.M. Bishop, B. Coppin, P. Dalton, G. Aamodt-Leeper, M. Bacarese-Hamilton, C. Creswell, R. McGurk, and P. A. Jacobs. 1997. Evidence from Turner 's syndrome of an imprinted X-linked locus affecting cognitive function. *Nature* 387: 705–708.

Smith, B. H. 1990. The cost of a large brain. *Behavioral and Brain Sciences* 13(2): 365–366.

————. 1991. Dental development and the evolution of life history in Hominidae. *American Journal of Physical Anthropology*. 86: 157–174.

————. 1992. Life history and the evolution of human maturation. *Evolutionary Anthropology* 1(4): 134–142.

Smith, E. A. 1992a. Human behavioral ecology. I. *Evolutionary Anthropology* 1(1): 20–25.

————. 1992b. Human behavioral ecology. II. *Evolutionary Anthropology* 1(2): 50–56.

————. 1995. Inuit sex ratio: A correction and an addendum. *Current Anthropology* 36: 658–659.

————. 1998a. Commentary on D. S. Wilson's "Hunting, sharing and multilevel selection." *Current Anthropology* 39: 90–91.

————. 1998b. Is Tibetan polyandry adaptive? Methodological and metatheoretical analyses. *Human Nature* 9: 225–261.

————. 1999. Three styles in the evolutionary analysis of human behavior. In L. Cronk, N. Chagnon, and W. Irons (eds.), *Evolutionary Biology and Human Social Behavior 20 Years Later*, 27–46. Hawthorne, N.Y.: Aldine de Gruyter.

————. 2004. Why do good hunters have higher reproductive success? *Human Nature* 15(4): 343–364.

Smith, E. A., and R. Bliege Bird. 2000. Turtle hunting and tombstone opening: Public generosity as costly signaling. *Evolution and Human Behavior* 21: 245–261.

————. 2005. Costly signaling and cooperative behavior. In H. Gintis, S. Bowles, R. Boyd, and E. Fehr (eds.), *Moral Sentiments and material Interests: On the Foundations of Cooperation in Economic Life*, 115–148. Cambridge, Mass.: MIT Press.

Smith, E. A., and R. Boyd. 1990. Risk and reciprocity: Hunter-gatherer socioecology and the problem of collective action. In E. Cashdan (ed.), *Risk and Uncertainty in Tribal and Peasant Economies*, 167–191. Boulder, Colo.: Westview Press.

Smith, E. A., and S. A. Smith. 1994. Inuit sex-ratio variation. *Current Anthropology* 35: 595–624.

Smith, E. A., and B. Winterhalder (eds.). 1992. *Evolutionary Ecology and Human Behavior*. New York: Aldine de Gruyter.

Smith, E. A., and M. Wishnie. 2000. Conservation and subsistence in small-scale societies. *Annual Review of Anthropology* 29: 493–524.

Smith, E. A., R. Bliege Bird, and D. Bird. 2003. The benefits of costly signaling: Merriam turtle hunters. *Behavioral Ecology* 14: 116–126.

Smith, E. A., M. Borgerhoff Mulder, S. Bowles, M. Gurven, T. Hertz, and M. K. Shenk. 2010a. Production systems, inheritance, and inequality in premodern societies: Conclusions. *Current Anthropology* 51(1): 85–94.

Smith, E. A., K. Hill, F. Marlowe, D. Nolin, P. Wiessner, M. Gurven, S. Bowles, M. Borgerhoff Mulder, T. Hertz, and A. Bell. 2010b. Wealth transmission and inequality among hunter-gatherers. *Current Anthropology* 51(1): 19–34.

Smith, J. E., and P. R. Kunz. 1976. Polygyny and fertility in nineteenth-century America. *Population Studies* 30: 465–480.

Smith, M. S., B. J. Kish, and C. B. Crawford. 1986. Inheritance of wealth as human kin investment. *Ethology and Sociobiology* 8(3): 171–182.

Smith, T., and T. Polacheck. 1981. Reexamination of the life table for northern fur seals with implications about population regulatory mechanisms. In C. W. Fowler and T. Smith (eds.), *Dynamics of Large Mammal Populations*, 99–120. New York: Wiley.

Smith, V. L. 1974. Economic theory and its discontents. *American Economic Review* 64(2): 320–322.

———. 1982. Microeconomic systems as an experimental science. *Amer. Econ. Rev.* 72: 923–955.

Smuts, B. B. 1985. *Sex and Friendship in Baboons*. New York: Aldine de Gruyter.

———. 1987a. Gender, aggression, and influence. In B. B. Smuts, D. L. Cheney, R. M. Seyfarth, R. W. Wrangham, and T. T. Strusaker (eds.), *Primate Societies*, 400–412. Chicago: University of Chicago Press.

———. 1987b. Sexual competition and mate choice. In B. B. Smuts, D. L. Cheney, R. M. Seyfarth, R. W. Wrangham, and T. T. Strusaker (eds.), *Primate Societies*, 385–399. Chicago: University of Chicago Press.

———. 1996. Male aggression against women: An evolutionary perspective. In D. Buss and N. Malamuth (eds.), *Sex, Power, Conflict: Evolutionary and Feminist Perspectives*, 231–268. New York: Oxford University Press.

Smuts, B. B., and D. J. Gubernick. 1992. Male-infant relationships in non-human primates: Paternal investment or mating effort: In B. Hewlett (ed.), *Father-Child Relations: Cultural and Biosocial Contexts*, 1–29. New York: Aldine de Gruyter.

Smuts, B. B., and R. Smuts. 1993. Male aggression and sexual coercion of females in nonhuman primates and other mammals: Evidence and theoretical implications. In P. Slater, J. Rosenblatt, M. Milinski, and C. Snowden (eds.), *Advances in the Study of Behavior*, 1–63. New York: Academic Press.

Smuts, B. B., D. Cheney, R. Seyfarth, R. Wrangham, and T. Struhsaker (eds.). 1987. *Primate Societies*. Chicago: University of Chicago Press.

Smuts, R. 1992. Fat, sex, class, adaptive flexibility, and cultural change. *Ethology and Sociobiology* 13: 521–542.

Snopkowski, K., and R. Sear. 2013. Kin influences on fertility in Thailand: Effects and mechanisms. *Evolution and Human Behavior* 34: 130–138.

Snowden, D. A., R. L. Kane, W. Beeson, G. Burke, J. Sprafka, J. Potter, D. Jacobs, and R. Phillips. 1989. Is early natural menopause a biologic marker of health and aging? *American Journal of Public Health* 79: 709–714.

Snyder, J., D. Fessler, L. Tiokhin, D. Frederick, S. Woo Lee, and C. Navarrete. 2011. Trade-offs in a dangerous world: Women's fear of crime predicts preferences for aggressive and formidable mates. *Evolution and Human Behavior* 32: 127–137.

Sober, E., and D. S. Wilson. 1998. *Do Unto Others: The Evolution and Psychology of Unselfish Behavior*. Cambridge, Mass.: Harvard University Press.

Solnick, S. 2001. Gender differences in the ultimatum game. *Economic Inquiry* 39(2): 189–200.

Sosis, R., and B. Ruffle. 2004. Ideology, religion, and the evolution of cooperation: Field tests on Israeli kibbutzim. *Research in Economic Anthropology* 23: 89–117.

Speck, F. G. 1931. Montagnais-Naskapi bands and early Eskimo distribution in the Labrador Peninsula. *American Anthropologist* 33: 557–600.

Spelke, E. 2005. Sex differences in intrinsic aptitude for mathematics and science? A critical review. *American Psychologist* 60: 950–958.

Sperber, D., F. Cara, and V. Girotto. 1995. Relevance theory explains the selection task. *Cognition* 57: 31–95.

Stanik, C., R. Kurzban, and P. Ellsworth. 2010. Rejection hurts: The effect of being dumped on subsequent mating efforts. *Evolutionary Psychology* 8(4): 682–694.

Starkweather, K., and R. Hames. 2012. A survey of non-classical polyandry. *Human Nature* 23(2): 149–172.

Stearns, S. C. 1992. *The Evolution of Life Histories*. Oxford: Oxford University Press.

Stearns, S., D. Govindaraju, D. Ewbank, and S. Byars. 2012. Constraints on the coevolution of contemporary human males and females. *Proceedings of the Royal Society B* 279: 4826–4844.

Stebbins, G. L. 1977. In defense of evolution: Tautology or theory? *American Naturalist* 111: 386–390.

Steiglitz, J., H. Kaplan, M. Gurven, J. Winking, and B. Vie Tayo. 2011. Spousal violence and paternal disinvestment among Tsimame forager-horticulturalists. *American Journal of Human Biology* 23: 445–457.

Sterck, E. 2012. The behavioral ecology of colobine monkeys. In J. Mitani, J. Call, P. Kappeler, R. Palombit, and J. Silk (eds.), *The Evolution of Primate Societies*, 65–90. Chicago: University of Chicago Press.

Stewart, K. J., and A. H. Harcourt. 1986. Gorillas: variation in female relationships. In B. B. Smuts, D. L. Cheney, R. M. Seyfarth, R. W. Wrangham, and T. T. Strusaker, (eds.), *Primate Societies*, 155–164. Chicago: University of Chicago Press.

Stiller, L. F. 1973. *The Rise of the House of Gorkha: A Study in the Unification of Nepal, 1768–1816*. New Delhi: Manjusri Publishing House.

Stoessinger, J. G. 1982. *Why Nations Go to War*. New York: St. Martin's Press.

Storr, W. 2011. The rape of men. *The Guardian*, July 16, 2011.

Strassmann, B. I. 1991. Sexual selection, paternal care, and concealed ovulation in humans. *Ethology and Sociobiology* 2: 31–40.

———. 1997. Polygyny as a risk factor for child mortality among the Dogon. *Current Anthropology* 38: 688–695.

———. 1999. Polygyny, family structure, and child mortality: A prospective study among the dogon of Mali. In L. Cronk, N. Changon, and W. Irons (eds.), *Evolution and Social Behavior*. New York: Aldine de Gruyter.

Strassmann, J., R. Page Jr., G. Robinson, and T. Seeley. 2011. Kin selection and eusociality. *Nature* 471: E5–E6.

Strate, J. 1985. The role of war in the evolution of political systems and the functional priority of defense. *Humboldt Journal of Social Relations* 12(2): 87–114.

Strier, K. 1996. Male reproductive strategies in New World primates. *Human Nature* 7: 105–123.

Strong, W. D. 1929. Cross-cousin marriage and the culture of the northeastern Algonkian. *American Anthropologist* 31: 277.

Struhsaker, T. T., and L. Leland. 1987. Colobines: Infanticide by adult males. In B. B. Smuts, D. L. Cheney, R. M. Seyfarth, R. W. Wrangham, and T. T. Struhsaker (eds.), *Primate Societies*, 83–98. Chicago: University of Chicago Press.

Sugiyama, Y., S. Kawamoto, O. Takenaka, K. Kumazaki, and N. Miwa. 1993. Paternity discrimination and inter-group relationships of chimpanzees at Bossou. *Primates* 34: 545–552.

Sundin, J. 1976. Theft and penury in Sweden, 1830–1920: A comparative study at the county level. *Scandinavian J. Hist.* 1: 265–292.

Sundin, J., and L.-G. Tedebrand. 1981. Mortality and morbidity in Swedish iron foundries 1750–1875. In Anders Brändström and Jan Sundin (eds.), *Tradition and Transition: Studies in Microdemography and Social Change*, 105–159. Demographic Database, Report no. 2. Umeå: University of Umeå.

———. 1984. Migratory workers and settled workers: A study of demographic behaviour at Swedish 19th-century iron foundries. Paper delivered at the Ninth Annual Meeting, Social Science History Association, October 25–28, 1984, Ontario, Canada.

Swallow, D. 2003. Genetics of lactase persistence and lactose intolerance. *Annual Review of Genetics* 37: 197–219.

Swanton, J. R. 1911. *Indian Tribes of the Lower Mississippi Valley and Adjacent Coast of the Gulf of Mexico.* Smithsonian Institution, Bureau of American Ethnology Bulletin 43. Washington, D.C.: U.S. Government Printing Office.

———. 1928a. Aboriginal culture of the Southeast. In *Forty-Second Annual Report of the Bureau of American Ethnology*, 673–727. Washington, D.C.: U.S. Government Printing Office.

———. 1928b. Social organization and social usages of the Indians of the Creek Confederacy. In *Forty-Second Annual Report of the Bureau of American Ethnology*, 23–473. Washington, D.C.: U.S. Government Printing Office.

Swartz, M., V. Turner, and A. Tuden (eds.). 1966. *Political Anthropology.* Chicago: Aldine de Gruyter.

Sweeney, T. A. 1992. *A Church Divided: The Vatican versus American Catholics.* New York: Prometheus Books.

Sybert, V. P., and E. McCauley. 2004. Turner's syndrome. *New England Journal of Medicine* 351: 1227–1238.

Sykes, B. 2002. *The Seven Daughters of Eve.* New York: Norton.

Szal, J. A. 1972. Sex differences in the cooperative and competitive behaviors of nursery school children. M.S. thesis, Stanford University, Stanford, Calif.

Takahata, Y., H. Ihobe, and G. Idani. 1996. Comparing copulations of chimpanzees and bonobos: Do females exhibit proceptivity or receptivity? In W. C. McGrew, L. Marchant, and T. Nishida (eds.), *Great Ape Societies*, 146–155. Cambridge: Cambridge University Press.

Tambiah, S. J. 1966. Polyandry in Ceylon. In C. von Fuhrer-Haimendorf (ed.), *Caste and Kin in Nepal, India, and Ceylon*, 264–358. London: Asia Publishing House.

Taşiran, A. C. 1995. *Fertility Dynamics: Spacing and Timing of Births in Sweden and the United States.* Amsterdam: Elsevier.

Taylor, P. D., and G. C. Williams. 1982. The lek paradox is not resolved. *Theoretical Population Biology* 22: 392–409.

Templeton, A. R. 1995. A cladistic analysis of phenotypic associations with haplotypes inferred from restriction endonuclease mapping or DNA sequencing. 5. Analysis of case/control sampling designs: Alzheimer's disease and the apoprotein E locus. *Genetics* 140: 403–409.

Terborgh, J., and A. W. Goldizen. 1985. On the mating system of the cooperatively breeding saddle-backed tamarin (*Saguinus fuscicollis*). *Behavioral Ecology and Sociobiology* 16: 293–299.

Thomas, D. S. 1941. *Social and Economic Aspects of Swedish Population Movements: 1750–1933*. New York: Macmillan.

Thompson, J., and M. Britton. 1980. Some socioeconomic differentials in fertility in England and Wales. In R. W. Hiorus (ed.), *Demographic Patterns in Developed Societies*, 1–10. London: Taylor and Francis.

Thornhill, R., and J. Alcock. 1983. *The Evolution of Insect Mating Systems*. Cambridge, Mass.: Harvard University Press.

Thornhill, R., and P. S. Burgoyne. 1993. A paternally imprinted X-chromosome retards the development of the early mouse embryo. *Development* 118: 171–174.

Thornhill, R., and S. W. Gangestad. 1994. Fluctuating asymmetry and human sexual behavior. *Psych. Science* 5: 297–302.

Thornhill, R., J. F. Chapman, and S. Gangestad. 2013. Women's preferences for men's scents associated with testosterone and cortisol levels: Patterns across the ovulatory cycle. *Evolution and Human Behavior* 34: 216–221.

Thornhill, R., C. Fincher, and D. Aran. 2009. Parasites, democratization, and the liberalization of values across contemporary countries. *Biological Reviews* 84: 113–131.

Thornhill, R., C. Fincher, D. Murray, and M Schaller. 2010. Zoonotic and non-zoonotic diseases in relation to human personality and social values: support for the parasite-stress model. *Evolutionary Psychology* 8(2): 151–160.

Thornhill, R., S. W. Gangestad, and R. Comer. 1995. Human female orgasm and mate fluctuating asymmetry. *Animal Behaviour* 50: 1601–1615.

Tiger, L., and R. Fox. 1971. *The Imperial Animal*. New York: Holt, Reinhart, and Winston.

Tilly, C. 1978. The historical study of vital processes. In C. Tilly (ed.), *Historical Studies of Changing Fertility*, 1–55. Princeton, N.J.: Princeton University Press.

Ting, T.-F. 1994. Shifts in reproductive patterns in China. *Population and Environment* 25(4): 299–317.

Timaeus, I. M., and A. Reynar. 1998. Polygynists and their wives in sub-Saharan Africa: An analysis of five demographic and health surveys. *Population Studies* 52: 145–162.

Tooby, J., and L. Cosmides. 1988. The evolution of war and its cognitive foundations. Institute for Evolutionary Studies Technical Report 88-1.

Torres, A., and J. D. Forrest. 1988. Why do women have abortions? *Family Planning Perspectives* 20(4): 169–176.

Tottenham, L., D. Saucier, L. Elias, and C. Gutwin. 2003. Female advantage for spatial location memory in both static and dynamic environments. *Brain and Cognition* 53(2): 381–383.

Trail, P. W. 1985. The intensity of sexual selection: Intersexual and interspecific comparisons require consistent measures. *American Naturalist* 126: 434–439.

Travisano, M., and R. E. Lenski. 1996. Long-term experimental evolution in *Escherichia coli*. IV. Targets of selection and the specificity of adaptation. *Genetics* 143: 15–26.

Travisano, M., J. A. Mongold, A. F. Bennett, and R. E. Lenski. 1995. Experimental tests of the roles of adaptation, chance, and history in evolution. *Science* 267: 87–90.

Trivers, R. L. 1971. The evolution of reciprocal altruism. *Quarterly Review of Biology* 46: 35–57.

———. 1972. Parental investment and sexual selection. In B. Campbell (ed.), *Sexual Selection and the Descent of Man*, 136–179. Chicago: Aldine de Gruyter.

———. 1974. Parent-offspring conflict. *American Zoologist* 14: 249–264.

Trivers, R. L. 1985. *Social Evolution*. Menlo Park, Calif.: Benjamin Cummings.

Trivers, R. L., and D. E. Willard. 1973. Natural selection of parental ability to vary the sex ratio. *Science* 179: 90–92.

Tronick, E. Z., R. B. Thomas, and M. Daltabuit. 1994. The Quechua manta pouch: A caretaking practice for buffering the multiple stressors of high altitude. *Child Development* 65: 1005–1013.

Trotter, M., and B. McCullough, 1984. Moas, men, and middens. In P. S. Martin and R. G. Klein (eds.), *Quaternary Extinctions: A Prehistoric Revolution*, 708–727. Tucson: University of Arizona Press.

Tuan, Y-F. 1968. Discrepancies between attitude and behavior: Examples from Europe and China. *Canadian Geographer* 12(3): 176–190.

———. 1970. Our treatment of the environment in ideal and actuality. *American Scientist* 58: 244–249.

Turke, P. W. 1988. Helpers at the nest: Childcare networks in Ifaluk. In L. Betzig, M. Borgerhoff Mulder, and P. Turke (eds.), *Human Reproductive Behaviour: A Darwinian Perspective*, 173–188. Cambridge: Cambridge University Press.

Turke, P. W. 1989. Evolution and the demand for children. *Population and Development Review* 15(1): 61–90.

———. 1990. Which humans behave adaptively, and why does it matter? *Ethology and Sociobiology* 11: 305–339.

———. 1992. Theory and evidence on wealth flows and old-age security: A reply to Fricke. *Population and Development Review* 17(4): 687–702.

Turke, P. W., and L. Betzig. 1985. Those who can do: Wealth, status, and reproductive success on Ifaluk. *Ethology and Sociobiology* 6: 79–87.

Tuschman, A. 2013. *Our Political Nature: The Evolutionary Nature of What Divides Us.* Amherst, N.Y.: Prometheus Books.

Tutin, C.E.G. 1979. Mating patterns and reproductive strategies in a community of wild chimpanzees (*Pan troglodytes schweinfurtii*). *Behavioral Ecology and Sociobiology* 6: 29–38.

United Nations. 1996. *World Population Prospects*, 1996 revision. http://esa.un.org/wpp/ (accessed May 22, 2014).

United Nation News Centre. 2009. UNICEF adviser says rape in Darfur continues with impunity. Originally published October 19, 2004.

Vale, M. 1981. *War and Chivalry: Warfare and Aristocratic Culture in England, France, and Burgundy at the End of the Middle Ages*. Athens: University of Georgia Press.

VanderLaan, D., D. Forrester, L. Petterson, and P. Vasey. 2012. Offspring production among the extended relatives of Samoan *fa'afine*. *PLoS ONE* 7(4):e36088. doi: 10.1371/journal.pone.0036088.

van Schaik, C. P. 1989. The ecology of social relationships among female primates. In V. Standen and R. Foley (eds.), *Comparative Socioecology: The Behavioural Ecology of Humans and Other Mammals*, 195–218. Oxford: Blackwell Scientific.

————. 1996. Social evolution in primates: The role of ecological factors and male behaviour. *Proceedings of the British Academy* 88: 9–31.

van Schaik, C. P., and R. I. Dunbar. 1990. The evolution of monogamy in large primates: A new hypothesis and some crucial tests. *Behaviour* 115: 30–61.

van Schaik, C. P., and K. Isler. 2012. Life history evolution in primates. In J. Mitani, J. Call, P. Kappeler, R. A. Palombit, and J. B. Silk (eds.), *The Evolution of Primate Societies*, 220–244. Chicago: University of Chicago Press.

Van Valen, L. 1973. A new evolutionary law. *Evolutionary Theory* 1: 1–30.

Vayda, A. P. 1960. *Maori Warfare*. Polynesian Society Maori Monographs No. 2. Wellington, New Zealand: Reed.

Vesa, T., P. Marteau, and R. Korpela. 2000. Lactose intolerance. *Journal of the American College of Nutrition* 19(2): 165S–175S.

Viazzo, P. P. 1990. *Upland Communities: Environment, Population, and Social Structure in the Alps since the Sixteenth Century*. Cambridge: Cambridge University Press.

Vinovskis, M. A. 1972. Mortality rates and trends in Massachusetts before 1860. *Journal of Economic History* 32: 184–218.

Voland, E. 1984. Human sex-ratio manipulation: Historical data from a German parish. *Journal of Human Evolution* 13: 99–107.

————. 1989. Differential parental investment: Some ideas on the contact area of European social history and evolutionary biology. In V. Standen and R. A. Foley (eds.), *Comparative Socioecology: The Behavioural Ecology of Humans and Other Mammals*, 391–402. Special Publication No. 8, British Ecological Society. Oxford: Blackwell Scientific.

————. 1990. Differential reproductive success within the Krummhörn population (Germany, 18th and 19th centuries). *Behavioral Ecology and Sociobiology* 26: 65–72.

Voland, E., and R.I.M. Dunbar. 1995. Resource competition and reproduction: The relationship between economic and parental strategies in the Krummhörn population (1720–1874). *Human Nature* 6: 33–49.

Voland, E., E. Siegelkow, and C. Engle. 1990. Cost/benefit oriented parental investment by high status families: The Krummhörn case. *Ethology and Sociobiology* 12: 105–118.

von Rueden, C., M. Gurven, and H. Kaplan. 2011. Why do men seek status? Payoffs to dominance and prestige. *Proceedings of the Royal Society B* 278: 2223–2232.

Voyer, D., and S. Voyer. 2014. Gender differences in scholastic achievement: A meta-analysis. *Psychological Bulletin*. doi: 10.1037/a0036620.

Voyer, D., A. Postma, B. Brake, and J. Imperato-McGinley. 2007. Gender differences in object location memory: A meta-analysis. *Psychonomic Bulletin and Review* 14(1): 23–38.

Wade, M. J. 1979. Sexual selection and variance in reproductive success. *American Naturalist* 114: 742–747.

Wade, M. J., and S. J. Arnold. 1980. The intensity of sexual selection in relation to male sexual behavior, female choice, and sperm precedence. *Animal Behaviour* 28: 446–461.

Walker, H. 2014. How women are underrepresented in politics all around the world. *Business Insider*, May 6. http://www.businessinsider.com/this-chart-shows-how-few-women-are-in-politics-2014-5#!Kh5kS (accessed April 7, 2014).

Walker, M. 1995. Menopause in female rhesus monkeys. *American Journal of Primatology* 35: 59–71.

Walker, R. S., K. Hill, M. Flinn, and R. Ellsworth. 2011. Evolutionary history of hunter-gatherer marriage practices. *PLoS ONE* 6(4): e19066. doi: 10.1371/journal.pone.0019066.

Wall, R. 1984. Real property, marriage, and children: The evidence from four pre-industrial communities. In R. M. Smith (ed.), *Land, Kinship, and the Life-Cycle*, 443–479. Cambridge, U.K.: Cambridge University Press.

Wallen, K. 1996. Nature needs nurture: The interaction of hormonal and social influences on the development of behavioral sex differences in rhesus monkeys. *Hormones and Behavior* 30: 364–378.

Waller, D. 2011. Special Operations funding quadruples. *Bloomberg News* June 7. http://www.bloomberg.com/news/2011-06-07/special-operations-spending-quadruples-with-commando-demand.html (accessed February 18, 2014).

Wang, F., J. Lee, and C. Campbell. 1995. Marital fertility control among the Qing nobility: Implications for two types of preventive checks. *Population Studies* 49: 383–400.

Warner, R. R., R. K. Harlan, and E. G. Leigh. 1975. Sex change and sexual selection. *Science* 190: 633–638.

Wasser, S. 1983. Reproductive competition and cooperation among female yellow baboons. In S. Wasser (ed.), *Social Behavior of Female Vertebrates*, 349–390. New York: Academic Press.

Watson, P. 1978. *War on the Mind*. New York: Basic Books.

Watts, D. 1989. Infanticide in mountain gorillas: New cases and a reconsideration of the evidence. *Ethology* 81: 1–18.

———. 1996. Comparative socio-ecology of gorillas. In W. C. McGrew, L. F. Marchant, and T. Nishida (eds.), *Great Ape Societies*, 16–28. Cambridge: Cambridge University Press.

———. 2012. The apes: Taxonomy, biogeography, life histories, and behavioral ecology. In J. Mitani, J. Call, P. Kappeler, R. A. Palombit and J. B. Silk (eds.), *The Evolution of Primate Societies*, 113–142. Chicago: University of Chicago Press.

Webster, D. 1975. Warfare and the evolution of the state: A reconsideration. *American Antiquity* 40(4): 464–470.

Webster, M., Jr., and S. Hyson. 1998. Creating status characteristics. *American Sociological Review* 63: 351–378.

Wen, X. 1993. Effect of son preference and population policy on sex ratios at birth in two provinces of China. *Journal of Biosocial Science* 25: 509–521.

West-Eberhard, M. J. 1975. The evolution of social behavior by kin selection. *Quarterly Review of Biology* 50: 1–33.

———. 1978. Temporary queens in Metapolybia wasps: Non-reproductive helpers without altruism? *Science* 200: 441–443.

———. 1989. Phenotypic plasticity and the origins of diversity. *Annual Review of Ecology and Systematics* 20: 249–278.

Whitam, F. L. 1983. Culturally invariable properties of male homosexuality: Tentative conclusions from cross-cultural research. *Archives of Sexual Behavior* 12(3): 207–226.

White, D. R. 1988. Rethinking polygyny: Co-wives, codes and cultural systems. *Current Anthropology* 29(4): 529–558.

White, D. R., and M. L. Burton. 1988. Causes of polygyny: Ecology, economy, kinship, and warfare. *American Anthropologist* 90(4): 871–887.

Whiting, B., and C. P. Edwards. 1973. A cross-cultural analysis of sex differences in the behavior of children aged three through eleven. *Journal of Social Psychology* 91: 171–188.

———. 1988. *Children of Different Worlds: The Formation of Social Behavior*. Cambridge, Mass.: Harvard University Press.

Whiting, B., and J. Whiting. 1975. *Children of Six Cultures: A Psychocultural Analysis*. Cambridge, Mass.: Harvard University Press.

Whiting, B., C. P. Edwards, and C. P. Edwards. 1992. *Children of Different Worlds: The Formation of Social Behavior*. Cambridge, Mass: Harvard University Press.

Whiting, J. 2009. The duration of maidenhood across cultures. In E. Chasdi (ed.), *Culture and Development: The Selected Papers of Joan Whiting*, 282–305. Cambridge: Cambridge University Press.

Whitten, P. L. 1987. Infants and adult males. In B. B. Smuts, D. L. Cheney, R. M. Seyfarth, R. W. Wrangham, and T. T. Struhsaker (eds.), *Primate Societies*, 343–357. Chicago: University of Chicago Press.

Whyte, M. K. 1978. Cross-cultural codes dealing with the relative status of women. *Ethnology* 17: 211–237.

———. 1979. *The Status of Women in Pre-industrial Society*. Princeton, N.J.: Princeton University Press.

Wickler, W., and U. Seibt. 1981. Monogamy in Crustacea and man. *Zeitschrift für Tierpsychologie* 57: 215–234.

Widdig, A., W. Jürgen Streich, and G. Tembrock. 2000. Coalition formation among Barbary macaques (*Macaca sylvanus*). *American Journal of Primatology* 50: 37–51.

Wilkinson, G. S. 1984. Reciprocal food sharing in the vampire bat. *Nature* 308: 181–184.

Wilkinson, P. 1986. Terrorism: international dimensions. In W. Gutteridge (ed.), *Contemporary Terrorism*, 29–56. New York: Facts on File.

Williams, C. L., and W. H. Meck. 1991. The organizational effects of gonadal steroids on sexually dimorphic spatial ability. *Psychoneuroendocrinology* 16: 155–176.

Williams, G. C. 1957. Pleiotropy, natural selection, and the evolution of senescence. *Evolution* 11: 398–411.

———. 1966. *Adaptation and Natural Selection*. Princeton, N.J.: Princeton University Press.

———. 1975. *Sex and Evolution*. Princeton, N.J.: Princeton University Press.

———. 1989. A sociobiological explanation of *Evolution and Ethics*. In J. Paradis and G. C. Williams (eds.), *Evolution and Ethics: T. H. Huxley's Evolution and Ethics with New Essays on Its Victorian and Sociobiological Context*, 179–214. Princeton, N.J.: Princeton University Press.

———. 1992a. Gaia, nature worship, and biocentric fallacies. *Quart. Rev. Biol.* 67: 479–486.

———. 1992b. *Natural Selection: Domains, Levels, and Challenges*. Oxford: Oxford University Press.

Wilson, D. S. 1975. New model for group selection. *Science* 189: 8701.

———. 1980. *The Natural Selection of Populations and Communities*. Menlo Park, Calif.: Benjamin Cummings.

———. 1998a. Game theory and human behavior. In L. Dugatkin and K. Reeves (eds.), *Game Theory and Animal Behavior*, 261–282. Oxford: Oxford University Press.

———. 1998b. Hunting, sharing and multilevel selection. *Current Anthropology* 39: 73–86 (comments on pp. 86–97).

Wilson, D. S., and E. Sober. 1994. Reintroducing group selection to the human behavioral sciences. *Behavioral and Brain Sciences* 17: 585–608.

Wilson, D. S., D. Near, and R. R. Miller. 1996. Machiavellianism: A synthesis of the evolutionary and psychological literatures. *Psychological Bulletin* 119: 285–299.

Wilson, D. S., D. O'Brien, D. T., and A. Sesma. 2009. Human prosociality from an evolutionary perspective: Variation and correlations on a city-wide scale. *Evolution and Human Behavior* 30: 190–200.

Wilson, E. O. 1975. *Sociobiology: The New Synthesis*. Cambridge, Mass.: Harvard University Press.

Wilson, J., J. Acheson, and T. Johnson. 2013. The cost of useful knowledge and collective action in three fisheries. *Ecological Economics* 96: 165–172.

Wilson, J., J. Hill, M. Kersula, C. L. Wilson, L. Whitsel, L. Yan, J. Acheson, Y. Chen, C. Clever, C. Congdon, A. Hayden, P. Hayes, T. Johnson, G. Morehead, R. Steneck, R. Turner, R. Vadas, C. J. Wilson. 2013. Costly information and the evolution of self-organization in a small, complex economy. *Journal of Economic Behavior and Organization* 905: 576–593.

Wilson, M., and M. Daly. 1997. Life expectancy, economic inequality, homicide, and reproductive timing in Chicago neighbourhoods. *British Medical Journal* 314: 1271–1274.

Wilson, M., M. Daly, and S. Gordon. 1998. The evolved psychological apparatus of decision-making is one source of environmental problems. In T. Caro (ed.), *Behavioral Ecology and Conservation Biology*, 501–523. New York: Oxford University Press.

Wilson, M., M. Daly, S. Gordon, and A. Pratt. 1996. Sex differences in valuation of the environment. *Population and Environment* 18(2): 143–159.

Winking, J., M. Gurven, H. Kaplan, and J. Stieglitz. 2009. The goals of direct paternal care among a South American population. *American Journal of Physical Anthropology* 139: 295–304.

Winter, J. 1989. Causes of war. In J. Groebel and R. Hinde (eds.), *Aggression and War: Their Biological and Social Bases*, 194–201. Cambridge: Cambridge University Press.

Winterhalder, B. 1996a. A marginal model of tolerated theft. *Ethology and Sociobiology* 17: 37–53.

———. 1996b. Social foraging and the behavioral ecology of intragroup transfers. *Evolutionary Anthropology* 5: 46–57.

———. 1996c. Gifts given, gifts taken: The behavioral ecology of non-market, intragroup exchange. *Journal of Archaeological Research* 5: 121–168.

———. 1997. Delayed reciprocity and tolerated theft. *Current Anthropology* 38: 74–75.

Wolf, A. P. 1981. Women, widowhood and fertility in pre-modern China. In J. Dupaquier, E. Helin, P. Laslett, M. Livi-Bacci, and S. Sogner (eds.), *Marriage and Remarriage in Populations of the Past*, 139–147. New York: Academic Press.

Wolf, D., R. Lee, T. Miller, G. Donnehower, and A. Genest. 2011. Fiscal externalities of becoming a parent. *Population and Development Review* 37(2): 241–266.

Wolfram, S. 2002. *A New Kind of Science*. Wolfram Media.

Wood, J. W. 1990. Fertility in anthropological populations. *Annual Review of Anthropology* 19: 211—242.

———. 1998. A theory of preindustrial population dynamics. *Current Anthropology* 39: 99–135.

Woolfenden, G. E., and J. Fitzpatrick. 1984. *The Florida Scrub Jay: Demography of a Cooperative Breeding Bird*. Princeton, N.J.: Princeton University Press.

Worldwatch Institute. 2013. *State of the World 2013: Is Sustainability Still Possible?* Washington, D.C.: Island Press.

Wrangham, R. 1979. Sex differences in chimpanzee dispersion. In D. Hamburg and E. Mc-Cown (eds.), *The Great Apes*, 481–490. Menlo Park, Calif: Benjamin Cummings.

———. 1980. An ecological model of female-bonded primate groups. *Behaviour* 75: 262–300.

———. 1987. Evolution of social structure. In B. Smuts, D. Cheney, R. Seyfarth, R. Wrangham, and T. Struhsaker (eds.), *Primate Societies*, 282–296. Chicago: University of Chicago Press.

Wrangham, R., and D. Peterson. 1996. *Demonic Males: Apes and the Origin of Human Violence*. Boston: Mariner Books.

Wright, L. 1997. *Twins and What They Tell Us about Who We Are*. New York: Wiley.

Wright, P. 1984. Biparental care in *Aotus trivirgatus* and *Callicebus moloch*. In M. Small (ed.), *Female Primates*, 59–75. New York: Alan Liss.

Wright, S. 1945. Tempo and mode in evolution: A critical review. *Ecology* 26: 415–419.

———. 1977. *Evolution and the Genetics of Populations*. Vol. 3, *Experimental Results and Evolutionary Deductions*. Chicago: University of Chicago Press.

Wrigley, E. A. 1983a. The growth of population in eighteenth century England: A conundrum resolved. *Past and Present*. 98: 121–150.

———. 1983b. Malthus's model of a pre-industrial economy. In J. Dupaquier, A. Fauve-Chamoux, and E. Grebenik (eds.), *Malthus Past and Present*, 111–124. New York: Academic Press.

Wrigley, E. A., and R. Schofield. 1981. *The Population History of England, 1541–1871*. Cambridge, Mass.: Harvard University Press.

Wrong, D. 1980. *Class Fertility Trends in Western Nations*. New York: Arno Press.

———. 1985. Trends in class fertility in Western nations. *Canadian Journal of Economics and Political Science*. 24: 216–210.

Wuethrich, B. 1998.Why sex? Putting theory to the test. *Science* 281: 1980–1982.

Wynne-Edwards, V. C. 1962. *Animal Dispersion in Relation to Social Behaviour*. Edinburgh: Oliver and Boyd

Yalman, N. 1967. *Under the Bo Tree*. Berkeley: University of California Press.

Yanca, C., and B. Low. 2004. Female allies and female power: A cross-cultural analysis. *Evolution and Human Behavior* 25: 9–23.

Yi, Z., T. Ping, G. Baochang, X. Yi, L. Bohua, and L. Yongping. 1993. Causes and implications of the recent increase in the reported sex ratio at birth in China. *Population and Development Review* 19(2): 283–302.

Young, C., S. Hähndel, B. Majolo, O. Schülke, and J. Ostner. 2013. Male coalitions and female behavior affect mating success independent of dominance rank and female receptive synchrony in wild Barbary macaques. *Behavioral Ecology and Sociobiology* 67(10): 1665–1677.

Yu, D., and G. Shepard. 1998. Is beauty in the eye of the beholder? *Nature* 396: 321–322.

Zahavi, A. 1975. Mate selection—a selection for the handicap principle. *Journal of Theoretical Biology* 53: 205–214.

Zahavi, A., and A. Zahavi. 1997. *The Handicap Principle: A Missing Piece of Darwin's Puzzle*. Oxford: Oxford University Press.

Author Index

Subject Index

Taxonomic Index

Society/Social Group Index